Networks and Services

Networks and Services

Carrier Ethernet, PBT, MPLS-TP, and VPLS

Mehmet Toy

A John Wiley & Sons, Inc., Publication

Published by John Wiley & Sons, Inc., Hoboken, New Jersey
Published simultaneously in Canada

For general information on our other products and services or for technical support, please contact our Customer Care Department within the United States at (800) 762-2974, outside the United States at (317) 572-3993 or fax (317) 572-4002.

Wiley also publishes its books in a variety of electronic formats. Some content that appears in print may not be available in electronic formats. For more information about Wiley products, visit our web site at www.wiley.com.

Library of Congress Cataloging-in-Publication Data:

Toy, Mehmet.
 Networks and Services : Carrier Ethernet, PBT, MPLS-TP, and VPLS / Mehmet Toy.
    pages cm.—(Information and communication technology series ; 95)
  Includes bibliographical references and index.
  ISBN 978-0-470-39119-8
  1. Ethernet (Local area network system) I. Title.
  TK5105.8.E83T69 2012
  621.39′81–dc23
                                    2012015783

Printed in the United States of America

10 9 8 7 6 5 4

To the memory of my parents
To my sister and brothers
To my wife, Fusun, and my sons, Onur, Ozan and Atilla

Contents

13 MPLS-TP(MPLS-Transport Profile)   303

14 Virtual Private LAN Services (VPLS)   333

Foreword

As Carrier Ethernet becomes the dominant delivery service and network for end-users (enterprises, mobile operators, wholesale providers, etc) worldwide, more telecommunications service providers are responding to end-user demand for Ethernet services to deliver efficient, high-bandwidth, and QoS-enabled networks.

In 2012, the MEF (the industry's defining body for Carrier Ethernet) advanced networks and services by a generation through introduction of Carrier Ethernet 2.0- CE 2.0; expanding the original 3 services to 8, and delivering three powerful, standardized features: Multi-CoS (Multiple Classes of Service), Manageability, Interconnect.

Multi-CoS defines performance objectives (based on Frame Delay, Inter-Frame Delay Variation, Frame Loss Ratio, Frame Delay Range, Mean Frame Delay, Availability, etc.) for 3 classes of service over 4 distance related "Performance Tiers" and maps more than 20 application types into these 3 classes. In other words, Multi-CoS standardizes each of 3 classes of service in terms of clearly defined performance objectives and it categorizes 20 applications into these 3 classes of service based upon required performance objectives of each application.

For the first time, Standardized Multi-CoS enables industry-wide adoption to deliver unprecedented network efficiency. It's unprecedented because the ability to quantify and predict network performance by application makes it viable to optimize bandwidth usage with resultant significant cost savings and while actually improving quality of service. This standardization reates a compelling case to implement Multi-CoS because the previous complexity, risks and costs of ad-hoc implementations are now minimized. Inherent in the MEF work is the ability to maintain this efficiency across interconnected networks ensuring end-user applications function properly in Multi-CoS implementations over well engineered regional, national and global networks.Collectively, these Multi-CoS features alone advance the state of the art for telecom and networking, demonstrating that Carrier Ethernet is so much more than a high speed, low cost version of TDM.

Next are a series of management definitions covering fault and performance monitoring to bring standardization, cost savings and scalability to network deployment. The third characteristic of CE 2.0 is Interconnect, which globalizes Carrier Ethernet across multiple provider networks and standardizes Ethernet wholesale services to significantly reduce costs and improve efficiency.

Those new to Carrier Ethernet could intuitively assume that the entire network requires all transport technologies, both access and backbone, to be IEEE 802.3 physical Ethernet. This is not at all the case. Only at the edge of the network and at the demarcation point between providers is the connection required to be via physical Ethernet. Carrier Ethernet services operate at what is defined at the "Eth Layer" in MEF parlance. The separation between the "App layer" (IP and above) and the "Tran Layer" (the physical transport layers below) means the Carrier Ethernet services are agnostic to transport technologies and networks. This has proven to be a key to the success of Carrier Ethernet since it decouples the migration from one technology to another, allows mixed technology infrastructures to deliver MEF-defined standardized Ethernet services.

Some examples of the various transport technologies in the access are standard IEEE 802.3z, ae, G.8031 fiber, G.8032 Ethernet rings, WDM fiber, GPON, Docsis, T1/E1, T3/E3, bonded copper, DSL, Ethernet over SDH/SONET, Ethernet over packet wireless, etc. In backbone networks this agnostic approach has enabled Carrier Ethernet to be delivered over PB/PBB/PBB-TE, MPLS VPWS/VPLS/TP, SONET/SDH OTN & WDM.

Therefore, telecom service providers must select and integrate one or more of transport technologies and networks to deliver well engineered Carrier Ethernet service offerings, potentially spanning several providers. Although the MEF augments technical specifications and implementation agreements with a series of best practices documents, implementation papers and the new successful Carrier Ethernet Professional Certification Program, further guidance in the form of a book that spans several technologies is much needed. *Networks and Services* explain current and future debated networking technologies and services supported by them. It is an invaluable reference for the industry.

NAN CHEN
Founding President, MEF

Preface

Networks may be grouped as access and backbone networks. Ethernet, Carrier Ethernet, TDM/SONET, IP, Passive Optical Networks (PON), wireless and cable technologies are common access network technologies while IP/MPLS and Dense Wavelength Division Multiplexing (DWDM) are common backbone network technologies. It is clear that using one technology at access and backbone will reduce equipment and operational costs. Both IP/MPLS and Carrier Ethernet are the two key candidates to be that technology. Various data, voice and video services offered by operators using point-to-point and multipoint topologies supported by these access and backbone technologies.

With the introduction of circuit concept over Ethernet by IEEE and MEF specifications, Ethernet is becoming a key technology for Metropolitan Area Networks and Wide Area Networks with deployment in more than 100 countries. Various Carrier Ethernet services have already been offered by telecommunications service providers. According to Infonetics Research, Carrier Ethernet equipment manufacturer revenue is to grow to $34 billion by 2013.

In addition to deployments of Carrier Ethernet Networks (CENs) by service providers, interconnectivity between CENs, External Network Network Interfaces (ENNIs), are growing rapidly as well. Support of OAM capabilities at ENNI for services crossing ENNIs are the key in providing end-to-end services. We expect OAM guidelines at ENNI will establish as service providers gain experience in working with each other. Currently there is little agreement between them for passing fault and performance management information such as alarms and measurements, and performing service monitoring.

Despite of large deployments of CENs and services over them, a limited set of features defined by MEF specifications have been offered to users today. We expect this to be changed in the coming years.

Carrier Ethernet services are offered end-to-end while Carrier Ethernet technology are often used at the access. Therefore, Carrier Ethernet needs to interwork with core network technologies such as IP/MPLS, and support pseudowires and circuit emulations, in supporting Ethernet Virtual Connections (EVCs) between customer locations.

The purpose of this book is to explain Carrier Ethernet and related technologies (i.e. Pseudowires, Transport Multi Protocol Label Switching, MPLS Transport Profile, Provider Backbone Bridging, and Virtual Private LAN Service) and services (i.e. Private Line, Virtual Private Line, LAN, Virtual LAN, Tree and

Virtual Tree) to those involved in their planning, deployment and operations. Some of the areas such as SLAs (Service Level Agreements), QoS, Availability Calculations, and Synchronization are still fertile areas for research. I am hoping that this book triggers further research in these areas.

The writing of this book took a long time. During that time, the various technologies including Carrier Ethernet covered in this book evolved and I tried to reflect the changes as much as possible.

I would like to thank MEF President Nan Chen for writing the foreword, my MEF colleagues for their contributions to Carrier Ethernet that became the key technology of L2 networking described here, and Diana Gialo and Kristen Parrish of John Wiley and Jayashree of Laserwords who helped with the publication.

MEHMET TOY
Allendale, NJ

Chapter 1

Introduction and Overview

1.1 INTRODUCTION

Ethernet has been the dominant technology for local area networks (LANs) for many years. With the introduction of circuit concept over Ethernet by the IEEE and MEF specifications, Ethernet is becoming a key technology for Metropolitan Area Networks (MANs) and Wide Area Networks (WANs).

This book makes an attempt to describe various aspects of data networks and services based on carrier Ethernet, pseudowires (PWs), transport multiprotocol label switching (T-MPLS), multiprotocol label switching transport profile (MPLS-TP), and virtual private local area network service (VPLS).

Chapter 2 describes the basics of Ethernet, such as protocol stack, bridges, switches, and hubs. Chapters 3, 4, and 5 cover the key techniques that are being used in building carrier class Ethernet networks and services, namely, synchronization, PWs, and protection, respectively.

Chapter 6 begins describing Carrier Ethernet network architecture and services that are currently deployed in the industry and evolving. It is followed by Chapters 7 and 8, describing traffic management and Ethernet Operations, Administrations, and Maintenance (OAM) capabilities of Carrier Ethernet[1], respectively.

Chapter 9 is devoted to circuit emulation services (CES) because of its complexity. For the same reason, Chapter 10 is devoted to Ethernet local management interface (ELMI).

Addressing scalability of Carrier Ethernet is being questioned, despite of the availability of S-Tag and C-Tag combination. Provider Backbone Bridges (PBBs) and Provider Backbone Transport (PBT) resolve this scalability issue. Chapter 11 describes these technologies in detail.

Three technologies, namely, T-MPLS, MPLS-TP, and VPLS can compete or work with Carrier Ethernet in forming data networks. Chapters 12, 13, and 14 describe them in detail.

---

[1]In this book, Carrier Ethernet and Metro Ethernet are used interchangeably.

Networks and Services: Carrier Ethernet, PBT, MPLS-TP, and VPLS, First Edition. Mehmet Toy.
© 2012 John Wiley & Sons, Inc. Published 2012 by John Wiley & Sons, Inc.

1.2 BASIC ETHERNET

Ethernet is physical layer LAN technology invented in 1973. Since its invention, Ethernet has become the most widely used LAN technology because of its speed, low cost, and relative ease of installation. Ethernet PHY supporting rates from 10 Mbps to 100 Gbps are available.

Chapter 2 describes CSMACD (Carrier Sense, Multiple Access, Collision Detect), protocol stack, bridges, switches, and hubs. Toward the end of the chapter, Ethernet frame formats defined by IEEE 802.3 and 802.2 are described.

1.3 SYNCHRONIZATION

The TDM (time-division multiplexing) networks transporting multiplexed voice and data traffic require highly accurate frequency information to be recoverable from their physical layers. Typically, a TDM circuit service provider (SP) will maintain a timing distribution network, providing synchronization traceable to a primary reference clock (PRC) that is compliant with clock quality. A reference timing signal of suitable quality is distributed to the network elements (NEs) processing the application. One approach is to follow a distributed PRC strategy. An alternative approach is based on a master–slave strategy.

Synchronization in Metro Ethernet Networks (MEN) is required mainly because of mobile backhaul and bandwidth explosion in mobile networks. Frequency, phase, and time synchronizations are needed. Chapter 3 describes two main synchronization techniques, Precision Time Protocol (1588v2) and Synchronous Ethernet (SyncE).

1.4 PSEUDOWIRES

Pseudowire Emulation Edge-to-Edge (PW3) is a mechanism used to emulate telecommunication services across packet-switched networks (PSNs), such as Ethernet, IP (internet protocol), or MPLS. The emulated services are mostly T1/T3 leased lines, frame relay, Ethernet, and minimize operating costs of service providers (SPs). PWs encapsulate cells, bit streams, and protocol data units (PDUs) and transport them across Carrier Ethernet Virtual Circuits (EVCs), IP, or MPLS tunnels. The transportation of encapsulated data usually require managing the sequence and timing of data units to emulate services such as T1/T3 leased lines and asynchronous transfer mode (ATM).

Chapter 4 describes various aspects of PWs, including its architecture, protocol stack, frame forwarding, control plane, resilience, and OAM.

1.5 PROTECTION

Services such as broadcast video, voice over IP, video on demand require five 9 availability. When failures occur, they are not supposed to be noticed by the subscriber. Automatic protection switching guarantees the availability of resources in the event of failure and ensures that switchover is achieved in less than 50 ms. The 50 ms is currently being debated in the industry. In fact, Metro Ethernet Forum mandates 500 msec protection switching at UNI (user network interface) or ENNI (external network–network interface). The smaller switching time increases the cost of the equipment.

Chapter 5 describes various protection techniques, including Linear Protection, Ring Protection, and Link Aggregation Group (LAG)/Link Aggregation Control Protocol (LACP).

1.6 CARRIER ETHERNET ARCHITECTURE AND SERVICES

Chapter 6 begins describing Carrier Ethernet network architecture and services that are developed by Metro Ethernet Forum and currently deployed in the industry.

Ethernet transport network is a two-layer network consisting of Ethernet MAC (ETH) layer network and Ethernet PHY (ETY) layer network. The ETY layer is the physical layer defined in IEEE 802.3. The ETH layer is the pure packet layer.

The ETH layer network is divided into ETH subnetworks that are also called ETH flow domain (EFD). An EFD is a set of all ETH flow points transferring information within a given administrative portion of the ETH layer network. EFDs may be partitioned into sets of nonoverlapping EFDs that are interconnected by ETH links. An IEEE 802.1D bridge represents the smallest instance of an EFD.

The termination of a Link is called a flow point pool (FPP). The FPP describes configuration information associated with an interface, such as an UNI or ENNI. Ethernet frame is exchanged over ETH layer that consists of preamble, start frame delimiter (SFD), destination MAC address (DA), source MAC address (SA), (Optional) 802.1QTag, Ethernet Length/Type (EtherType), user data, padding if required, frame check sequence (FCS), and extension field, which is required only for 100 Mbp half-duplex operation. A service frame can be a unicast, multicast, broadcast, and Layer 2 Control Protocol (L2CP) frame.

Two interfaces are defined for Carrier Ethernet, UNI, and ENNI. A connection between UNIs is called Ethernet Virtual Connection (EVC), while a connection between an UNI and an ENNI is called Operator Virtual Connection (OVC). An EVC or an OVC can be point-to-point, point-to-multipoint, or multipoint-to-multipoint. The following services are built on top of these EVCs and OVCs:

- Ethernet line (E-Line) services consisting of Ethernet private line (EPL) and Ethernet virtual private line (EVPL) services
- Ethernet local area network (E-LAN) services consisting of EPL and EVPL services

- Ethernet Tree (E-Tree) services consisting of Ethernet Private Tree (EP-Tree) and Ethernet Private Virtual Tree (EVP-Tree) services
- Ethernet Access services offered over ENNI

The operator responsible from the end-to-end EVC may order a tunnel between its remote user that is connected to another operator's network and its ENNI gateway. This access of the remote user to the SP's network is called User Network Interface Tunnel Access (UTA). The remote user end of the tunnel is called Remote User Network Interface (RUNI), while the SP end of this tunnel is called Virtual User Network Interface (VUNI). Attributes of these services are described in detail in Chapter 6.

Chapter 6 is followed by Chapters 7 and 8, describing traffic management and OAM capabilities, respectively.

1.7 CARRIER ETHERNET TRAFFIC MANAGEMENT

Traffic Management, which is also called packet conditioning, is queuing, scheduling, and policing frames. The conditioning function includes the following:

1. Counting bytes and packets that pass and drop
2. Policing packet flow according to predetermined burst/flow rate (includes both color aware and color unaware)
3. Setting color
4. Setting 1P/Differentiated Services Code Point (DSCP)/type of service (TOS) priority value
5. Re-marking/remapping 1P/DSCP/TOS based on a predefined re-marking table
6. Shaping packets to conformance to predetermined flow rate
7. Sending packets to a particular queue

Once frames are classified into a Class of Service (CoS) flow, ingress frames are then policed according to the CoS bandwidth profile assigned to the flow, consisting of Committed Information Rate (CIR), Excess Information Rate (EIR), Committed Burst Size (CBS), Excess Burst Size (EBS), Coupling Flag (CF), and Color Mode (CM).

Frames that are marked "Green" by the policer are always queued, and frames that are marked "Yellow" are only queued if the fill level of the queue is less than a defined threshold. This ensures that "Green" frames are always forwarded.

The MEF defines three classes as H, M, and L. Their priorities are indicated by Priority Code Point (PCP), DSCP, or TOS bytes. Each CoS label has its own performance parameters.

- Class H is intended for real-time applications with tight delay/jitter constraints such as VoIP
- Class M is intended for time critical
- Class L is intended for non-time-critical data such as e-mail

Service-level agreements (SLAs) between an SP and subscriber for a given EVC are defined in terms of delay, jitter, loss, and availability performances. SLAs are further defined for the following four performance tiers (PTs):

1. PT1 (Metro PT) for typical Metro distances (<250 km, 2 ms propagation delay),
2. PT2 (Regional PT) for typical Regional distances (<1200 km, 8 ms propagation delay),
3. PT3 (Continental PT) for typical National/Continental distances (<7000 km, 44 ms propagation delay),
4. PT4 (Global PT) for typical Global/Intercontinental distances (<27,500 km, 172 ms propagation delay).

Chapter 7 also describes SLA-application mapping.

1.8   ETHERNET OPERATIONS, ADMINISTRATIONS, AND MAINTENANCE (OAM)

Ethernet operations, administrations, and maintenance (OAM) consists of fault management, performance management, testing, and service-monitoring capabilities to support Ethernet services described in Chapter 6. Measurements; events/alarms; and provisioning attributes for interfaces (i.e., UNI, VUNI, ENNI), EVC, and OVC are defined.

Ethernet OAM can be categorized into the following:

- Link layer OAM
- Service layer OAM
- OAM via ELMI.

Link OAM provides mechanisms to monitor link operation and health, and improves fault isolation. The major features covered by this protocol are as follows:

- Discovery of MAC address of the next hop
- Remote failure indication for link failures
- Power failure reporting via Dying Gasp
- Remote loopback.

Ethernet (Alarm Indication Signal) AIS/RDI (Remote Defect Indication) is used to suppress downstream alarms and eliminate alarm storms from a single failure. This is similar to AIS/RDI in legacy services such as SONET (synchronous optical network), TDM, and frame relay.

The service OAM consists of continuity check, loopback, link trace, delay/jitter measurements, loss measurements, and in-service and out-of-service testing protocols to monitor SLAs, identify service-level issues, and debug them.

Connectivity Fault Management (CFM) creates a maintenance hierarchy by defining maintenance domain and maintenance-domain levels, where maintenance end points (MEPs) determine domain boundaries. Maintenance points (MPs) that are between these two boundary points, MEPs, are called Maintenance Intermediate Points (MIPs).

Continuity check messages (CCM) are issued periodically by MEPs. They are used for proactive OAM to detect loss of continuity (LOC) among MEPs and discovery of each other in the same domain. MIPs will discover MEPs that are in the same domain using CCMs as well. In addition, CCM can be used for loss measurements and triggering protection switching.

Ethernet loopback (ETH-LB) function is an on-demand OAM function that is used to verify connectivity of an MEP with a peer MP(s). Loopback is transmitted by an MEP on the request of the administrator to verify connectivity to a particular MP.

Ethernet link trace (ETH-LT) function is an on-demand OAM function initiated in an MEP on the request of the administrator to track the path to a destination MEP. They allow the transmitting node to discover connectivity data about the path. The PDU used for ETH-LT request information is called the link trace message (LTM), and the PDU used for ETH-LT reply information is called the link trace reply (LTR).

Performance measurements can be periodic and on-demand. They are available in 15-min bins. They can be in-service measurements to monitor health of the network as well as user SLAs, and out-of-service measurements before turning up the service, for isolating troubles and identifying failed components during failures.

Delay measurement can be performed using the 1DM (one-way delay measurement) or DMM/DMR (delay measurement message/delay measurement reply) PDUs. Loss measurement can be performed by counting service frames using the LMM/LMR (loss measurement message/loss measurement reply) PDUs as well as by counting synthetic frames via the SLM/SLR (synthetic loss message/synthetic loss reply) PDUs.

In addition to the above, Chapter 8 describes availability and testing based on RFC2544 and ITU-T Y.1731.

1.9 CIRCUIT EMULATION

The CES allows Carrier Ethernet networks to be able to support legacy equipment of users by supporting legacy interfaces such as TDM. As a result, the users benefit from Carrier Ethernet capabilities without replacing their equipment.

In the CES, data streams are converted into frames for transmission over Ethernet. At the destination site, the original bit stream is reconstructed when the headers are removed, payload is concatenated, and clock is regenerated, while ensuring very low latency. The MEN behaves as a virtual wire.

The TDM data is delivered at a constant rate over a dedicated channel. The TDM PW operates in various modes, Circuit Emulation over PSN CESoPSN, Structure-Agnostic TDM over Packet (SAToP), TDM over IP (TDMoIP), and HDLCoPSN (High-Level Data Link Control (HDLC) emulation over PSN).

In SAToP, the TDM is typically unframed. When it is framed or even channelized, the framing and channelization structure are completely disregarded by the transport mechanisms. In such cases, all structural overhead must be transparently transported along with the payload data, and the encapsulation method employed provides no mechanisms for its location or utilization. On the other hand, structure-aware TDM transport may explicitly safeguard TDM structure. The unstructured emulation mode is suitable for leased line.

Ethernet CES provides emulation of TDM services, such as N × 64 kbps, T1, T3, and OC-n, across a MEN, but transfers the data across MEN. From the customer perspective, this TDM service is the same as any other TDM service.

Circuit emulation applications interconnected over a Circuit Emulation Services over Ethernet (CESoETH) service may exchange signaling in addition to TDM data. With structure-agnostic emulation, it is not required to intercept or process CE (customer-edge) signaling. Signaling is embedded in the TDM data stream, and hence it is carried end-to-end across the emulated circuit. With structure-aware emulation, transport of common channel signaling (CCS) may be achieved by carrying the signaling channel with the emulated service, such as channel 23 for DS1.

In addition, Chapter 9 describes performance monitoring, provisioning, and fault management of CES over Ethernet.

1.10   ETHERNET LOCAL MANAGEMENT INTERFACE (ELMI)

Chapter 10 describes ELMI protocol that operates between the CE device and network element (NE) of SP.

The ELMI protocol includes the following procedures:

- Notification to the CE device of the addition of an EVC
- Notification to the CE device of the deletion of an EVC
- Notification to the CE device of the availability (active/partially active) or unavailability (inactive) state of a configured EVC
- Notification to the CE device of the availability of the RUNI
- Communication of UNI and EVC attributes to the CE device.

In order to transfer ELMI messages between the UNI-C and UNI-N, a framing or encapsulation mechanism is needed. The ELMI frame structure is based on the IEEE 802.3 untagged MAC-frame format, where the ELMI messages are encapsulated inside Ethernet frames.

At ELMI, STATUS Message is sent by the UNI-N to UNI-C in response to a STATUS ENQUIRY message to indicate the status of EVCs or for the exchange of sequence numbers. STATUS ENQUIRY is sent by the UNI-C to request status or to verify sequence numbers.

The ELMI procedures are characterized by a set of ELMI messages that will be exchanged at the UNI. These message exchanges can be asynchronous or periodic. Periodic message exchanges are governed by timers, status counters, and sequence numbers.

1.11 PBT

Addressing scalability of Carrier Ethernet is being questioned. PBBs and PBT described in Chapter 11 attempts to resolve the scalability issue. These extensions to the Ethernet protocols are developed to transform Ethernet to a technology ready for use in MANs/WANs.

Services supported in LAN/MEN such as E-LAN and E-LINE will be supported end-to-end. This results in no changes to the customer's LAN equipment, providing end-to-end usage of the technology, contributing to wider interoperability and low cost. SLAs provide end-to-end performance, based on rate, frame loss, delay, and jitter, and enable traffic engineering (TE) to fine-tune the network flows.

Underlying protocols for the PBT are as follows:

- IEEE 802.1AB link layer discovery protocol, which is used to discover the network layout and forwarding this information to the control plane or management layer
- The IEEE 802.1ag protocol to monitor the links and trunks in the PBT layout
- Expansion of the PBB protocol defined in IEEE 802.1ah
- IEEE 802.1Qay, PBBs with TE or PBT protocol.

The PBB, namely, MAC-in-MAC encapsulation, supports complete isolation of individual client-addressing fields and isolation from address fields used in the operator backbone. Client provider bridge (PB) frames are encapsulated and forwarded in the backbone network, based on new B-DA (backbone destination address), B-SA (backbone source address), and B-VID (backbone VLAN-ID).

Although Q-in-Q supports a tiered hierarchy (i.e., no tag, C-Tag/C-VLAN ID, and S-Tag/S-VLAN ID), the SP can create 4094 customer VLANs, which is insufficient for large metropolitan and regional networks. The 802.1 ah introduces a new 24-bit tag field (I-SID), service instance identifier, to overcome 12-bit S-VID (S-VLAN ID) defined in PB. This 24-bit tag field is proposed as a solution to the scalability limitations encountered with the 12-bit S-VID defined in PBs.

PBBs operate the same way as traditional Ethernet bridges. CFM addresses the end-to-end OAM, such as loopback at specific MAC, link trace, and continuity check.

The PBB located at the backbone of the PBT network is called backbone core bridge (BCB). The bridge located at the edge of PBT network is called backbone edge bridge (BEB). The BCB is an S-VLAN bridge used within the core of a PBBN. The BEB is a system that encapsulates customer frames for transmission across a Provider Backbone Bridge Network (PBBN).

The BEB is of three types: I type Backbone Edge Bridge (I-BEB), B type Backbone Edge Bridge (B-BEB), and IB type Backbone Edge Bridge (IB-BEB).

I-component is responsible for encapsulating frames received from customers, assigning each to a backbone service instance and destination identified by a backbone destination address, backbone source address, and a service instance identifier (I-SID).

B-component is responsible for relaying encapsulated customer frames to and from I-components or other B-components when multiple domains interconnect, either within the same BEB or externally connected, checking that ingress/egress is permitted for frames with that I-SID, translating the I-SID (if necessary) and using it to assign the supporting connection parameters (backbone addresses if necessary and VLAN identifiers) for the PBBN, and relaying the frame to and from the Provider Network Port(s).

The PBBN provides multipoint tunnels between provider bridged networks (PBNs), where each B-VLAN carries many S-VLANs.

Traffic engineered provider backbone bridging (PBB-TE), PBT, is intended to bring connection-oriented characteristics and deterministic behavior to Ethernet. It turns off Ethernet's spanning tree and media-access-control address-flooding and learning characteristics. That lets Ethernet behave more similar to a traditional carrier transport technology.

The frame format of PBT is exactly as the format used for implementing the PBB. The difference is in the meaning of the frame's fields. The VID and the B-DA fields together form a 60-bit globally unique identifier.

The control plane is used to manage the forwarding tables of the switches. To create PBT tunnels, all switches need to be controlled from one (PBT) domain. This technique enables circuit switching on an Ethernet

Chapter 11 further describes PBT networks and PBT-MPLS interworking.

1.12 T-MPLS AND MPLS-TP

Chapters 12, 13, and 14 describe three technologies that are competing or working with Carrier Ethernet in forming data networks, T-MPLS, MPLS-TP, and VPLS. Although MPLS-TP is supposed to replace T-MPLS, T-MPLS is already deployed. Therefore, Chapter 12 is devoted to it.

T-MPLS offers packet-based alternatives to SONET circuits and promises much greater flexibility in how packet traffic is transported through their metro and core optical networks. In T-MPLS, a new profile for MPLS is created, so that MPLS label switched paths (LSPs) and PWs can be engineered to behave similar to TDM circuits or Layer 2 virtual connections.

The T-MPLS is intended to be a separate layer network with respect to MPLS. However, the T-MPLS will use the same data-link protocol ID (e.g., EtherType), frame format, and forwarding semantics as defined for MPLS frames. Unlike MPLS, it does not support a connectionless mode and is intended to be simpler in scope, less complex in operation, and more easily managed. Layer 3 features have been eliminated and the control plane uses a minimum of IP to lead to low-cost equipment implementations.

As an MPLS subset, T-MPLS abandons the control protocol family, which the Internet Engineering Task Force (IETF) defines for MPLS. It simplifies the data plane, removes unnecessary forwarding processes, and adds ITU-T transport style protection switching and OAM functions (e.g., connectivity verification, alarm suppression, RDI).

The key differences of T-MPLS when compared with MPLS include the following:

- Use of bidirectional LSPs
- No PHP (Penultimate Hop Popping) option
- No ECMP (Equal Cost Multiple Path) option.

The T-MPLS, similar to MPLS, defines UNI interface, which is the interface between a client and service node, and NNI (Network–Network Interface), which is between two service nodes.

In a typical T-MPLS network, a primary LSP and backup LSP are provided. The switching between the primary and secondary LSP tunnels can take place within 50 ms. These T-MPLS tunnels can support both Layer 3 IP/MPLS traffic flows and Layer 2 traffic flows via PWs. The T-MPLS protection can be linear or ring.

The MPLS-TP is a continuation of T-MPLS. A Joint Working Group (JWT) was formed between the IETF and ITU-T to achieve mutual alignment of requirements and protocols and come up with another approach. The T-MPLS is renamed as MPLS-TP to produce a converged set of standards for MPLS-TP. The MPLS-TP is a packet-based transport technology based on the multiprotocol label switching traffic engineering (MPLS-TE) and PW data plane architecture.

The objective is to achieve the transport characteristics of SONET/SDH (synchronous digital hierarchy) that are connection oriented; a high level of availability; quality of service; and extensive operations, administration, and maintenance (OAM) capabilities.

With the MPLS-TP, network provisioning can be achieved via a centralized Network Management System (NMS) and/or a distributed control plane. The generalized multiprotocol label switching (GMPLS) can be used as a control plane that provides a common approach for management and control of multilayer transport networks.

Networks are typically operated from a network operation center (NOC) using an NMS that communicates with the network elements (NEs). The NMS provides FCAPS management functions (i.e., fault, configuration, accounting, performance, and security management).

For MPLS-TP, the NMS can be used for static provisioning while the GMPLS can be used for dynamic provisioning of transport paths. The control plane is mainly used to provide restoration functions for improved network survivability in the presence of failures and facilitates end-to-end path provisioning across network or operator domains.

Similar to T-MPLS, MPLS-TP uses a subset of IP/MPLS standards, where features that are not required in transport networks such as IP forwarding, PHP, and ECMP are not supported or made optional. On the other hand, MPLS-TP defines extensions to existing IP/MPLS standards and introduces established requirements from transport networks. Among the key new features are comprehensive OAM capable of fast detection, localization, troubleshooting, and end-to-end SLA verification; linear and ring protection with sub-50 ms recovery; separation of control and data plane; and fully automated operation without control plane using NMS.

Static and dynamic provisioning models are possible. The static provisioning model is the simplified version commonly known as static MPLS-TP. This version does not implement even the basic MPLS functions, such as label distribution protocol (LDP) and Resource Reservation Protocol–Traffic Engineering (RSVP-TE), since the signaling is static. It does, however, implement support for GAL (Generic Associated Channel Label) and G-ACh (Generic Associated Channel), which is used in supporting OAM functions.

An MPLS-TP label switching router (LSR) is either an MPLS-TP provider-edge (PE) router or an MPLS-TP provider (P) router for a given LSP. An MPLS-TP PE router is an MPLS-TP LSR that adapts client traffic and encapsulates it to be transported over an MPLS-TP LSP by pushing a label or using a PW. An MPLS-TP PE exists at the interface between a pair of layer networks. An MPLS-TP label edge router (LER) is an LSR that exists at the endpoints of an LSP and therefore pushes or pops the LSP label.

An MPLS-TP PE node can support UNI providing the interface between a CE and the MPLS-TP network, and NNI providing the interface between two MPLS-TP PEs in different administrative domains.

The details of the MPLS-TP architecture, OAM, and security are described in Chapter 13.

1.13 VIRTUAL PRIVATE LAN SERVICES (VPLS)

The MPLS facilitates the deployment and management of Virtual Private Networks (VPNs). The MPLS-based VPN can be classified as follows:

- Layer 3 multipoint VPNs or IP VPNs that are often referred to as Virtual Private Routed Networks (VPRNs)
- Layer 2 point-to-point VPNs, which basically consist of a collection of separate Virtual Leased Lines (VLL) or PW
- Layer 2 multipoint VPNs or VPLS.

The VPLS is a multipoint service, but unlike IP VPNs, it can transport non-IP traffic and leverages advantages of Ethernet.

Two VPLS solutions are proposed as follows:

1. VPLS using Border Gateway Protocol (BGP) that uses BGP for signaling and discovery
2. VPLS using label distribution that uses LDP signaling and basically an extension to the Martini draft.

Both approaches assume tunnel LSPs between PEs. PWs (PWE3s) are set up over tunnel LSPs (i.e., virtual connection (VC) LSPs).

In order to establish MPLS LSPs, Open Shortest Path First (OSPF-TE) and RSVP-TE can be used, where OSPF-TE will take bandwidth availability into account when calculating the shortest path, while RSVP-TE allows reservation of bandwidth.

There are two key components of VPLS, PE discovery and signaling. PE discovery can be via Provisioning Application, BGP, and RADIUS. Signaling can be targeted via LDP and BGP.

In order to offer different classes of service within a VPLS, IEEE 802.1P bits in a customer Ethernet frame with a VLAN tag is mapped to EXP bits in the PW and/or tunnel label.

VPLS is a multipoint service, therefore, the entire SP network appears as a single logical learning bridge for each VPLS. The logical ports of this SP bridge are the customer ports as well as the PWs on a virtual private local area network service edge (VE). The SP bridge learns MAC addresses, at its VEs while a learning bridge learns MAC addresses on its ports. Source MAC addresses of packets with the logical ports on which they arrive are associated in the Forwarding Information Base (FIB) to forward packets.

In LDP-based VPLS, an interface participating in a VPLS must be able to flood, forward, and filter Ethernet frames. Each PE will form remote MAC address to PW associations and associate directly attached MAC addresses to local customer facing ports. Connectivity between PEs can be via MPLS transport tunnels as well as other tunnels over PWs such as GRE, L2TP, and IPSec. The PE runs the LDP signaling protocol and/or routing protocols to set up PWs, setting up transport tunnels to other PEs and delivering traffic over PWs.

A full mesh of LDP sessions is used to establish the mesh of PWs. Once an LDP session has been formed between two PEs, all PWs between these two PEs are signaled over this session. A hierarchical topology can be used in order to minimize the size of the VPLS full mesh when there is a large number of PWs.

Hierarchical virtual private local area network service (H-VPLS) was designed to address scalability issues in VPLS. In VPLS, all PE nodes are interconnected in a full mesh to ensure that all destinations can be reached. In H-VPLS, a new type of node is introduced called the multitenant unit (MTU), which aggregates multiple CE connections into a single PE, to reduce the number of PE-to-PE connections.

In BGP approach, VPLS control plane functions mainly autodiscovery and provisioning of PWs are accomplished with a single-BGP update advertisement. In the autodiscovery, each PE discovers other PEs that are part of a given VPLS instance via BGP. When a PE joins or leaves a VPLS instance, only the affected PE's configuration changes, while other PEs automatically find out about the change and adapt.

The BGP route target community (or extended communities) is used to identify members of a VPLS. A PE announces usually via I-BGP that it belongs to a specific VPLS instance by annotating its Network Layer Reach-ability Information (NLRI) for that VPLS instance with route target RT and acts on this by accepting NLRIs from other PEs that have route target RT.

When a new PE is added by the SP, a single-BGP session is established between the new PE and a route reflector. The new PE then joins a VPLS domain when the VPLS instance is configured on that PE. Once discovery is done, each pair of PEs in a VPLS establishes PWs to each other, transmit certain characteristics of the PWs that a PE sets up for a given VPLS, and tear down the PWs when they are no longer needed. This mechanism is called signaling.

Autodiscovery and signaling functions are typically announced via I-BGP. This assumes that all the sites in a VPLS are connected to PEs in a single-autonomous system (AS). However, sites in a VPLS may connect to PEs in different ASs. In this case, I-BGP connection between PEs and PE-to-PE tunnels between ASs are established.

Hierarchical BGP VPLS is used to scale the VPLS control plane when using the BGP.

The advantages of VPLS may be summarized as follows:

- Complete customer control over their routing, where there is a clear demarcation of functionality between the SP and customer that makes troubleshooting easier
- Ability to add a new site without configuration of the SP's equipment or the customer equipment at existing sites
  - Minimize MAC address exposure, improving scaling by having one MAC address per site (i.e., one MAC per router) or per service
  - Improve customer separation by having CE router to block unnecessary broadcast or multicast traffic from customer LANs
  - MPLS core network emulates a flat LAN segment that overcomes distance limitations of Ethernet-switched networks and extends Ethernet broadcast capability across WAN
  - Point-to-multipoint connectivity connects each customer site to many customer sites
- A single CE–PE link transmits Ethernet packets to multiple remote CE routers
- Fewer connections required to get full connectivity among customer sites.

Adding, removing, or relocating a CE router requires configuring only the directly attached PE router. This results in substantial OpEx savings.

Chapter 2

Basic Ethernet

2.1 INTRODUCTION

Ethernet is a physical layer local area network (LAN) technology invented in 1973. Since its invention, Ethernet has become the most widely used LAN technology because of its speed, low cost, and relative ease of installation. This is combined with wide computer-market acceptance and the ability to support most of the network protocols.

Ethernet network system is invented by Robert Metcalfe at Xerox. Metcalfe's Ethernet was modeled after the Aloha network developed in the 1960s at the University of Hawaii. The Aloha system used two distinct frequencies in a hub/star configuration, where the hub broadcasts packets to everyone on the "outbound" channel while clients sending data to the hub on the "inbound" channel. Data received was immediately resent, allowing clients to determine whether or not their data had been received. Any machine noticing corrupted data would wait a short time and then resend the packet. Aloha was important because it used a shared medium for transmission.

Similarly, Metcalfe's system used a shared medium for transmission, however, detected collisions between simultaneously transmitted frames and included a listening process before frames were transmitted, thereby greatly reducing collisions.

The first Metcalfe system ran at 2.94 Mbps. DEC, Intel, and Xerox (DIX) issued a DIX Ethernet standard for 10 Mbps Ethernet systems by 1980. Also, the Institute of Electrical and Electronics Engineers (IEEE) started project 802 to standardize LANs. In 1985, IEEE published the portion of the standard pertaining to Ethernet based on the DIX standard—IEEE 802.3 Carrier Sense Multiple Access with Collision Detection (CSMA/CD) Access Method and Physical Layer Specifications. The IEEE standards have been adopted by the American National Standards Institute (ANSI) and by the International Organization of Standards (ISO).

In 1995, 100 Mbps Ethernet over wire or fiber-optic cable, which is called Fast Ethernet, was standardized by IEEE, IEEE 802.3u. This standard

Networks and Services: Carrier Ethernet, PBT, MPLS-TP, and VPLS, First Edition. Mehmet Toy.
© 2012 John Wiley & Sons, Inc. Published 2012 by John Wiley & Sons, Inc.

also allowed for equipment that could autonegotiate the two speeds such that Ethernet device can automatically switch from 10 to 100 Mbps, and vice versa.

2.2 CSMA/CD

Ethernet uses the CSMACD (Carrier Sense, Multiple Access, Collision Detect). The "Multiple Access" means that every station is connected to a single copper wire or a set of wires that are connected together to form a single data path. The "Carrier Sense" means that before transmitting data, a station checks the wire to see if any other station is already sending data. If the LAN appears to be idle, then the station can begin to send data.

To illustrate how collision takes place, let us consider two Ethernet stations that are 800 feet apart from each other and send data at a rate of 10 Mbps. Given light and electricity travel about 1 ft in a nanosecond, after the electric signal for the first bit has traveled about 100 ft down the wire, the same station has begun to send the second bit. If both stations begin transmitting signals at the same time, then they will be in the middle of the third bit before the signal from each reaches the other station. Therefore, their signals will "collide" nanoseconds later. When such a collision occurs, the two stations stop transmitting, "back off," and try again later after a randomly chosen delay period.

For example, in Figure 2.1, when PC1 begins sending data, the signal passes down the wire and just before it reaches PC2, PC2, which hears nothing and thinks that the LAN is idle, begins to transmit its own data. A collision occurs. The second station recognizes this immediately, but the first station will not detect it until the collision signal retraces the first path all the way back through the LAN to its starting point.

Any system based on collision detect must control the time required for the worst round trip through the LAN. For Ethernet, this round trip is limited to 50 ms. At 10 Mbps rate, this is enough time to transmit 500 bits. At 8 bits per byte, this is slightly less than 64 bytes. To make sure that the collision is recognized, Ethernet requires that a station must continue transmitting until the 50 ms period has ended. If the station has less than 64 bytes of data to send, then it must pad the data by adding zeros at the end.

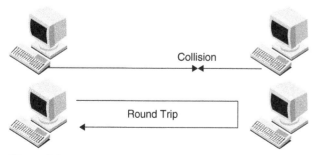

Figure 2.1 Example for collision in LAN.

Besides the constant listening, there is an enforced minimum quiet time of 9.6 ms between frame transmissions, to give other devices a chance. If a collision occurs, retransmission is determined by an algorithm that chooses unique time intervals for resending the frames. The Ethernet interface backs off or waits for the chosen time interval and then retransmits if no activity is detected. The process is repeated until the frame collides up to 16 times and then it is discarded.

The CSMA/CD protocol is designed to allow fair access by all transmission devices to shared network channels. If deferred and retransmitted traffic is less than 5% of the total network traffic, the Ethernet network is considered to be healthy.

The total distance between an Ethernet transmitter and receiver is limited because of the timing of the Ethernet signals on the cable as described above. Ethernet requires a repeater or bridge or switch to go beyond these hard limits. They are discussed in the following sections.

2.3 FULL DUPLEX, PAUSE, AUTONEGOTIATION

Reference 1 describes full-duplex Ethernet operation between a pair of stations, which is simultaneous transmit and receive over twisted-pair or fiber-optic cables that support two unidirectional paths. Ethernet devices need support from simultaneous transmit and receive functions, in addition to cabling.

Also Reference 1 introduces traffic flow control called MAC control protocol and PAUSE. If traffic gets too heavy, the control protocol can pause the flow of traffic for a brief time period.

When two network interfaces are connected to each other to autonegotiate the best possible shared mode of operation [2], the autonegotiation mechanism detects the speed but not the duplex setting of an Ethernet peer that did not use autonegotiation. An autonegotiating device defaults to half duplex, when the remote does not negotiate.

2.4 REPEATERS AND HUBS

Repeaters are used when the distance between computers is great enough to cause signal degradation. Repeater provides signal amplifications and retiming that enable the collision domain to extend beyond signal attenuation limitations. It takes the signal from one Ethernet cable and repeats it to another cable. Repeater has no memory and does not depend on any particular protocol. It duplicates everything, including the collisions. The repeaters, too, must listen and pass their signals only when the line is clear.

If a collision was detected, the repeater can transmit a jam signal onto all ports to ensure collision detection. Repeaters could be used to connect segments such that there were up to five Ethernet segments between any two hosts. Repeaters could detect an improperly terminated link from the continuous collisions and stop forwarding data from it.

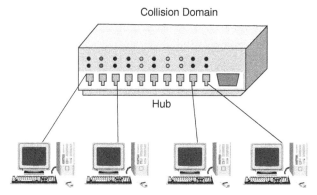

Figure 2.2 Hub configuration.

Active hubs and repeaters connect LAN segments. Active hubs (Fig. 2.2) are described as the central device in a star topology, and while it connects multiple cable segments at one point, it also repeats the incoming signals before transmitting them onto their destinations. Hubs can be designed to support Simple Network Management Protocol (SNMP) to administer and configure the hub.

2.5 BRIDGES

Bridges transfer MAC-layer packets from one network to another (Fig. 2.3). They serve as gatekeepers between networks. Bridges control traffic by checking source and destination addresses and forwarding only network-specific traffic. Bridges also check for errors and drop traffic that is corrupted, misaligned, and redundant. Bridges help prevent collisions and create separate collision domains by holding and examining entire Ethernet packets before forwarding them on. This allows the network to cover greater distances and adds more repeaters onto the network.

Most bridges can learn and store Ethernet addresses by building tables of addresses with information from traffic passing through them. This is a great advantage for users who move from place to place, but it can cause some problems when multiple bridges start network loops.

While repeaters could isolate some aspects of Ethernet segments, such as cable-related issues, they still forward all traffic to all Ethernet devices. This created practical limits on how many machines could communicate on an Ethernet network. Also, as the entire network was a one-collision domain and all hosts had to be able to detect collisions anywhere on the network. Segments joined by repeaters had to operate at the same speed, making phased-in upgrades impossible.

To alleviate these problems, bridging was created to communicate at the data link layer while isolating the physical layer. Bridges learn where devices are: by watching MAC addresses. They do not forward packets across segments when

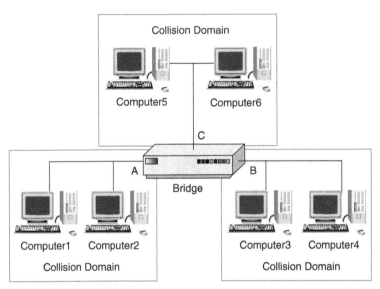

Figure 2.3   Bridge configuration.

they know that the destination address is not located in that direction. Broadcast traffic is still forwarded to all network segments.

When handling many ports at the same time, a bridge reads the entire Ethernet frame into a buffer and compares the destination address with an internal table of known MAC addresses and make a decision as to whether to drop the packet or forward it to another or all segments.

2.6 SWITCHES

Ethernet switches (Fig. 2.4) allow interconnection of bridge ports of different rates and break a network up into multiple networks consisting of a number of smaller collision domains and links. The following two types of switches are commonly used:

- *Cut through Switches.* They read the destination address on the frame header and immediately forward the frame to the switch port attached to the destination MAC address. With this, they improve performance because they only read the header information and transmit the rest of the frame without inspection.
- *Store and Forward Switches.* They hold the frame until the whole packet is inspected for proper length and CRC integrity, similar to a bridge. This type of switch makes sure that the frame is fit to travel before transmitting it. Valid frames are processed for filtering and forwarding and corrupt frames are prevented from forwarding onto an output port.

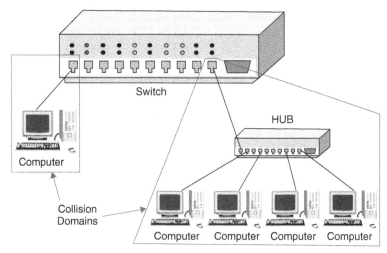

Figure 2.4 Switch configuration.

Unicast bandwidth, multicast bandwidth, and size of MAC address database are the key parameters of the switches. The QoS procedures partition bandwidth among multiple subscribers and protect unicast and multicast bandwidths. The number of MAC addresses learned from any one customer can be limited to the limit amount of traffic and hide subscribers from malicious or inadvertent attacks against this shared resource.

2.7 PHYSICAL LAYER

The Ethernet system consists of the following three basic elements:

- physical medium carrying Ethernet signals between computers,
- a set of medium access control rules embedded in each Ethernet interface that allow multiple computers to fairly arbitrate access to the shared Ethernet channel, and
- an Ethernet frame carrying data over the system.

The Ethernet physical layer evolved over time considerably. The speed ranges from 1 Mbps to 100 Gbps, while the physical medium can range from bulky coaxial cable to twisted pair to optical fiber. In general, network protocol stack software will work identically on different medium.

Many Ethernet adapters and switch ports support multiple speeds, using autonegotiation to set the speed and duplex for the best values supported by both connected devices.

The first Ethernet networks, 10BASE5, used thick yellow cable with vampire taps as a shared medium that uses CSMA/CD. Later, 10BASE2 Ethernet used thinner coaxial cable with BNC connectors as the shared CSMA/CD medium. The StarLAN 1BASE5 and 10BASE-T used twisted pairs.

Currently, 10BASE-T, 100BASE-TX, and 1000BASE-T utilize twisted-pair cables and RJ45 connectors. They run at 10 Mbps, 100 Mbps, and 1 Gbps, respectively.

A data packet on the wire is called a frame. A frame has preamble and start frame delimiter. After a frame has been sent, transmitters are required to transmit 12 octets of idle characters before transmitting the next frame. This takes 96 ns for 100 Mbps rate. The maximum net bit rate of 10 Mbps Ethernet is approximately 9.75 Mbps, assuming a continuous stream of 1500 bytes payload.

Ethernet over fibers are commonly used in structured cabling applications. Fiber has advantages in performance, electrical isolation, and distance, up to tens of kilometers with some versions.

There are two different types of optical fibers: multimode and single-mode. A multimode fiber allows many light propagation paths, while a single-mode fiber allows only one light path.

In multimode fiber, the time it takes for the light to travel through a fiber is different for each mode, resulting in a spreading of the pulse at the output of the fiber referred to as intermodal dispersion. The difference in the time delay between the modes is called differential mode delay (DMD). Intermodal dispersion limits multimode fiber bandwidth.

A single-mode fiber has higher bandwidth than multimode fiber, as it has no intermodal dispersion. This allows for higher data rates over much longer distances than achievable with multimode fiber.

Although single-mode fiber has higher bandwidth, multimode fiber supports high data rates at short distances. Relaxed tolerances on optical coupling requirements allow lower cost transceivers or lasers. As a result, multimode fiber has dominated in shorter distance and cost-sensitive LAN applications.

The bandwidth achievable on twisted-pair wiring is limited by the quality, distance, and interference associated with each individual pair. Therefore, it is important to utilize lower overhead protocols to transport data for effective utilization of every pair.

Copper access network mainly uses either TDM or ATM (asynchronous transfer mode) IMA (Inverse Multiplexing over ATM). ATM segments data into 48-byte cells, where each segment is encapsulated with a 5-byte header. When carrying Ethernet frames over the ATM, there is a partial cell fill at the end of each frame. For example, a 64-byte Ethernet frame must be transported using two 48-byte ATM cells (for a total of 96 bytes used for data), with an additional 10 bytes of overhead for the ATM cell headers. This results in an efficiency of 60% in this example (only 64 out of 106 bytes carry actual data).

IEEE 802.3ah introduced a new encapsulation scheme called 64/65-octet encapsulation, which is used on all standard 2BASE-TL and 10PASS-TS Ethernet interfaces. With 64/65-octet encapsulation, at most 5 out of every 69 octets (7%) are required for overhead. The efficiency is much higher for frame sizes greater than 64 octets.

2BASE-TL of IEEE 802.3ah also improves efficiency for multipair technologies such as IMA by introducing an aggregation multiplexing and demultiplexing layer into the Ethernet stack that is responsible for taking an Ethernet frame and partitioning it over multiple variable speed links in a manner that best utilizes the speed of each pair. For example, an implementation could partition a frame into variable size fragments, where the size of the fragments depends on the speed of the link, with the faster links carrying the larger fragments.

The multiple pairs provide protection against facility failures. The pairs in an aggregated group can be scattered across a large system (i.e., distributed bonding).

With distributed bonding, pairs can be aggregated across multiple system components instead of being restricted to a single element or line card. Mid-band Ethernet supports native Ethernet at rates from 2 Mbps to 45 Mbps over bonded copper pairs. The services are exactly the same as fiber-based Metro Ethernet and conforms to the IEEE 802.3ah standard.

Mid-band Ethernet services push the Ethernet service edge well beyond the fiber footprint.

2.7.1 10 Mbps Ethernet

For many years 10BASE2, which is also called ThinNet or Cheapernet, was the dominant 10 Mbps Ethernet standards. It is a 50 Ω coaxial cable that connects machines, where each machine uses a T-adaptor to connect to its network interface card (NIC). It requires terminators at each end.

10BASE-T runs over four wires (i.e., two twisted pairs) on a category 3 or category 5 cable. A hub or switch sits in the middle and has a port for each node. This is also the configuration used for 100BASE-T and gigabit Ethernet.

For Ethernet over fiber, 10BASE-F is the standards for 10 Mbps.

2.7.2 Fast Ethernet

100BASE-T is used for any of the three standards for 100 Mbps Ethernet over twisted-pair cable, 100BASE-TX, 100BASE-T4, and 100BASE-T2. The 100BASE-TX has dominated the market and is synonymous with 100BASE-T.

The 100BASE-TX is 100-Mbps Ethernet over category 5 cable that uses two out of four pairs. There are three types of it as follows:

- 100BASE-T4: 100-Mbps Ethernet over category 3 cable as used for 10BASE-T installations. It uses all four pairs in the cable and is limited to half duplex. It is now obsolete, as category 5 cables are common.
- 100BASE-T2: 100-Mbps Ethernet over category 3 cable uses only two pairs and supports full duplex. It is functionally equivalent to 100BASE-TX, but supports old cable. No products supporting this standard were ever manufactured.

On the other hand, 100-Mbps Ethernet over fiber is supported by 100BASE-FX.

2.7.3 Gigabit Ethernet

Gigabit Ethernet has the same IEEE 802.3 frame format, full duplex, and flow control methods as 10 Mbps and 100 Mbps Ethernet, only faster. In addition, it takes advantage of CSMA/CD when in half-duplex mode supports SNMP and takes advantage of jumbo frames to reduce the frame rate to the end host.

Gigabit Ethernet can be transmitted over CAT 5 cable and optical fiber as 1000Base-CX for short distance (i.e., up to 25 m) transport over copper, 1000BASE-T over unshielded twisted-pair copper cabling (category 5/5e), 1000Base-SX with 850 nm wavelength over fiber, and 1000Base-LX with 1300 nm wavelength that is optimized for longer distances over single-mode fiber.

Gigabit Ethernet types defined in IEEE 802.3 and their descriptions are given in Table 2.1.

2.7.4 10, 40, and 100 G Ethernet

The 10 G Ethernet family of standards encompasses media types for single-mode fiber for long haul, multimode fiber for distances up to 300 m, copper backplane for distances up to 1 m, and copper twisted pair for distances up to 100 m. It was first standardized as IEEE Std 802.3ae-2002, and later it was included in IEEE Std 802.3-2008.

10GBASE-SR is designed to support short distances over deployed multimode fiber cabling; it has a range of between 26 and 82 m depending on cable type.

10GBASE-LX4 uses wavelength division multiplexing to support ranges of between 240 and 300 m over deployed multimode cabling. It also supports 10 km over single-mode fiber.

The 10GBASE-LR and 10GBASE-ER support 10 and 40 km, respectively, over single-mode fiber.

The 10GBASE-T is designed to support copper twisted pair as specified by the IEEE Std 802.3-2008.

The 10GBASE-SW, 10GBASE-LW, and 10GBASE-EW use the WAN PHY, designed to interoperate with OC-192/STM-64 SONET/SDH equipment. They correspond at the physical layer to 10GBASE-SR, 10GBASE-LR, and 10GBASE-ER, respectively, and hence use the same types of fiber and support the same distances.

Ethernet of 10 G over fiber interfaces are listed in Table 2.2. The operation of 10 G Ethernet is similar to that of lower speed Ethernets as well. It maintains the IEEE 802.3 Ethernet frame size and format that preserves layer 3 and greater protocols. However, 10 G Ethernet only operates over point-to-point links in full-duplex mode. In addition, it uses only multimode and single-mode optical fibers for transporting Ethernet frames. Operation in full-duplex mode eliminates the need for CSMA/CD.

Table 2.1 Gigabit per Second Interface Types, Related Standards, and Their
Descriptions

Type	Description
1000BASE-T	PAM-5-coded signaling, CAT5/CAT5e/CAT6 copper cabling with four twisted pairs
1000BASE-SX	8B10B NRZ-coded signaling, multimode fiber for distances up to 550 m
1000BASE-LX	8B10B NRZ-coded signaling, multimode fiber for distance up to 550 m or single-mode fiber for distances up to 10 km.
1000BASE-LH	It is a long haul solution and supports distances up to 100 km over single-mode fiber
1000BASE-CX	8B10B NRZ-coded signaling, balanced shielded twisted pair over special copper cable, for distances up to 25 m
1000BASE-BX10	Bidirectional over single strand of single-mode fiber for distances up to 10 km
1000BASE-LX10	Over a pair of single-mode fibers for distances up to 10 km
1000BASE-PX10-D	Downstream, from head end to tail end, over single-mode fiber using point-to-multipoint topology for distances of at least 10 km
1000BASE-PX10-U	Upstream, from a tail end to the head end, over single-mode fiber using point-to-multipoint topology for distance of at least 10 km
1000BASE-PX20-D	Downstream, from head end to tail ends, over single-mode fiber using point-to-multipoint topology for distances of at least 20 km
1000BASE-PX20-U	Upstream, from a tail end to the head end, over single-mode fiber using point-to-multipoint topology for distance of at least 20 km
1000BASE-ZX	Up to 100 km over single-mode fiber
1000BASE-KX	One meter over backplane

The growth of bandwidth intense applications such as video on demand, high performance computing, and virtualization drives 40G and 100G Ethernet [3–5] usage in long-haul networks as well as in regional and metro networks. 40G has been developed as an intermediate solution before deploying 100G that can reduce the number of optical wavelengths required. 1 Terabit Ethernet implementation is also being considered.

Tables 2.3 and 2.4 list 40 Gbps and 100 Gbps interfaces defined by IEEE 802.3.

2.7.5 LAN PHY

The IEEE 802.3ae defines two broad physical layer network applications, LAN PHY and wide area network (WAN) PHY.

Table 2.2   10GBASE-X Fiber Interfaces

Interface	PHY	Optics	Description
10GBASE-SR	LAN	850 nm serial	Designed to support short distances over deployed multimode fiber cabling and has a range between 26 and 82 m depending on cable type. It also supports 300 m operation over a new 2000 MHz km multimode fiber.
10GBASE-LR	LAN	1310 nm serial	Supports 10 km over single-mode fiber
10GBASE-ER	LAN	1550 nm serial	Supports 40 km over single-mode fiber
10GBASE-LX4	LAN	4 nm × 1310 nm CWDM	Uses wavelength division multiplexing to support ranges between 240 and 300 m over deployed multimode cabling. Also supports 10 km over single-mode fiber
10GBASE-SW	WAN	850 nm serial	A variation of 10GBASE-SR using the WAN PHY, designed to interoperate with OC-192/STM-64 SONET/SDH equipment
10GBASE-LW	WAN	1310 nm serial	A variation of 10GBASE-LR using the WAN PHY, designed to interoperate with OC-192/STM-64 SONET/SDH equipment
10GBASE-EW	WAN	1550 nm serial	A variation of 10GBASE-ER using the WAN PHY, designed to interoperate with OC-192/STM-64 SONET/SDH equipment

Table 2.3   40GBASE-X Interfaces

Type	Description
40GBASE-SR4	100 m operation over multimode fiber
40GBASE-LR4	10 km operation over single-mode fiber
40GBASE-CR4	10 m operation copper cable assembly
40GBASE-KR4	1 m operation over backplane

Table 2.4   100GBASE-X Interfaces

Name	Description
100GBASE-SR10	100 m operation over multimode fiber
100GBASE-LR4	10 km operation over single-mode fiber
100GBASE-ER4	40 km operation over single-mode fiber
100GBASE-CR10	10 m operation copper cable assembly

The LAN PHY operates at close to the 10 G Ethernet rate to maximize throughput over short distances. Two versions of LAN PHY are standardized, serial 10GBASE-R and 4-channel course wave division multiplexing (CWDM) 10GBASE-X.

The 10GBASE-R uses a 64B/66B encoding system that raises the 10 G Ethernet line rate from a nonencoded 9.58–10.313 Gbps. However, the 10GBASE-X still uses 8B/10B encoding because all of the 2.5 Gbps CWDM channels it employs are parallel and run at 3.125 Gbps after encoding.

The MAC to PHY data rate for both LAN PHY versions is 10 Gbps. Encoding is used to reduce clocking and data errors caused by long runs of ones and zeros.

The WAN PHY supports connections to circuit-switched SONET networks. Besides the sublayers added to the LAN PHY, the WAN PHY adds another element called the WAN interface sublayer (WIS). The WIS takes data payload and puts it into a 9.58464 Gbps frame that can be transported at a rate of 9.95328 Gbps. The WIS does not support every SONET feature, but it carries out enough overhead functions, including timing and framing, to make the Ethernet frames recognizable and manageable by the SONET equipment they pass through.

2.7.6 LAN PHY/WAN PHY Sublayers

The physical medium is defined by physical layer, while media access layer is defined by ISO data link layer that is divided into two IEEE 802 sublayers, the media access control (MAC) sublayer and the MAC-client sublayer. The IEEE 802.3 physical layer corresponds to the ISO physical layer.

The MAC sublayer has the following two primary responsibilities:

- Data encapsulation, including frame assembly before transmission, and frame parsing/error detection during and after reception, and
- Media access control, including initiation of frame transmission and recovery from transmission failure.

The MAC-client sublayer is the logical link control (LLC) layer defined by IEEE 802.2, which provides the interface between the Ethernet MAC and the upper layers in the protocol stack of the end station. The LLC protocol operates on top of the MAC protocol.

Bridge entity defined by IEEE 802.1 provides LAN-to-LAN interfaces between LANs that use the same protocol (e.g., Ethernet to Ethernet) and also between different protocols (e.g., Ethernet to token ring).

The protocol headers comprises two parts, an LLC protocol header and a Subnetwork Access Protocol (SNAP) header.

The LLC protocol is based on the HDLC link protocol and uses an extended 2-byte address. The first address byte indicates a destination service access point (DSAP) and the second address a source service access point (SSAP). These identify the network protocol entities that use the link layer service.

A control field is also provided, which may support a number of HDLC modes. These include Type 1 (connectionless link protocol), Type 2 (connection-oriented protocol), and Type 3 (connectionless acknowledged protocol).

The SNAP header is used when the LLC protocol carries IP packets and contains the information that would otherwise have been carried in the 2-byte MAC frame type field.

The physical (PHY) layer includes the following sublayers:

- Physical coding sublayer (PCS) that encodes and decodes the data stream between the MAC and PHY layer. There are three categories for the PCS:
  - 10GBASE-R that is serially encoded (64B/66B) (LAN PHY);
  - 10GBASE-X that is serially encoded (8B/10B) and used for wavelength division multiplexing (WDM) transmissions (LAN PHY);
  - 10GBASE-W that is serially encoded (64B/66B) and compatible with SONET standards for a 10 Gbps WAN (WAN PHY);
- Physical medium attachment (PMA), which is an optional interface for connection to optical modules. There are two PMA interfaces, 10 G Ethernet attachment unit interface (XAUI) and 10 G Ethernet 16-bit interface (XSBI). The XAUI is an interface to specialized 10 G Ethernet optical modules and system backplanes. The XSBI is a serial optics interface for LAN and WAN PHY applications. Intermediate and long reach optical modules use this interface. It requires more power and more pins than an XAUI.
- Physical medium-dependent (PMD) sublayer supports distance objectives of the physical medium-dependent sublayer. Four different PMDs are defined to support single- and multimode optical fibers:
  - 850 nm serial—MMF, up to 65 m
  - 1310 nm serial—SMF, up to 10 km
  - 1550 nm serial—SMF, up to 40 km
  - 1310 nm CWDM—MMF, up to 300 m
  - 1310 nm CWDM—SMF, 10 km.

2.8 TEMPERATURE HARDENING

Because of the inherent reach limitations of copper, when compared to fiber, equipment must be deployable in controlled environment such as in the central office (CO) and remote terminals (RTs), and nonenvironmentally controlled basements and wiring terminals.

In harsh environmental conditions, equipment must be environmentally hardened to withstand a diverse range of temperatures from -40 to $85°C$ and resist various environmental elements such as dust, rain, snow, and sleet. For outside plant deployments, equipment must not only be hardened but also fit the physical characteristics of the enclosure. In particular, it must adhere to the depth limitations of an RT and have only frontal access to all connectors so that it can fit flush against the wall.

To fully leverage Ethernet, copper access equipment needs to support deployments, where the equipment can be located at a considerable distance from the next switch in the network. Network modularity along with long-distance optics, short-distance optics, or CAT5 cabling supports the most flexible deployments.

2.9   STANDARDS

The early development of Ethernet was done by Xerox. Later, it was refined and its second generation is called Ethernet II. It is also called DIX after its corporate sponsors Digital, Intel, and Xerox [6]. As the holder of the trademark, Xerox established and published the standards.

The IEEE formed the 802 committee to look at Ethernet, token ring, fiber optic, and other LAN technology and published the following:

- IEEE 802.3—hardware standards for Ethernet cards and cables,
- IEEE 802.2—hardware standards for token ring cards and cables, and
- IEEE 802.5—new message format for data on any LAN.

The 802.3 standard [7] further refined the electrical connection to the Ethernet. It was immediately adopted by all the hardware vendors.

Standard Ethernet frame sizes are between 64 and 1518 bytes. Jumbo frames are between 64 and 9215 bytes. Because larger frames translate to lower frame rates, using jumbo frames on Gigabit Ethernet links greatly reduces the number of packets that are received and processed by the end host.

Some of the Ethernet-related IEEE standards are given in Table 2.5.

2.10   ETHERNET FRAME TYPES AND THE ETHERTYPE FIELD

There are several types of Ethernet frames as follows:

- The Ethernet Version 2 or Ethernet II frame, which is also called DIX frame (named after DEC, Intel, and Xerox)
- Novell's nonstandard variation of IEEE 802.3 ("raw 802.3 frame") without an IEEE 802.2 LLC header
- IEEE 802.2 LLC frame
- IEEE 802.2 LLC/SNAP frame.

All four types of Ethernet frames may optionally contain a IEEE 802.1Q tag to identify the VLAN it belongs to and its IEEE 802.1p priority.

2.10.1   Ethernet Frames

A block of data transmitted on the Ethernet is called a frame. An Ethernet frame has a header (Preamble—Length), payload (LLC—Pad), and a trailer (Frame Check Sequence).

Table 2.5   List of IEEE Standards for Ethernet

Name	Description
IEEE 802.3a	10Base-2 (thin Ethernet)
IEEE 802.3c	10 Mbps repeater specifications (clause 9)
IEEE 802.3d	FOIRL (fiber link)
IEEE 802.3i	10Base-T (twisted pair)
IEEE 802.3j	10Base-F (fiber optic)
IEEE 802.3u	100Base-T (Fast Ethernet and autonegotiations)
IEEE 802.3x	Full duplex
IEEE 802.3z	1000Base-X (gigabit Ethernet)
802.3ab	1000Base-T (gigabit Ethernet over twisted pair)
IEEE 802.3ac	VLAN tag (frame size extension to 1522 bytes)
IEEE 802.3ad	Parallel links (link aggregation)
IEEE 802.3ae	10 G Ethernet
IEEE 802.3ah	Ethernet in the first mile
802.3as	Frame expansion
IEEE 802.3at	Power over Ethernet Plus
802.1aj-TPMAR	Architecture
IEEE 802.1ad	QinQ
IEEE 802.3ad	Link aggregation/protection
IEEE 802.1 w	Testing
802.1ag	Connectivity fault management
IEEE 1588v2	Synchronization

The first 12 bytes of every frame contains the 6-byte destination address and a 6-byte source address. Each Ethernet adapter card comes with a unique factory installed address. Use of this hardware address guarantees a unique identity to each card.

In normal operation, an Ethernet adapter will receive only frames with a destination address that matches its unique address, or destination addresses that represent a multicast message. In "promiscuous" mode, they receive all frames that appear on the LAN.

2.10.2   Ethernet Address

Each Ethernet NIC has a unique identifier called an MAC address. Each MAC address assigned by the IEEE Registration Authority for the card manufacturers is a 48-bit number, of which the first 24 bits identify the manufacturer, manufacturer ID, or organizational unique identifier (OUI). The second half of the address (extension of board ID) is assigned by the manufacturer. The number is usually programmed into the hardware so that it cannot be changed.

At the physical layer, the Destination field is preceded by a 7-byte preamble and a 1-byte start of frame delimiter. At the end of the data field is a 4-byte checksum.

There are the following three common frame formats that are described in the following sections:

1. Ethernet II or DIX
2. IEEE 802.3 and 802.2
3. SNAP.

2.10.3 Ethernet II or DIX

The most common Ethernet frame format, type II Versions 1.0 and 2.0 of the DIX Ethernet specification have a 16-bit subprotocol label field called the Ether-Type. IEEE 802.3 replaced that with a 16-bit length field, with the MAC header followed by an IEEE 802.2 LLC header. The maximum length of a frame was 1518 bytes for untagged IEEE802.3 frames and 1522 for 802.1Q tagged. The two formats were eventually unified by the convention that values of that field between 64 and 1522 indicated the use of the new 802.3 Ethernet format with a length field and that values of 1536 decimal (0600 hexadecimal) and greater indicated the use of the original DIX or Ethernet II frame format with an EtherType subprotocol identifier. This convention allows software to determine whether a frame is an Ethernet II frame or an IEEE 802.3 frame, allowing the coexistence of both standards on the same physical medium.

A 2-byte Type code (Fig. 2.5) was assigned by Xerox to identify each vendor protocol. To allow collision detect, the 10 Mb Ethernet requires a minimum packet size of 64 bytes. Any shorter message must be padded with zeros. Since short Ethernet frames must be padded with zeros to a length of 64 bytes, each of these higher level protocols required either a larger minimum message size or an internal length field that can be used to distinguish data from padding.

The preamble is a 64-bit (8-byte) field that contains a synchronization pattern consisting of alternating ones and zeros and ending with two consecutive ones. After synchronization is established, the preamble is used to locate the first bit of the packet. The preamble is generated by the LAN interface card.

Pre-Amble (7 bytes)	SFD (1 byte)	Destination Address (DA) (6 bytes)	Source Address (SA) (6 bytes)	Type (2 bytes)	Data (46–1500 bytes)	CRC (4 bytes)

SFD - Start Frame Delimiter (10101011)
Ethernet II Frame (64–1518 bytes)

Type for Decnet 0×60 0×03
Type for Novell IPX 0×81 0×37
Type for TCP/IP 0×80 0×00
Type for XNS of Xerox 0×06 0×00

Figure 2.5 Frame format of Ethernet II.

In a header, Preamble sets bit timing and signals that a frame is being sent (10 Mbps Ethernet). For 100 and 1000 Mbps Ethernet, systems signal constantly and do not need preamble or start frame delimiter fields.

2.10.4 IEEE 802.3

The 802.3 format is depicted in Figure 2.6.

To allow collision detect, the 10 Mb Ethernet requires a minimum packet size of 64 bytes. Any shorter message must be padded with zeros. In order for the Ethernet to be interchangeable with other types of LANs, it would have to provide a length field to distinguish significant data from padding.

After a frame has been sent, transmitters are required to transmit 12 octets of idle characters before transmitting the next frame. For 10M, 100M, and 1000M, this takes 9600, 960, and 96 ns, respectively.

The DIX standard did not need a length field because the vendor protocols that used it (XNS, DECNET, IPX, IP) all had their own length fields. The 802.3 standard replaced the 2-byte type field with a 2-byte length field.

Xerox had not assigned any important types to have a decimal value below 1500. Since the maximum size of a packet on Ethernet is 1500 bytes, there was no conflict or overlap between DIX and 802 standards. Any Ethernet packet with a type/length field less than 1500 is in 802.3 format with a length, while any packet in which the field value is greater than 1500 must be in DIX format with a type.

2.10.5 IEEE 802.2

The 802.2 variants of Ethernet are not in widespread use on common networks, currently. 802.2 Ethernet is used to support transparent translating bridges between Ethernet and IEEE 802.5 token ring or FDDI networks. The most common framing type used today is Ethernet Version 2, with its EtherType set to 0x0800 for IPv4 and 0x86DD for IPv6.

IP traffic can be encapsulated in IEEE 802.2 LLC frames with SNAP. IP Version 6 can also be transmitted over Ethernet using IEEE 802.2 with LLC/SNAP (Fig. 2.7).

The IEEE 802.1Q tag, if present, is placed between the source address and the EtherType or length fields. The first 2 bytes of the tag are the Tag Protocol

Pre-Amble (7 bytes)	SFD (1 byte)	Destination Address (DA) (6 bytes)	Source Address (SA) (6 bytes)	Length (2 bytes)	Data (46–1500 bytes)	CRC (4 bytes)

SFD - Start Frame Delimiter (10101011)

Figure 2.6 Frame format of IEEE 802.3.

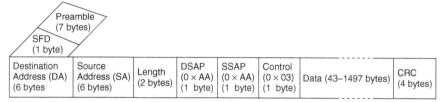

DSAP - Destination Service Access Point
SSAP - Source Service Access Point
SFD - Start Frame Delimiter (10101011)

Figure 2.7   Frame format of IEEE 802.2.

Identifier (TPID) value of 0x8100. This is located in the same place as the Ether-Type/length field in untagged frames; therefore, an EtherType value of 0x8100 means that the frame is tagged. The true EtherType/length is located after the Q-tag. The TPID is followed by 2 bytes containing the tag control information (TCI), which is the IEEE 802.1p priority and VLAN ID. The Q-tag is followed by the rest of the frame, using one of the types described above.

The 802.2 header follows the 802.3 header and substitutes Type with a new field. The 802.2 header is 3 bytes long for control packets or the kind of connectionless data sent by all the old DIX protocols. A 4-byte header is defined for connection-oriented data, which refers primarily to SNA and NETBEUI. The first 2 bytes identify the SAP. The two SAP fields are set to 0x0404 for SNA and 0xF0F0 for NETBEUI.

2.10.6 SNAP

Some protocols operate directly on top of 802.2 LLC, which provides both datagram and connection-oriented network services. The LLC header includes the additional eight-bit address fields (Fig. 2.8), called service access points (SAPs) when both source and destination SAP are set to the value 0xAA, the SNAP service is requested. The SNAP header allows EtherType values to be used with all IEEE 802 protocols and supporting private protocol ID spaces (Fig. 2.8). In IEEE 802.3x-1997, the IEEE Ethernet standard was changed to explicitly allow the use of the 16-bit field after the MAC addresses.

Under SNAP, the 802.2 header appears to be a datagram message (control field 0x03) between SAP ID 0xAA. The first 5 bytes of what 802.2 considers data are actually a subheader ending in the 2-byte DIX type value. Any of the old DIX protocols can convert their existing logic to legal 802 SNAP by simply moving the DIX type field back 8 bytes from its original location.

Ethernet II uses one bit to indicate multicast addresses, while 802.3 uses two bits. On 802.3, the first bit is similar to the multicast bit in that it indicates whether the address is for an individual or for a group, and the second bit indicates whether the address is locally or universally assigned. The second bit is rarely used on Ethernet (CSMA/CD) networks.

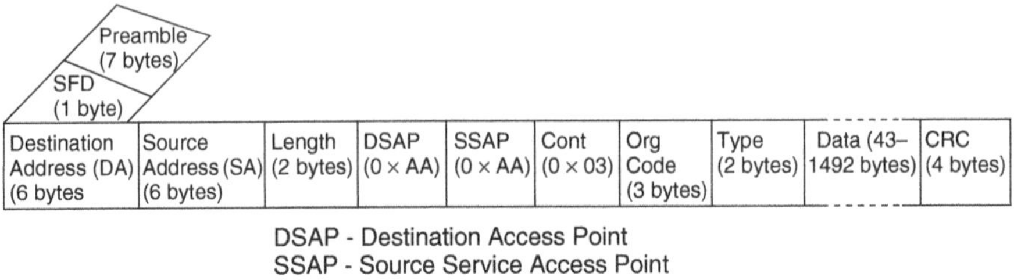

Destination Address (DA) (6 bytes	Source Address (SA) (6 bytes)	Length (2 bytes)	DSAP (0 × AA)	SSAP (0 × AA)	Cont (0 × 03)	Org Code (3 bytes)	Type (2 bytes)	Data (43– 1492 bytes)	CRC (4 bytes)

DSAP - Destination Access Point
SSAP - Source Service Access Point

Figure 2.8   Frame format of SNAP.

Adding IEEE 802.2 LLC information to an 802.3 physical packet format requires three additional fields at the beginning of the data field: a 1-byte DSAP field, a 1-byte SSAP field, and a 1-byte control field.

2.11 CONCLUSION

Ethernet is physical layer LAN technology invented in 1973. Since its invention, Ethernet has become the most widely used LAN technology. At present, 1 and 10 Gbps LAN interfaces are commonly used. The demand for 40 and 100 Gbps Ethernet equipment is expected to grow in the near future.

REFERENCES

1. IEEE 802.3x. 1997. IEEE Standards for Local and Metropolitan Area Networks: Specification for 802.3 Full Duplex Operation.
2. IEEE standard 802.3u. 1995. IEEE Local and Metropolitan Area Networks-Supplement - Media Access Control (MAC) Parameters, Physical Layer, Medium Attachment Units and Repeater for 100Mb/s Operation, Type 100BASE-T.
3. Duelk M. "Consideration for 40 gigabit Ethernet", May 2007. http://grouper.ieee.org/groups/802/3/hssg/public/may07/duelk_01_0507.pdf.
4. "Solving 100Gb/s Transmission Challenges", Ciena, 2010.
5. Metcalf B. "Bob Metcalfe on the Terabit Ethernet", Feb. 2008. http://www.lightreading.com/video.asp?doc_id=146223.
6. Digital Equipment Corporation, Intel Corporation, Xerox Corporation. 1980. The Ethernet: A Local Area Network. Available at http://portal.acm.org/citation.cfm?id=1015591.1015594. —Version 1.0 of the DIX specification.
7. IEEE 802.3-2008/Cor 1–2009 - IEEE Standard for Information technology–Local and metropolitan area networks–Specific requirements–Part 3: Carrier Sense Multiple Access with Collision Detection (CSMA/CD) Access Method and Physical Layer Specifications Corrigendum 1: Timing Considerations for PAUSE Operation.

Chapter 3

Synchronization

3.1 INTRODUCTION

Depending on the application, timing, phase, and/or frequency, synchronization is necessary among systems and networks. If one system is ahead of the others, the others are behind that particular one. For a user, switching between these systems would cause time to jump back and forward. This would be very confusing and results in various errors.

Imagine two isolated networks running their own wrong clocks connect to the *Internet* where some email messages arrive 5 min before they were sent and replies arrive 2 min before the messages were sent. We all will be very confused.

On the other hand, time division multiplexed (TDM) networks optimized for constant rate signals that transport multiplexed voice and data traffic require highly accurate frequency information to be recoverable from their physical layers. Isolated physical clocks, such as piezoelectric crystals, cannot be expected to agree with each other over time. Various effects, such as temperature changes and aging, cause even reliable frequency sources to wander. Frequency discrepancies lead to valid bits being lost and invalid bits inserted, which is called *bit slips*. In order to avoid bit slips, somewhere in every TDM network, there is an extremely accurate primary reference clock (PRC) from which all other TDM clocks in the network directly or indirectly derive their timing. Clocks derived in this manner are said to be traceable to a PRC.

The heart of a synchronous network is a stable and reliable clock source. The PRC is commonly used to describe a very accurate clock. Global positioning system (GPS) receivers are typically used to drive the reference clock where GPS is a satellite-based sourced clocking system. Fixed, earth-mounted reference station signals are added in the calculations of the GPS reception to increase the accuracy/clock quality in differential global positioning system (DGPS) systems.

Some countries such as Germany, Japan, and France also maintain a radio-based distribution with atomic clock sources with 2 MW power and 3000 km radius reach, providing high quality clocking and low variances.

The primary clock signal has to be distributed "downwards" to the edges of the network to sync up all necessary devices. Special-purpose T1 or E1 circuits

Networks and Services: Carrier Ethernet, PBT, MPLS-TP, and VPLS, First Edition. Mehmet Toy.
© 2012 John Wiley & Sons, Inc. Published 2012 by John Wiley & Sons, Inc.

are typically used to deliver clocking to the relevant boxes. At the tail end, a cellular base station controller could be the main device enabling cell sites to align with other cell sites within a network, allowing users to tax the network by roaming at speeds of up to 200 km/h (130 miles/h). High speed roaming and the handoff of calls between cell towers truly tax the infrastructure and require the utmost in-clock synchronization. One major requirement for synchronous networks is their ability to keep clocks in sync once they lose their connection from the master source(s), called *holdover time*. Holdover time depends solely on the accuracy of the built-in oscillator. The longer the devices can run autonomously, the more expensive those components become. A holdover time of 24 h is essential, while more than a week is commercially infeasible for deployments within every leaf node in the network.

Ethernet is the low cost technology of the choice for enterprise and residential networks, and expected to be used in WAN as well, with the introduction of Carrier Ethernet capabilities. Migration from TDM networks to Ethernet-based packet technology introduces new challenges. Packet switching was originally introduced to handle asynchronous data. However, applications such as the transport of TDM service and the distribution of synchronization over packet networks require the strict synchronization requirements of those applications. An acceptable level of quality, such as limited slip rate, must be maintained.

Synchronization in TDM networks is well understood and implemented. Typically, a TDM circuit service provider will maintain a timing distribution network, providing synchronization traceable to a PRC that is compliant with clock quality [1].

In some private network applications involving circuit emulation, it may be sufficient to distribute a non-PRC quality level (QL) common clock toward Circuit Emulation Services (CES) Interworking Function (IWF). However, the use of synchronization timing below PRC QL could result in internetworking difficulties between different network domains.

In order to achieve synchronization, network operators distribute a reference timing signal of suitable quality to the network elements processing the application. One approach is to follow a distributed PRC strategy. An alternative approach is based on a master–slave strategy.

Different performance can be requested in case the packet network is part of an access network or is the underlying layer of the core network. The distribution of a synchronization reference over a portion of a core network may be requested to comply with strict jitter and wander requirements. On the other hand, in the access network, requirements may be relaxed to allow a distribution of a timing reference signal with a performance level lower than PRC quality to support the timing requirements of the end node.

The nodes involved in packet-oriented transmission technology, such as ATM nodes, do not require any synchronization for the implementation of the packet switching function. In fact, at any entrance point of a packet switch, an individual device provides packet timing adaptation of the incoming signal to the internal timing. For example, in ATM networks, idle cell stuffing is used to deal with frequency differences. Transmission links do not need to synchronize with each other.

However, synchronization functions in packet networks, especially on the boundary of the packet networks, are dependent on the services carried over the network. As the packet network supports TDM-based applications, it needs to provide correct timing at the traffic interfaces. For TDM-based services, the interworking function (IWF) may require network-synchronous operation in order to provide acceptable performance.

The transport of TDM signals through packet networks requires that the signals at the output of the packet network comply with TDM timing requirements in order for TDM equipment to interwork. This is independent of the application whether it is voice or data.

The adaptation of TDM signals into the packet network is called *circuit emulation services* (*CES*) [38–40]. Jitter and wander limits at traffic and/or synchronization interfaces, long-term frequency accuracy that can influence the slip performance, and total delay that is critical for real-time services such as the voice service must be satisfied.

The PDH timing requirements for traffic interfaces are mainly related to jitter, wander, and slip performance where timing information is only present at the physical layer. These values are specified in Reference 5 for the network based on 2048 kbps hierarchy and in Reference 14 for the network based on 1544 kbps hierarchy. In addition, Reference 21 specifies the applicable slip rate objectives.

Time distribution relates to the transfer of time rather than frequency. Frequency is a relative measure but is generally assumed to be measured relative to a frequency standard. Time differs from frequency in that it represents an absolute, monotonically increasing value that can be generally traced to the rotation of the earth (e.g., year, day, hour, minute, second). Mechanisms to distribution time are significantly different than what is used to distribute frequency.

Time stamps may be used in some network applications to support generation of frequency. The notion of time carried by these time stamps compared with the time generated by the local oscillator can be used to recover a frequency reference for the local oscillator. Differential methods can also be used to recover timing from packets. In this case, the time stamp needs only to be a relative and can be used as an estimate of phase. As phase and frequency are related, it is possible to use this relative information to re-create a frequency reference. This is known as *differential timing* [10]. As an example, synchronous residual time stamp (SRTS) [11] is a well-known method standardized for use in ATM AAL1 that allows relative phase to be signaled as a time stamp, to be sent across a packet network to be used to re-create the frequency of a PDH signal.

3.2 APPLICATION REQUIREMENTS

While accurate clocking is necessary for running services such as cell towers with mobile applications, each network operator has its own clocking mechanism with varying levels of accuracy. Accuracy is highly dependent on the resources available to the operator.

An American National Standards Institute (ANSI) standard entitled "Synchronization Interface Standards for Digital Networks" (ANSI/T1.101) [40] defines the stratum levels and minimum performance requirements for digital network synchronization. The requirements for the stratum levels are shown in Table 3.1, which provides a comparison and summary of the drift and slip rates for the strata clock systems. Stratum 0 is defined as a completely autonomous source of timing, which has no other input, other than perhaps a yearly calibration. The usual source of stratum 0 timing is an atomic standard (Caesium Beam or Hydrogen Maser). The minimum adjustable range and maximum drift is defined as a fractional frequency (f/f) offset of 1×10^{-15} or less.

Clock QLs are defined by the industry standards organizations to maintain clock quality in the network for time-sensitive services that need synchronization, to avoid overflow or underflow of slip buffers, bit errors, and other adverse effects. Reference 5 provides criteria for controlled slip rate.

Synchronization in Metro Ethernet Networks is required mainly owing to Mobile Backhaul and bandwidth explosion in Mobile networks, and replacement of synchronous physical layer with Ethernet. The need for transport and recovery of clocking information has two main aspects:

Table 3.1 Accuracy Requirements for Stratum Clocking

Stratum	Accuracy/ Adjust Range	Pull-In Range	Stability	Time to First Frame Slip
1	1×10^{-11}	N/A	N/A	72 d
2	1.6×10^{-8}	Must be capable of synchronizing to clock with accuracy of $\pm 1.6 \times 10^{-8}$	1×10^{-10}/d	7 d
3E	1.0×10^{-6}	Must be capable of synchronizing to clock with accuracy of $\pm 4.6 \times 10^{-6}$	1×10^{-8}/d	3.5 h
3	4.6×10^{-6}	Must be capable of synchronizing to clock with accuracy of $\pm 4.6 \times 10^{-6}$	3.7×10^{-7}/d	6 min (255 in 24 h)
4E	32×10^{-6}	Must be capable of synchronizing to clock with accuracy of $\pm 32 \times 10^{-6}$	Same as accuracy	Not yet specified
4	32×10^{-6}	Must be capable of synchronizing to clock with accuracy of $\pm 32 \times 10^{-6}$	Same as accuracy	N/A

- Time-of-day (ToD) for correct logging, access control, time-stamping, and so on.
- Clock phase, where high precision is needed to sync up mobile cells or to drive TDM circuits.

Three types of synchronization are needed: frequency, phase, and time. Tables 3.2 and 3.3 list some of the applications and required synchronization types.

Reference 14 and Reference 37 set limits on the magnitude of jitter and wander at network interfaces. The wander may not exceed given values anywhere in the network. Thus, a circuit emulation link, for example, may consume only part of the wander budget.

- GSM, WCDMA, and CDMA2000 require frequency accuracy of 0.05 ppm at air interface.
- CDMA2000 requires time synchronization at ± 3 μs level (± 10 μs worst case).
- WCDMA time division duplex (TDD) mode requires 2.5 μs time accuracy between neighboring base stations (i.e., ± 1.25 μs of coordinated universal time, UTC).

These requirements are too difficult to achieve without good transparent clocks or boundary clocks (BCs) in each intermediate node. Some cellular operators do have control over the transport network so they could use IEEE1588 compliant switches for achieving time synchronization.

In order to emulate TDM service, two things have to interwork:

- The emulated service has to transport the phase information or has to follow a certain clock discipline.
- The payload has to be transparently carried through the network. The lowest layer must set the line speed and structure of the circuit. A selected number of timeslots are then used for signaling, payload, and clocking information.

Typical speeds for TDM circuit emulations start at 64 kbps ISDN-B channels and can go up to OC-3/STM-1 levels running at 155 Mbps. In order to recover potential jitter that might occur during transmission, it is common to use buffering to compensate. Yet, buffering will always add latency to the overall service performance. Values over 100 ms are quite long and might interfere with VoIP services. As the payload type may be unknown, a fully transparent service must forward every bit.

Payload-aware emulations are far more effective on bandwidth utilization, but they are also more costly on component complexity and software management. In order to recover clock information, three typical methods are used:

- Internal Clocking by using its own onboard oscillator where the device clocks a line directly from its own independent source.

Table 3.2   Synchronization Applications

Synchronization Type	Application	Required or Targeted Quality
Frequency	TDM support (CES)	PRC traceability
	Third-Generation Partnership Project (3GPP2) base stations (including long-term evolution, LTE)	Frequency assignment shall be better than ±0.05 ppm for frequency division duplex (FDD), and micro- to femtocells ±0.25 ppm
	IEEE 802.16 (WiMAX)	Unsynchronized orthogonal frequency division multiple access (OFDMA): frequency accuracy shall be better than ±2 ppm
	Digital Video Broadcasting Terrestrial/Handheld (DVB-T/H)	Frequency accuracy depends on radio frequency, down to a few ppb
Time Phase (relative time) Time-of-day ("wall-clock," absolute time)	3GPP2 Code Division Multiple Access (CDMA) 3GPP Universal Multiple Telecommunications Service (UMTS) TDD	Time alignment error should be <3 μs for phased and shall be less than <10 μs for ToD Intercell synchronization accuracy must be better than ±2.5 μs between base stations (or <±1.25 μs from common source)
	DVB-T/H single-frequency network (SFN)	All transmitters within a SFN must broadcast an identical signal to within 1 μs accuracy
	3GPP LTE multi-MEDI Broadcast over a single-frequency network (MBSFN)	Cell synchronization accuracy should be better than or equal to 3 μs for SFN support
	802.16D/e TDD	Requirements depends on mode, modulation, application, implementation, and option used; it likely would have to be better than 5 μs be better than 5 microseconds; ≤±1/16× cyclic prefix
	SLA monitoring and correlation of logs	Y.1731 [7] Delay and Jitter measurements. The short-term goal is to improve precision to <1 ms. The target is a few orders of magnitude below average delay (i.e., ~10–100 μs). For correlation, the finer the time-stamping, the faster the correlation.

Table 3.3 Time-Phase Requirements

Application	Time/Phase Synchronization Accuracy
CDMA2000	±3 ms with respect to UTC (during normal conditions) ±10 ms of UTC (when the time sync reference is disconnected)
WCDMA (TDD mode)	2.5 ms phase difference between Base Stations
TD-SCDMA (TDD mode)	3 ms phase difference between Base Stations
LTE (TDD)	3 ms time difference between Base Stations (small cell) 10 ms time difference between Base Stations (large cell)
MBSFN	<±1 ms with respect to a common time reference (continuous timescale)
WiMAX (TDD mode)	Depends on several parameters As an example ±0.5 ms and ±5 ms have been mentioned for a couple of typical cases
IP network delay monitoring	Depends on the level of quality that shall be monitored As an example ±100 ms with respect to a common time reference (e.g., UTC) may be required. ±1 ms has also been mentioned
Billing and alarms	±100 ms with respect to a common time reference (e.g., UTC)

- External Clocking that takes the clock information from an external connection and distributes it among all connected services belonging to this phase system. Typically, a primary reference input such as GPS or building integrated timing supplies/sync supply units (BITS/SSU), also known as standalone synchronization equipment (SASE), is used as a Grandmaster Clock.
- Loop-timed/Line-Derived Clocking where the clock source is taken from a selected TDM circuit.

Separate clock disciplines can be handled within one device, potentially supporting diverse applications that do not have the same clock source.

Table 3.4 describes timing methods for Mobile Backhaul as an example.

3.3 SYNCHRONIZATION STANDARDS

In order to provide clocking across Ethernet, the following options are available:

- **Synchronous Ethernet-**Layer-1-embedded timing information on the physical layer requiring full hardware support to link up. This approach provides only phase, not ToD.
- **IEEE 1588v2 (PTPv2)-**Layer-2-embedded operation, administration, and maintenance (OAM) frames with highest priority to ship clock/phase *and* ToD information across the packet network. Hardware-NIC (network interface card) support is advised for higher accuracy, but the packets remain standard Ethernet frames.

- **IETF RFC4330 (NTPv4)-**Layer-3-ToD information with high precision, but slow sync-up times and no phase lock support to aid TDM-emulation services.

Most of the standards relevant to the options above are from three different standards organizations, IETF [4–8], ITU-T [10, 12–19], and IEEE [20, 22].

Emulation of a native service is very different than transporting of the payload of a TDM line carrying a known protocol type. The difference is described as "content-aware" or "payload-agnostic," where any type of payload can be carried along with the source clock information. The main difference is related to the amount of bandwidth consumed when transporting the emulated service. If a payload-aware service detects no activity within the line, little to no bandwidth is being used to transport the service, whereas in case of payload-agnostic mode, every data bit from the TDM-port has to be transported and reconstituted across the network. After reconstruction, the TDM service with a clock recovery process called *phase locked loop* (PLL) ensures that a stable clock will be delivered despite packet delay variances (jitter) or other interfering factors.

There are two main standards currently being used to run pseudowire services on most platforms: circuit emulation over PSN (CESoPSN) [39] and structure agnostic TDM over packet (SAToP) [23]. With these two standards, it is possible to emulate a clock-sensitive service across a variable packet core transport.

3.4 NTP/SNTP

Network Time Protocol (NTP) is an *Internet* protocol used to synchronize the clocks of computers to a time reference. A full implementation of the NTP protocol is too complicated for many systems. A simplified version of the protocol, namely, SNTP had been defined. SNTP (simple network time protocol) is also basically an NTP, but lacks some internal algorithms that are not needed for all types of servers.

Table 3.4   Timing Technologies for CES

Technology	Frequency Accuracy	Frequency (MTIE/TDEV)	Phase/Time (Time-of-Day)
Internal XO	Yes	No	No
GPS	Yes	Yes	Yes
BITS/SSU (SASE)	Yes	Yes	Yes
Synchronous Ethernet (physical layer)	Yes	Yes	No
IEEE 1588v2 (protocol layer)	Yes	Yes	Yes
NTP v4 (protocol layer)	Yes	Yes	Yes
T1/E1 (leased line)	Yes	Yes	No

MTIE/TDEV: Maximum Time Interval Error/Time Deviation.

NTP needs a reference clock that defines the *true time* to operate. All clocks are set toward that true time. NTP uses UTC as reference time.

NTP is a fault-tolerant protocol that will automatically select the best of several available time sources to synchronize to. Multiple candidates can be combined to minimize the accumulated error.

NTP is highly scalable. A synchronization network may consist of several reference clocks. Each node of such a network can exchange time information either bidirectionally or unidirectionally. Propagating time from one node to another forms a hierarchical graph with *reference clocks* at the top.

Having available several time sources, NTP can select the best candidates to build its estimate of the current time. The protocol is highly accurate, using a resolution of less than a nanosecond (about 2^{-32} s). Even when a network connection is temporarily unavailable, NTP can use measurements from the past to estimate current time and error.

NTP implementations are supported by most of the popular UNIX and Windows operating systems. Among those are AIX, FreeBSD, HP-UX, Irix, Linux, NetBSD, SCO UNIX, OpenBSD, OSF/1, Solaris, System V.4, Windows Vista, Windows NT 4.0, Windows 2000, Windows XP, and Windows.NET Server 2003.

According to A Survey of the NTP Network [10], there were at least 175,000 hosts running NTP in the *Internet*. Among these there were over 300 valid *stratum-1* servers. In addition there were over 20,000 servers at stratum 2, and over 80,000 servers at stratum 3 (Fig. 3.1).

Currently, there are version three and version four implementations of NTP available. NTPv4 [24] algorithms can deal with high delay variations a bit better than version three. On the other hand, NTPv4 [24] uses floating-point operations where NTPv3 [25] used integer arithmetic. This might be an issue for older systems without a floating-point unit.

The new features of version four as compared to version three are as follows:

- use of floating-point arithmetic instead of fixed-point arithmetic;
- redesigned clock discipline algorithm that improves accuracy, handling of network jitter, and polling intervals;
- support for the *nanokernel* implementation that provides nanosecond precision as well as improved algorithms;
- public-key cryptography known as *autokey* that avoids having common secret keys;
- automatic server discovery (*manycast* mode);
- fast synchronization at start-up and after network failures (*burst mode*);
- new and revised drivers for reference clocks;
- support for new platforms and operating systems.

On the other hand, SNTP Version 4 [26] accommodates Internet Protocol Version 6 (IPv6) and Open Systems Interconnection (OSI) addressing. However, certain optional extensions to the basic Version 3 model, including an anycast mode and an authentication scheme, are designed specifically for multicast and anycast modes.

NTP and SNTP services are defined on UDP/IP-Port: 123. SNTP can talk to NTP servers. NTP is more complex than SNTP and can handle multiple sources. SNTP accuracy today is typically between 1 and 50 ms. NTPv4 claims to be able to be close to nx10 μs—depending on the network and hardware. GPS receivers provide typically <10–200 ns—stratum-1.

With (S)NTP, the transmission time stamp is being transmitted in the synchronization packet itself compared to Precision Time Protocol (PTP; IEEE1588), where it is transmitted in a following packet. In this way measurement of transmission, reception, and transmission of measured time stamps can be decoupled.

Systems use NTP to synchronize clocks to TOD master (NTP time server). Clock sync accuracy is affected by frame jitter in time sync request/response messages (Figs. 3.2 and 3.3).

Slave clock accuracy depends on packet delay and jitter. Packet delay can be accounted for in computation, but jitter affects tracking accuracy. In one-way delay measurement, jitter effect is additive. For example, if there is 1 ms jitter between Slave 1 and the Master, and 1 ms between Slave 2 and the Master, we may end up with 2 ms jitter between 2 ms jitter between Slaves 1 and 2. In order to minimize jitter impact on the accuracy, various proprietary algorithms are developed by various vendors.

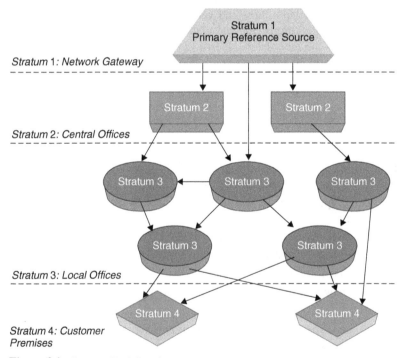

Figure 3.1 Stratum Clock Levels.

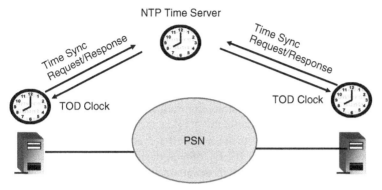

Figure 3.2 TOD clock synchronization based on an NTP time server.

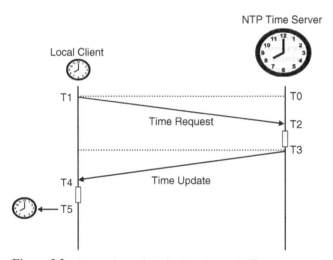

Figure 3.3 Current time calculation based on an NTP server.

RTP (Real Time Protocol) v1 [26] and RTPv2 [28] are among them. RTCP (Real Time Transport Control Protocol) [28] provides feedback on the quality of the data distribution and carries a persistent transport-level identifier for an RTP source. All participants send RTCP packets, therefore the rate must be controlled in order for RTP to scale up to a large number of participants. By having each participant send its control packets to all the others, each can independently observe the number of participants. The number of participants is used to calculate the rate at which the packets are sent.

3.5 PRECISION TIME PROTOCOL (IEEE 1588)

In recent years an increasing number of systems utilize a more real-time clocks and distributed architecture, all of which are synchronized to each other within the system. These clocks are used to manage distributed file systems, backup, and recovery systems, and many other similar activities. These computers typically interact via LANs and the Internet. In this environment, the most widely used technique for synchronizing the clocks is the NTP or the related SNTP.

IEEE 1588 is designed to fill a niche not well served by either of the two dominant protocols, NTP and GPS. IEEE 1588 is designed for local systems requiring very high accuracies beyond those attainable using NTP. It is also designed for applications that cannot bear the cost of a GPS receiver at each node, or for which GPS signals are inaccessible.

IEEE 1588 enables submicrosecond synchronization of clocks by having a master clock send multicast synchronization message frames containing time stamps. All IEEE 1588 receivers correct their local time on the basis of the received time stamp and an estimation of the one-way delay from transmitter to receiver.

IEEE 1588 is a protocol designed to synchronize real-time clocks in the nodes of a distributed system that communicate using a network. IEEE 1588 defines a protocol enabling precise synchronization of clocks applicable to systems communicating by local area networks supporting multicast messaging, including, but not limited to, Ethernet. The protocol will enable heterogeneous systems that include clocks of various inherent precision, resolution, and stability to synchronize.

The PTP is a high PTP for synchronization used in measurement and control systems residing on a local area network. Accuracy in the submicrosecond range may be achieved with low cost implementations.

PTP was originally defined in the IEEE 1588–2002 [29]. In 2008, a revised standard, IEEE 1588–2008 [22] was released, which is also known as PTPv2. The PTPv2 improves accuracy, precision, and robustness but is not backward compatible with the original 2002 version [31].

PTP is based on IP multicasting and can be used on any network that supports multicasting. It can be scaled for a large number of PTP nodes. Precision is typically in the range of 100 ns–100 μs depending on real-time capabilities of end-system.

The master cyclically transmits a unique synchronization (SYNC) message to the related slave clocks at defined intervals, by default every 2s.

The master clock measures the exact time of transmission. The slave clocks measure the exact times of reception. The master then sends in a second message, the follow-up message, and the exact time of transmission of the corresponding sync message to the slave clocks.

On reception of the sync message and, for increased accuracy, on reception of the corresponding follow-up message, the slave clock calculates the correction (offset) in relation to the master clock taking into account the reception time stamp of the sync message. The slave clock must then be corrected by this offset. If there were to be no delay over the transmission path, both clocks would now be synchronous.

For highly accurate synchronization, the time of transmission and reception of PTP messages can be determined as precisely and as closely as possible to the hardware.

Data format for PTP over Ethernet is depicted in Figure 3.4.

Although PTP, which is 1588 v2, can be implemented over any packet network, the major focus has been on the development of PTP over UDP/IPv4. The protocol stack is depicted in Figure 3.5.

The protocol will support system-wide synchronization accuracy in the sub-microsecond range with minimal network and local clock computing resources.

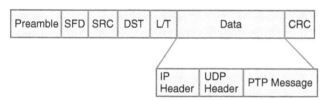

Figure 3.4 Data format for PTP over Ethernet.

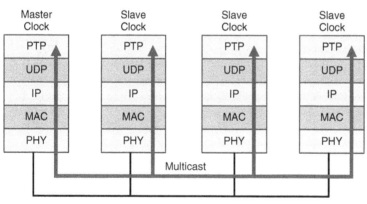

Figure 3.5 IEEE 1588 protocol stack.

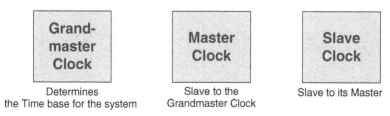

Figure 3.6 Clock hierarchy.

The default behavior of the protocol will allow simple systems to be installed and operated without administrator's involvement.

The IEEE 1588 organizes the clocks into a master–slave hierarchy based on observing the clock property information contained in multicastSync messages (Fig. 3.6). Each slave synchronizes to its master based on Sync, Delay_Req, Follow_Up, and Delay_Resp messages exchanged between master and its slave (Fig. 3.7).

The time base in an IEEE 1588 system is the time base of the Grandmaster Clock such as a stratum-1 source. The Grandmaster Clock time base is implementation dependent. All other clocks synchronize to the grand master. If the Grandmaster Clock maintains a UTC time base, the 1588 protocol distributes the appropriate leap second information to the slaves. Figure 3.8 depicts the use of the 1588 clocking in CES application.

In the synchronization process, Master Clock sends **Sync** and **Follow_up** messages (Fig. 3.9). Sync message contains an estimate of the sending time (~t1). When received by a slave clock the receipt time is noted. Follow_Up message is always associated with the preceding Sync message and contain the "precise sending time = (t1)" as measured as close as possible to the physical layer of the network. When received by a slave clock the "precise sending time" is used in computations rather than the estimated sending time contained in the Sync message.

On the other hand, Slave clock sends **Delay_Req** messages. The slave measures and records the sending time (t3). When **Delay_Req** is received by the master clock the receipt time is noted (t4). In response, Master sends **Delay_Resp** message containing the receipt time of the associated **Delay_Req** message (t4). When received by a slave clock the receipt time is noted and used in conjunction with the sending time of the associated Delay_Req message as part of the latency calculation.

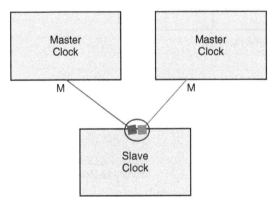

Figure 3.7 Slave receiving inputs from multiple masters.

Synchronization computation in the Slave clock is:

offset = receipt time − precise sending time − one way delay
       (for a Sync message)
one_way_delay = {MS Delay + SM Delay}/2 (assumes symmetric delay)
MS Delay = receipt time − precise sending time (for a Sync message)
SM Delay = Delay_Req receipt time − precise sending time
            (of a Delay_Req message)

From this offset the slave corrects its local clock.

In order to synchronize a pair of clocks using 1588, a Sync message is sent from master to slave (Fig. 3.9). The apparent time difference between the two clocks is measured:

MS_Time_Difference = slave's receipt time − master's sending time
                     = t2 − t1
MS_Time_Difference = offset + MS Delay

where MS Delay is the time for Sync message to travel from master to slave, which is a combination of propagation and queuing delay.

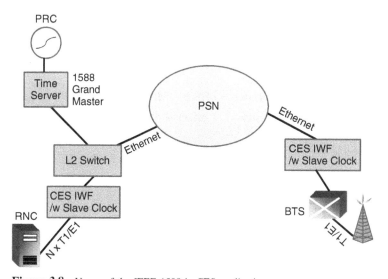

Figure 3.8   Usage of the IEEE 1588 in CES application.

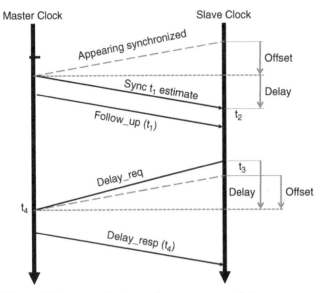

Figure 3.9 Message Exchanges between master and slave.

After that, a Delay_Req message from slave is sent to master and the apparent time difference between the two clocks is measured:

SM_Time_Difference = master's receipt time − slave's sending time

$$= t4 − t3$$

SM_Time_Difference = − offset + SM Delay

where SM Delay is the time for Delay_Req message to travel from slave to master, which is a combination of propagation and queuing delay.

From the above, we have the following two equations:

$$\text{MS\_Time\_Difference = offset + MS Delay} \qquad (3.1)$$

$$\text{SM\_Time\_Difference = − offset + SM Delay} \qquad (3.2)$$

From Equations 3.1 and 3.2,

offset = {(MS_Time_Difference − SM_Time_Difference) − (MS Delay − SM Delay}/2

MS Delay + SM Delay = {MS_Time_Difference + SM_Time_Difference}

if we assume

MS Delay = SM Delay = One_Way_Delay

then,

offset = {MS_Time_Difference − SM_Time_Difference}/2

one_way_delay = {MS_Time_Difference + SM_Time_Difference}/2

For **MS_Time_Difference = 10 minutes** and **SM_Time_Difference = −5 minutes**, we can have **offset = {10 − (− 5)}/2 = 7.5 minutes** and **one_way_delay = {10 + (-5)}/2 = 2.5 minutes**.

IEEE 1588v2 employs a two-way methodology, where packets are sent back and forth from the clock master to the clock slaves. This overcomes high amplitude, ultralow frequency wander that defeats other methods such as adaptive clock recovery techniques. Also, the standard is virtually independent of the physical media and can flow over low speed twisted-pair, high speed optical fiber, wireless, or even satellite links without requiring equipment design modifications.

It is not limited to TDM circuit emulation like the in-band solutions, but it can support CES better than adaptive clocking by distributing a precise network clock to every IWF node in the system. It can also be used for any pure, packet-based network, providing synchronization for future backhaul networks to be deployed by mobile operators.

The standard can distribute time/phase, frequency, or both. Telecom operators can use it to sell a synchronization service to customers (residential, wireless operators, etc.). It is resilient because a failed network node can be routed around. It is also resilient because the synchronization can come from one or more grandmaster clock nodes.

IEEE 1588v2 packets fully comply with Ethernet and IP standards and are backward compatible with all existing Ethernet and IP routing and switching equipment. There is no requirement for intermediate switches or routers to be IEEE 1588v2 aware. They see these timing packets as normal packet data.

The protocol calls for synchronization packets with time stamps to be sent from master clocks to all slave clocks and for individual slave clocks to send time-stamped packets to the master. The clock grandmaster maintains a time base locked to a PRC and establishes a separate synchronization session with each of the slaves it serves. The master and slave exchange timing packets according to the syntax of the IEEE 1588v2 protocol.

This process provides the timing clock recovery algorithm with the time stamps it needs to precisely re-create the master time base. From this time base, the synchronization signals used by the network equipment are synthesized. The timing clock recovery algorithm filters most noise, packet queuing delay, and propagation delay created by the transport network.

Accuracy of Time Stamping is influenced by protocol stack and execution environment. Messages generated or received by the PTP application are delayed within sending and receiving clocks. Delay components are cumulative and consist of constant and varying parts, called *transit delay and delay variation* (Jitter).

Almost all of these effects can be bypassed if time stamps are taken as near as possible to the physical layer with hardware assistance.

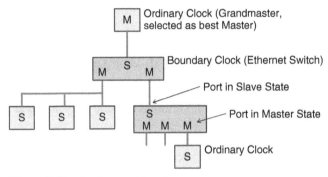

Figure 3.10 Synchronous hierarchy.

To cope up with the heavily fluctuating delay of storing and forwarding network elements such as switches/routers, two approaches are envisaged:

- BCs are used (Fig. 3.10). BCs forward frames as switches are used to, but they also contain a clock. PTP frames are processed by the BC in order to synchronize the internal clock and to synchronize other clocks connected to the BC. This leads to a hierarchical topology of clocks. A BC's synchronization performance is independent of the network load.
- Drift and offset are estimated by statistical methods. The methods require a long history of measured data. This requires reasonably stable oscillators and results in slow reaction to topological or environmental changes.

In a network using IEEE 1588, PTP configures and segments the network automatically. Each node employs a grand master clock (GMC) algorithm—also known as a *best master clock* algorithm—to determine the most accurate clock in the segment. Each node advertises its clock properties and features to other nodes using "Sync" messages. Once the GMC is selected, all nodes synchronize to it by adjusting their clocks accordingly. The GMC algorithm runs periodically, providing resynchronization as nodes are added or removed. BCs and transparent clocks both provide accurate distribution of the PTP protocol across network devices such as switches, routers, and repeaters. A transparent clock forwards PTP event messages (but does not act as a master or slave) and provides correction for the delay time across the device. A BC synchronizes other IEEE 1588 clocks across the subnets defined by a router, switch, or other devices that could block the transmission of all IEEE 1588 messages. A BC also eliminates the jitter typically generated by network devices owing to internal buffering. In addition, a BC typically acts as a master for all connected subnets, and appropriately retransmits 1588 management messages.

EEE 1588 BCs or transparent switches reduce the effect of jitter in Ethernet-based IEEE 1588 networks. A switch acting as a BC runs the PTP protocol, and is synchronized to an attached master clock. The BC in turn acts as a master clock to all attached slaves.

Even with hardware assistance, some fluctuations can still be observed due to quantization effects resulted from time stamp resolution and jitter in the data path and oscillator instabilities. Stochastic fluctuations may be removed by filtering and averaging algorithms. Long-term averaging requires a reasonable oscillator stability. If a topology change occurs such as a fast reconfiguration in ring configuration, then filtering and averaging slower the convergence. If reconfiguration can reliably detected, filtering and averaging should be bypassed to accelerate convergence.

Issues with IEEE 1588 may be expressed as follows:

- IEEE-1588 only allows the values of sync interval to be 1, 2, 8, 16, and 64 s. It is difficult to maintain performance in a loaded network with sync packet rate of 1 pps and an inexpensive oscillator.
- IEEE 1588 relies on a symmetric network.
- IEEE 1588 does not have provision for redundancy support.
- IEEE 1588 relies on BCs topology. BCs are not available in legacy telecom networks.
- IEEE 1588 only supports multicast.
- Long PTP messages consuming too much bandwidth.

3.6 SYNCHRONOUS-ETHERNET NETWORKS (SyncE)

There are two methods for clocking distribution:

- Plesiochronous and network-synchronous methods, called *Synchronous Ethernet*.
- Packet-based methods such as IEEE 1588 (PTP) as described in previous sections.

Synchronous Ethernet is defined by ITU a means of using Ethernet to transfer timing (frequency) via the Ethernet PHY layer. This is a general case of Layer 1 timing and was introduced in Reference 10.

SyncE uses the PHY clock transmissions and generates the clock signal from "bit stream" similar to traditional SONET/SDH/PDH PLLs. Each node in the Packet Network recovers the clock (Fig. 3.11).

The IEEE 802.3 standards require that the line rate of Ethernet operate within a specific rate (±100 ppm) relative to an absolute reference. According to the Synchronous Ethernet, this rate must be traceable to an external reference. As a result, Ethernet devices requiring frequency recovery via synchronous Ethernet need to support a synchronization status message. Synchronous-Ethernet ports nominally operate within a frequency tolerance range of ±4.6 ppm. However, in order to operate with nonsynchronous interfaces and to maintain data continuity, synchronous-Ethernet receivers must also operate at ±100 ppm.

Ethernet networks are free-running (±100 ppm). However, in the case of synchronous Ethernet, a master–slave synchronization architecture at the physical layer is used to provide reference timing signal distribution over packet networks, from backbone level to access level. This method can be used to provide timing recovery at the IWFs for constant bit rate (CBR) services transported over packet networks. It could also be used to provide a reference timing signal down to edge access equipment in a pure Ethernet network supporting synchronous Ethernet.

Within existing Ethernet technology, the service is effectively asynchronous. In synchronous Ethernet, existing Ethernet services will continue to be mapped into and out of the Ethernet physical layer at the appropriate rates.

Hierarchical timing distribution is recommended for synchronous-Ethernet networks. Timing should not be passed from a synchronous Ethernet in free-run/holdover mode to a higher quality clock because the higher quality clock should not follow the synchronous-Ethernet signal during fault conditions.

In general, a reference timing signal traceable to a PRC is injected into the Ethernet switch using an external clock port to deliver a physical layer clock from the Ethernet switch to the IWF (Fig. 3.12). This signal is extracted and processed via a synchronization function before injecting timing onto the Ethernet bit stream. The synchronization function provides filtering and may require holdover.

Clearly, there may be a number of Ethernet switches between the element where the reference timing signal is injected and the IWF. In such cases the synchronization function within the Ethernet switch must be able to recover synchronization "line timing" from the incoming bit stream.

The network clock is the clock used to discipline the synchronization function within the Ethernet switch. The clock injected into the synchronization function will be synchronous, that is, locked to the network clock.

The packet-based methods rely on timing information carried by the packets (e.g., sending dedicated Time Stamp messages as shown in Fig. 3.13). Methods using two-way transfer of timing information are also possible such as NTP or similar protocols. In some cases, this is the only alternative to a PRC-distributed approach.

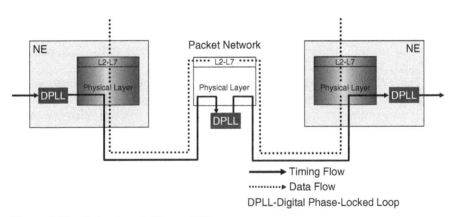

Figure 3.11 Timing through Ethernet PHY.

Timing recovery for CBR services such as TDM CES require that the timing of the signal is similar on both ends of the packet network and is handled by the IWFs. The service clock is preserved in such a way that the incoming service clock frequency is replicated as the outgoing service clock frequency.

Functional blocks defined for synchronization-related functions, including clock functions, time distribution functions, clock selection functions, and IWF, are necessary to implement circuit emulation. These functional blocks such as ITU-T G.8262 [12] clocking may be embedded into network equipment or into other equipment such as an SSU/SASE.

Clocks within the network have been categorized based on performance, and a master–slave synchronization scheme is employed. As in GPS, in some cases, the satellite is used for the distribution of frequency.

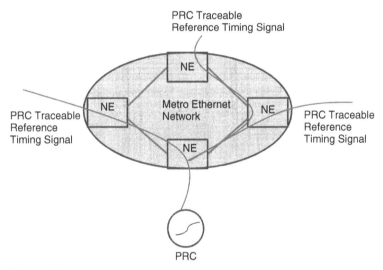

Figure 3.12 Master–slave synchronization network over synchronous Ethernet.

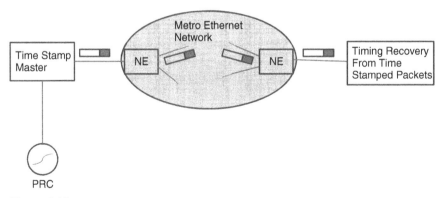

Figure 3.13 Timing distribution via time stamps in packet systems.

Four clocking methods are used for synchronization of packet networks [10]:

- Network-synchronous operation,
- differential methods,
- adaptive methods, and
- reference clock available at the TDM end systems.

3.6.1 Clocking Methods for Synchronization

Network-synchronous operation method refers to the fully network-synchronous operation by using a PRC-traceable network-derived clock or a local PRC such as GPS as the service clock. This method does not preserve the service timing.

According to the differential methods, the difference between the service clock and the reference clock is encoded and transmitted across the packet network (Fig. 3.14). The service clock is recovered on the far end of the packet network making use of a common reference clock (i.e., service timing is preserved). The SRTS method [11] is an example of this family of methods.

Differential methods may work with IWF reference clocks that are not PRC traceable.

In the adaptive methods, the timing can be recovered based on the interarrival time of the packets or on the fill level of the jitter buffer. It should be highlighted that the method preserves the service timing (Fig. 3.15).

When the reference clock is available at the TDM end systems, there is no need to recover the timing because both the end systems have direct access to the timing reference, and will retime the signal leaving the IWF.

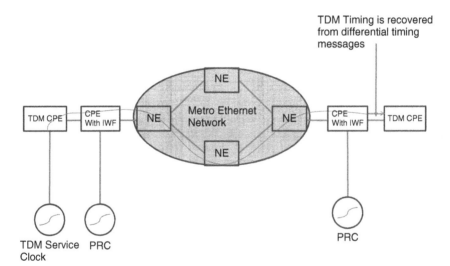

Figure 3.14 Example of timing recovery operation based on differential methods.

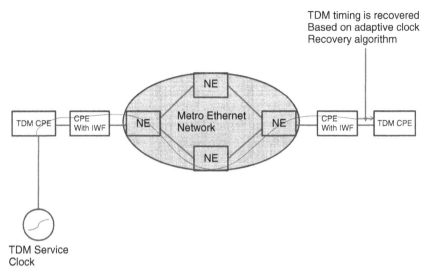

TDM timing is recovered
Based on adaptive clock
Recovery algorithm

Figure 3.15   Adaptive method.

3.6.2   Impact of Packet Network Impairments on Synchronization

Fundamentally, network synchronization is required in Layer 1 networks to manage buffers. PDH, SDH, and OTN networks have Layer 1 buffers to accommodate changes in rates. These buffers are managed by mechanisms such as stuff bytes and pointers, together with system clocks to accommodate different clocking domains.

In packet networks, the data is delivered over the network in packets or frames, rather than being carried as a continuous stream of CBR. Packets may be statistically multiplexed. Within a single switch, multiple packet streams may have to converge onto a single output buffer, resulting in a buffer contention. Packets can be delayed or dropped.

As packets may traverse different routes, a stream of packets from ingress to egress may exhibit significant packet delay variation. Additionally, packets may be misordered resulting in additional buffering. Therefore, large buffers are required to perform packet-level processing.

The following performance parameters affect timing distribution and clock recovery:

- Packet Transfer Delay and Delay Variation—In Differential Methods, packet transfer delay and delay variation should not affect clock recovery performance when a network reference clock is available at both ends and Differential Methods are used.

  In Adaptive Methods, adaptive recovery of the service clock from a packet stream containing CBR data is usually achieved by some computing function of the arrival rate or arrival times of the packets at the destination node.

   If the delay through the packet network is constant, the frequency of arrival of packets at the destination node is not affected by the network. There may be a phase lag in the recovered clock due to the delay through the network, but there should be no frequency or phase wander. If the delay varies, it may be perceived by a clock recovery process as a change in phase or frequency of the original service clock.

- Packet Loss Ratio—TDM circuits carried over packet networks may be extremely vulnerable to bit errors caused by packet loss. A single bit error in the packet leads to the whole packet being discarded, yielding a burst of consecutive bit errors in the recovered TDM stream. Hence, even moderate levels of packet loss (from the viewpoint of conventional packet network) may cause unavailability of a TDM circuit.

- Packet Severe Loss Block Outcomes—References 30 and 36 define a severe loss block outcome as occurring when, for a block of packets observed at an ingress interface during time interval T, the ratio of lost packets to total packets exceeds a threshold. Similar effects are expected in Ethernet networks. During these impairments the timing recovery mechanism has to handle the total loss of packets.

- Packet Error Ratio—In order to meet the network requirements, limiting the jitter and wander production of the synchronous-Ethernet solution in a wide area network environment will be necessary.

The synchronization function within the synchronous-Ethernet switch should be based on the performance characteristics of an embedded clock. Such a clock will ensure that proper network operation occurs when the clock is synchronized from another similar synchronous-Ethernet clock or a higher quality clock. For consistency with existing synchronization networks, the embedded clock may be based on Reference 34.

3.6.3 Stabilization Period

The stabilization period is an important parameter during the start-up phase and when switching between timing references. When equipment operates in holdover mode for hours, the phase error when selecting a new clock reference is largely dominated by the phase error caused by the frequency error of the clock in holdover. When the adaptive method is used, the requirement on the stabilization period may depend on the actual phase noise in the packet network. A large packet delay variation may require a long period before the clock can lock to the timing reference.

   The filter implementation and the characteristics of the internal oscillator are important as well. In fact, depending on the holdover characteristics, longer time could be accepted when switching from a reference to a second reference, as a good holdover can allow longer locking periods.

The stabilization period is not well defined. The period of 900 s is used for adaptive methods.

3.6.4 IWF Synchronization Function

The IWF provides the necessary adaptations between TDM and packet streams. The possible supported timing options for the Tx clock are as follows:

- timing from recovered source clock carried by the TDM input (loop timing or line timing);
- timing from the network clock (the network clock can be derived either from the physical layer of the traffic links from the packet network or through an external physical timing interface, e.g., 2048 kHz);
- timing from a free-running clock (it shall provide an accuracy according to relevant TDM/CBR service interface, e.g., 2048 kbps shall comply with ITU-T Rec. G.703, ± 50 ppm);
- differential methods;
- adaptive timing (including clock recovery using dedicated time stamps).

Depending on the services to be provided, a suitable subset of the listed timing options shall be supported. Slip control in the TDM Tx direction is necessary to control possible over/underflow in the playout buffer. Slips shall be performed on $n \times 125$ µs frames. When TDM transmitter and/or receiver clocks are in holdover or are traceable to clocks in holdover, and a synchronous clock recovery technique (differential method or network-synchronous operation) is used, slips (most likely uncontrolled) will occur.

When selecting a new timing source, the output wander may temporarily exceed the output wander limit. However, the output wander must be within the output wander limit by the end of a period called the *stabilization period*.

Another characteristic that is relevant for the IWF is the latency. The latency requirements are normally defined on the network level specifying the total latency in the end-to-end connection.

Network-synchronous operation may be used when a signal traceable to PRC is available at the IWFs and it is not required to preserve the service clock. Differential methods maybe used when a PRC-traceable reference is available at the IWF. With this method it is possible to preserve the service clock. Adaptive methods may be used when the delay variation in the network can be controlled. With this method it is possible to preserve the service clock.

3.6.5 PRC

A typical synchronous-Ethernet architecture will have a PRC located in one of three locations depending on the overall architecture:

- Core-located—The PRC will be located at the core node, centrally located with some form of distribution to the IWF.

- Access-located—The PRC will be located at some point further back within the network, typically at the multiservice access point. This architecture might use more PRC nodes compared to the core-located architecture.
- IWF-located—The PRC will be located geographically with the IWF and there will be a direct synchronization connection to the IWF; therefore, there will be one PRC per IWF.

Synchronization status messaging (SSM) provides a mechanism for downstream Ethernet switches to determine the traceability of the synchronization distribution scheme back to the PRC or highest quality clock that is available. The synchronization function processes the SSM. Synchronization status messages for Ethernet implement the SSM channel using an IEEE 802.3 organizational-specific slow protocol (OSSP).

For upstream network failure conditions, the synchronization function takes appropriate action based on the SSM and preset priorities, and selects an alternate synchronization feed from the network or an external interface.

SSM messages represent the QL of the system clocks located in the various network elements. *Quality level* refers to the holdover performance of a clock. A "heart-beat" message is used to provide a continuous indication of the clock QL. A message period of one second meets the message rate requirements of IEEE 802.3 slow protocols. To minimize the effects of wander that may occur during holdover, an event type message with a new SSM QL is generated.

To protect against possible failure, the lack of the messages is considered to be a failure condition.

An ESMC information PDU, containing the current QL used by the system clock selection algorithm, is generated once per second.

For synchronous Ethernet, the slow protocol used for the transmission of the SSM code relies on the use of a "heart-beat" timer. ESMC information PDUs are sent periodically at a rate of one PDU per second. Lack of reception of an ESMC information PDU within a five-second period results in the QL being set to do-not-use (DNU).

Synchronous-Ethernet equipment will require a reference source selection mechanism to provide traceability to upstream elements and ultimately the PRC with respect to frequency. The selection mechanism controls the physical timing flows within the equipment. The selection mechanism must be able to select an appropriate external reference source, traffic reference source, and internal clock (i.e., local oscillator).

3.6.6 Operation Modes

The operation mode can be asynchronous or synchronous. Ethernet equipment that is not synchronous-Ethernet capable work in an asynchronous mode, where each input interface gets its timing from its input signal, which is within a frequency range of ±100 ppm (±20 ppm for 10G WAN). The output interfaces might each have a free-running oscillator generating timing within a frequency rate of ±100 ppm (±20 ppm for 10G WAN).

Synchronous-Ethernet equipment is equipped with a system clock. Synchronous-Ethernet interfaces are able to extract the received clock and pass it to a system clock. This equipment clock may work in several modes, QL or priority modes [32]. Each interface of a synchronous-Ethernet equipment might be configured to work in either nonsynchronous or synchronous operation mode.

A synchronous-Ethernet interface configured in asynchronous mode is an interface that, for the receive side, does not pass the recovered clock to the system clock and is therefore not a candidate reference to the synchronization selection process. It does not process the Ethernet synchronization messaging channel (ESMC) that may be present and therefore cannot extract the QL value.

On the transmit side, its output frequency might be synchronized to the embedded Ethernet equipment clock (EEC), but this remains unknown to the receive interface at the other termination of the link. As an asynchronous interface does not generate an ESMC and therefore does not transmit a QL, this interface does not participate in the synchronization network. It is functionally identical to an asynchronous interface.

A synchronous-Ethernet interface can be configured in synchronous operation mode. Its receive side is able to extract the frequency of its input signal and passes it to a system clock that can be an EEC or better quality clock. It processes the ESMC and extracts the QL value that are required for a frequency reference.

The transmit part of the interface is locked to the output timing of the system clock and generates the ESMC to transport a QL. 1G copper Ethernet interfaces perform link autonegotiation to determine the master and slave clocks for the link.

3.6.7 Frequency Accuracy of Slave Clock

For Option 1, under free-running conditions, the EEC output frequency accuracy should not be greater than 4.6 ppm with regard to a reference traceable to a [ITU-T G.811] clock. 1 month and 1 year for the time interval for this accuracy have been proposed.

For Option 2, under prolonged holdover conditions, the output frequency accuracy of the different types of node clocks should not exceed 4.6 ppm with regard to a reference traceable to a PRC, over a time period of 1 year (Table 3.5).

3.6.8 EEC

Reference 10 outlines minimum requirements for timing devices used in synchronous Ethernet for synchronizing the equipment. It supports clock distribution based on network-synchronous line-code methods as in synchronous Ethernet. It is called *embedded Ethernet equipment clock (ECC)*.

EEC has Option 1 and Option 2. The first option, referred to as *EEC-Option 1*, applies to synchronous-Ethernet equipment that is designed to interwork with networks optimized for the 2048 kbps hierarchy. These networks allow the worst-case synchronization reference as specified in Reference 36. The second option, referred to as *EEC-Option 2*, applies to synchronous-Ethernet equipment that is designed to interwork with networks optimized for the 1544 kbps hierarchy [34].

A synchronous-Ethernet equipment slave clock needs to support all of the requirements specific to one option and not mix requirements between EEC-Options 1 and 2.

The noise generation of an EEC represents the amount of phase noise produced at the output when there is an ideal input reference signal or the clock is in holdover state. Performance level for a suitable reference is at least 10 times more stable than the output level.

The clock ability to limit this noise is described by its frequency stability. The maximum time interval error (MTIE) and time deviation (TDEV) characterize noise generation performance. MTIE and TDEV are measured through an equivalent 10 Hz, first-order, low pass measurement filter, at a maximum sampling time of 1/30 s. The minimum measurement period for TDEV is 12 times the integration period.

Wander generation bounds for locked mode for both options are given in Reference 14.

When a clock is not locked to a synchronization reference, the random noise components are negligible compared to deterministic effects such as initial frequency offset.

On the other hand, jitter generation has different limits for different interface rates. The peak-to-peak amplitude measured over a 60 s interval should not exceed 0.5 for 1 Gbps interface with measuring filter bandwidth of 2.5 kHz–10 MHz and 10 Gbps interface with measuring filter bandwidth of 20 kHz–0 MHz [12].

Bounds for noise tolerance, noise transfer, transit response and holdover performance of slave clocks should be also satisfied [12]. In order to determine the EEC noise tolerance, the worst-case network limit is used. The tolerance of an EEC indicates the minimum phase noise level at the input of the clock to be accommodated while maintaining the clock within performance limits, without causing any alarms or switch reference or going into holdover mode.

Table 3.5 Option 1 and Option 2 Functionalities

Functionality	Option 1	Option 2
Minimum pull-in range	±4.6 ppm (independent from internal oscillator frequency offset)	±4.6 ppm (independent from internal oscillator frequency offset)
Hold-in range	Not required	±4.6 ppm (independent from internal oscillator frequency offset)
Pull-out range	±4.6 ppm	Not applicable

The noise transfer characteristics determine EEC properties related to the transfer of excursions of the input phase relative to the carrier phase.

Transient response and holdover performance defines the ability to withstand disturbances and avoid transmission defects or failures. The short- and long-term phase transient response (holdover) requirements are given in Reference 12.

3.7 CONCLUSION

Depending on the application, timing, phase, and/or frequency synchronization is necessary among systems and networks. For example, TDM networks require highly accurate frequency information to be recoverable from their physical layers. In every TDM network there is at an extremely accurate PRC from which all other TDM clocks in the network derive their timing.

Clocking distribution is performed either via synchronous Ethernet or packet-based methods such as IEEE-1588 (PTP). The 1588v2 is picking up momentum in the industry to provide time, phase, and frequency synchronization owing to the fact that it does not require to change the existing hardware.

REFERENCES

1. ITU G.811, Timing characteristics of primary reference clocks, 09/97.
2. ITU-T G.823, The control of jitter and wander within digital networks which are based on the 2048 kbit/s hierarchy, 2000.
3. ITU-T G.824, The control of jitter and wander within digital networks which are based on the 1544 kbit/s hierarchy.
4. ITU-T G.822, Controlled slip rate objectives on an international digital connection, 1993.
5. ITU G.8261, Timing and Synchronization aspects in Packet Networks, 05/2006.
6. ITU-T I.363.1, B-ISDN ATM Adaptation Layer Specification: Type 1 AAL, 1996.
7. ANSI/T1.101–1987, Synchronization Interface Standard for Digital Networks.
8. Lewis Jeremy, Hybrid Mode Synchtronous Ethernet and IEEE 1588 in Wireless TDD Applications, 2010. http://timing.zarlink.com/zarlink/Hybrid_Mode_Synchronous_Ethernet_IEEE_1588-_WSTS.pdf.
9. ITU-T Y.1731, OAM functions and mechanisms for Ethernet based networks, 2008.
10. RFC 4330, Network Time Protocol - NTPv4 -(Draft 8) -SNTPv4, 2006.
11. RFC 3985, Pseudowire Emulation Edge-to-Edge (PWE3) Architecture, 2005.
12. RFC 4197, Requirements for Edge-to-Edge Emulation of Time Division Multiplexed (TDM) Circuits over Packet Switching Networks, 2005.
13. RFC 4553, Structure-Agnostic Time Division Multiplexing (TDM) over Packet (SAToP), 2006.
14. ITU G.8262, ITU-T G.8262, Timing characteristics of synchronous Embedded Ethernet Equipment Clock (EEC), 7/2010.
15. ITU-T G.8260, Definitions and terminology for synchronization in packet networks, 2010.
16. ITU-T G. 8271, Basic of time and phase synchronization, 2011.
17. ITU-T G.8261.1, Network requirements for frequency synchronization, 2011.
18. ITU-T G.8264, Distribution of timing information through packet networks, 2008.
19. ITU-T G.8265, Architecture and requirements for packet based frequency delivery, 2010.
20. ITU-T G.8265.1, Precision time protocol telecom profile for frequency synchronization, 2010.

21. IEEE 1588v2, Standard for Local and Metropolitan Area Networks - Timing and Synchronization for Time-Sensitive Applications in Bridged Local Area Networks - 802.1as (Draft 7.6, Nov 2010).

22. IEEE 1588–2008, Standard for a Precision Clock Synchronization Protocol for Networked Measurement and Control Systems.

23. RFC 5086, Structure-Aware Time Division Multiplied (TDM) Circuit Emulation Service over Packet Switched Network (CESoPSN), 2007.

24. RFC 4553, Structure-Agnostic Time Division Multiplexing (TDM) over Packet (SAToP); June 2006.

25. RFC 5905, Network Time Protocol Version 4: Protocol and Algorithms Specification, June 2010.

26. RFC 1305, Network Time Protocol (Version 3) Specification, Implementation and Analysis, March 1992.

27. RFC 2030, Simple Network Time Protocol (SNTP) Version 4 for IPv4, IPv6 and OSI, Oct 1996.

28. RFC 1889, RTP: A Transport Protocol for Real-Time Applications, Jan 1996.

29. RFC 3550, RTP: A Transport Protocol for Real-Time Applications, July 2003.

30. IEEE 1588–2002, "Standard for a Precision Clock Synchronization Protocol for Networked Measurement and Control Systems".

31. National Institute of Standards and Technology (NIST), "IEEE 1588 Systems", 2011. Available at http://ieee1588.nist.gov/.

32. ITU-T Y.1540, Internet protocol data communication service - IP packet transfer and availability performance parameters, 2003.

33. ITU-T Y.1561, Performance and availability parameters for MPLS networks, 2005.

34. ITU-T G.813, Timing characteristics of SDH equipment slave clocks (SEC), 2003.

35. ITU-T G.781, Synchronization layer functions, 1999.

36. ITU-T G.803, Architecture of transport networks based on the synchronous digital hierarchy (SDH), 2003.

37. ITU-T G.812, Timing requirements of slave clocks suitable for use as node clocks in synchronization networks, 2006.

38. MEF 18, Abstract Test Suite for Circuit Emulation Services over Ethernet, 2007.

39. MEF 3, Circuit Emulation Service Definitions, Framework and Requirements in Metro Ethernet Networks, 2004.

40. MEF 8, Implementation Agreement for the Emulation of PDH Circuits over Metro Ethernet Networks. 2004.

Chapter 4

Pseudowires

4.1 INTRODUCTION

Pseudowire Emulation Edge to Edge (PW3) is a mechanism used to emulate telecommunications services across packet-switched networks such as Ethernet, IP, or MPLS. The emulated services are mostly T1/T3 leased lines, frame relay, Ethernet, and ATM to maximize return on existing assets and minimize operating costs of the service providers (SPs).

Pseudowires (PWs) encapsulate cells, bit streams, and protocol data units (PDUs) and transport them across IP or MPLS tunnels. The transportation of encapsulated data usually require managing the sequence and timing of data units to emulate services such as T1/T3 leased lines and ATM.

For a customer equipment (CE), a PW appears to be a dedicated circuit for the emulated service. This provides simplicity to implement various services over packet networks (PSNs, packet-switched network).

4.2 PROTOCOL LAYERS

The logical protocol layer structure is needed to support a PW as shown in Figure 4.1.

The payload is transported over the encapsulation layer. The encapsulation layer carries any information, not already present within the payload itself including sequencing, and the interface to the PW demultiplexer layer.

The PW demultiplexer layer provides the ability to deliver multiple PWs over a single PSN tunnel.

The PSN convergence layer provides the enhancements needed to make the PSN conform to a consistent interface to the PW, making the PW independent of the PSN type. A detailed protocol stack reference model for PWs is given in Figure 4.2 [1].

The PW provides the CE with an emulated physical or virtual connection to its peer at the far end. Native service PDUs from the CE are passed through an encapsulation layer at the sending PE (provider edge) and then sent over the

Networks and Services: Carrier Ethernet, PBT, MPLS-TP, and VPLS, First Edition. Mehmet Toy.
© 2012 John Wiley & Sons, Inc. Published 2012 by John Wiley & Sons, Inc.

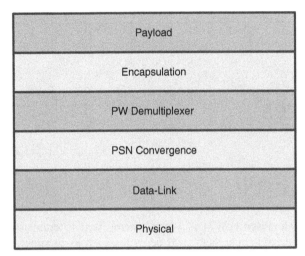

Figure 4.1 Logical Protocol Layering Model.

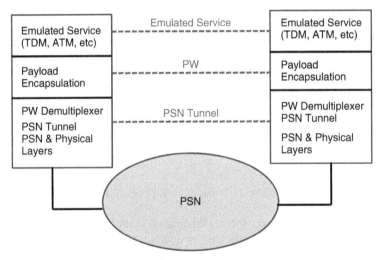

Figure 4.2 PWE3 protocol stack reference model.

PSN. The receiving PE removes the encapsulation and restores the payload to its native format for transmission to the destination CE.

4.3 PAYLOAD TYPES

The native data units (payloads) can be packet, cell, bit stream, and structured bit stream. The PW services for these payloads are given in Table 4.1.

A packet payload is a variable-size data unit delivered to the PE via an attachment circuit (AC) and relayed across the PW as a single unit or in multiple

Table 4.1 Payload Types and Corresponding PW Services

Payload Type	PW Service
Structured bit stream	SONET/SDH (e.g., S-PE, VT, NxDS0)
Bit stream	Unstructured E1, T1, E3, T3
Packet	Ethernet, HDLC framing, frame relay, ATM AAL5 PDU
Cell	ATM

segments, where the combined size of packet payload and its PWE3 and PSN headers is larger than the PSN path MTU. A packet payload may need sequencing and real-time support.

In some applications such as frame relay, the packet payload is selected based on an identifier, part of the forwarder function. For example, frame relay PDUs will be selected based on the frame delay data link connection identifier (DLCI).

A cell payload is created by capturing, transporting, and replaying groups of octets presented on the wire in a fixed-size format. For example, the cell payloads for ATM and MPEG transport stream packets (DVB) are 53-octets and 188-octets, respectively.

To reduce per-PSN packet overhead, multiple cells may be concatenated into a single payload. The cell payload service will need sequence numbering and may also need time synchronization.

A bit stream payload is created by capturing, transporting, and replaying the bit pattern on the emulated wire. For example, E1 and T1 send "all-ones" to indicate failure that can be detected without any knowledge of the structure of the bit stream. Sequencing and time synchronization are needed.

A structured bit stream payload is created by some knowledge of the under-lying structure of the bit stream to capture, transport, and replay the bit pattern on the emulated wire. Sequencing and time synchronization are needed.

Some parts of the original bit stream may be stripped in the PSN-bound direction. For example, in structured SONET, the section and line overhead may be stripped. The stripped information may appear in the encapsulation layer to facilitate the reconstitution.

4.4 PSEUDOWIRE ARCHITECTURE

A PW is a connection between two PE equipment connecting two ACs. An AC can be a frame delay DLCI, an ATM VCI/VPI, an Ethernet port, a VLAN, an MPLS LSP, etc. Figure 4.3 illustrates the network reference model for point-to-point PWs [2].

The two PEs (Provider Edge 1 and Provider Edge 2) have to provide one or more PWs on behalf of their client CEs (CE1 and CE2) to enable the client CEs to communicate over the PSN. A PSN tunnel is established to provide a data path for the PW. The PW traffic is invisible to the core network, and the core

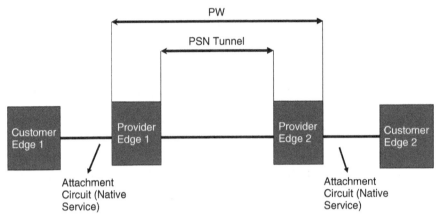

Figure 4.3   PWE3 network reference model.

network is transparent to the CEs. Native data units (bits, cells, or packets) arrive via the AC, are encapsulated in a PW-PDU, and are carried across the underlying network via the PSN tunnel. The PEs perform the necessary encapsulation and decapsulation of PW-PDUs and handle any other functions required by the PW service, such as sequencing or timing.

4.4.1   PWE3 Preprocessing

Applications such as frame relay have to perform operations on the native data units received from the CE (including both payload and signaling traffic) before they are transmitted across the PW by the PE. Examples include Ethernet bridging, SONET cross-connect, translation of locally significant identifiers such as VCI/VPI, or translation to another service type. These operations could be carried out in an external equipment or within the PE, where processed data is then presented to the PW via a virtual interface within the PE.

Both PEs or one of the PEs can support preprocessing (PREP) functionality. Figure 4.4 shows the interworking of one PE with preprocessing and a second without this functionality. The functional interface between the PREP and the PW is that represented by a physical interface carrying the service. This reference point effectively defines the necessary interworking specification.

Figure 4.5 illustrates how the protocol stack reference model is extended to include the provision of preprocessing. The required preprocessing can be divided into two components as Forwarder Service Processing (FWRD) and Native Service Processing (NSP). Forwarders can be single input and single output, as well as multiple inputs and multiple outputs as depicted in Figures 4.6 and 4.7.

Figure 4.6 shows a simple forwarder (i.e., single input and single output) that performs some type of filtering operation. Figure 4.7 shows forwarding between multiple ACs and multiple PWs, where payloads are extracted from one or more ACs and directed to one or more PWs.

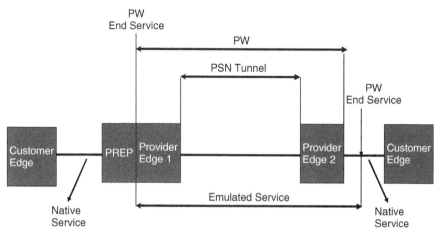

Figure 4.4 Preprocessing within the PWE3 network reference model.

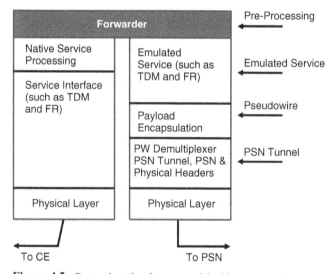

Figure 4.5 Protocol stack reference model with preprocessing.

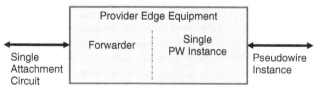

Figure 4.6 Simple point-to-point service.

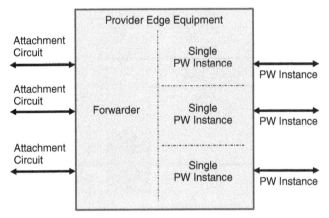

Figure 4.7   Multiple AC to multiple PW forwarding.

Some applications required some form of data or address translation, or some other operation requiring knowledge of the semantics of the payload. This is the function of the native service processor (NSP). The use of the NSP approach simplifies the design of the PW by restricting a PW to homogeneous operation.

Figure 4.8 illustrates the relationship between NSP, forwarder, and PWs in a PE. The NSP function may apply any transformation operation on the payloads, as they pass between the physical interface to the CE and the virtual interface to the forwarder. These transformation operations will be limited to those that have been implemented in the data path and that are enabled by the PE configuration. A PE device may contain more than one forwarder.

4.4.2   Payload Convergence Layer

The primary task of the payload convergence layer is the encapsulation of the native data units containing an L2 or L1 header (payload) in PW-PDUs. The

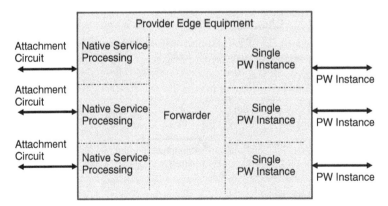

Figure 4.8   NSP in a multiple AC to multiple PW forwarding PE.

payload convergence header carries the additional information. This information may be used to replay the native data units at the CE-bound physical interface.

From its underlying PW demultiplexer and PSN layers, the PW encapsulation layer and its associated signaling requires a reliable control channel for signaling line events, status indications, and CE–CE events that are translated and sent between PEs.

In addition, the encapsulation layer may require the following depending on applications:

1. A high priority indicated via DSCP or EXP bit or a bit in the tunnel header, unreliable, sequenced channel. A typical use is for CE-to-CE signaling.
2. A sequenced channel for data traffic that is sensitive to packet reordering.
3. An unsequenced channel for data traffic insensitive to packet order.

The PW encapsulation layer provides the necessary infrastructure to adapt the specific payload type being transported over the PW to the PW demultiplexer layer that is used to carry the PW over the PSN. It consists of payload convergence, timing, and sequencing sublayers. The sublayering and its context with the protocol stack are shown in Figure 4.9.

The payload convergence sublayer provides the provision of per-packet signaling and other out-of-band information, while the timing and sequencing layers provide timing and sequencing services to the payload convergence layer.

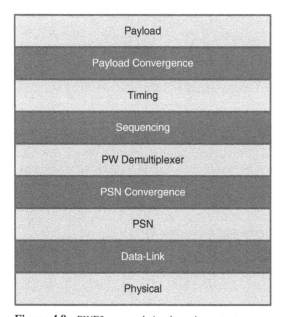

Figure 4.9 PWE3 encapsulation layer in context.

The sequencing function provides frame ordering, frame duplication detection, and frame loss detection that are keys to the emulation of the invariant properties of a physical wire.

The size of the sequence-number space depends on the speed of the emulated service and on the maximum time of the transient conditions in the PSN. A sequence-number space greater than 216 may be needed to prevent the sequence-number space from wrapping during the transient.

When packets carrying the PW-PDUs traverse a PSN, they may arrive out of order at the destination PE. For some services, the frames (control frames, data frames, or both) must be delivered in order. Providing a sequence number in the sequence sublayer header for each packet is one possible approach. Alternatively, it is possible to drop misordered PW-PDUs instead of trying to sort PW-PDUs into the correct order.

For some native services, the receiving PE has to play out the native traffic, as it was received at the sending PE. The timing information is either sent between the two PEs or, in some cases, received from an external reference.

Some applications such as CES (Chapter 9) require clock recovery and timed payload delivery. Clock recovery is the extraction of output transmission bit timing information from the delivered packet stream. Timed delivery is the delivery of noncontiguous PW-PDUs to the PW output interface with a constant phase relative to the input interface. The timing of the delivery may be relative to a clock derived from the packet stream received over the PSN clock recovery, or to an external clock.

There will be cases where the combined size of the payload and its associated PWE3 and PSN headers may exceed the PSN path MTU. Then, fragmentation and reassembly have to be performed for the packet to be delivered.

A PE implementation may not to support fragmentation and will drop packets that exceed the PSN MTU.

4.4.3   PW Demultiplexer Layer and PSN

The purpose of the PW demultiplexer layer is to allow multiple PWs to be carried in a single tunnel. This minimizes complexity and conserves resources. Some types of native service are capable of grouping multiple circuits into a "trunk"; for example, multiple ATM VCs in a VP, multiple Ethernet VLANs on a physical media, or multiple DS0 services within a T1 or E1. A PW may interconnect two end trunks. That trunk would have a single multiplexing identifier.

The demultiplexer layer provides three main functions, multiplexing, fragmentation, and identifying PDU length and delivering PDU.

If the PSN provides a fragmentation and reassembly service of adequate performance, it may be used to obtain an effective MTU that is large enough to transport the PW-PDUs.

PDU delivery to the egress PE is the function of the PSN Layer. If the underlying PSN does not provide all the information necessary to determine the length of a PW-PDU, the encapsulation layer must provide it.

It is a common practice to use an error detection mechanism such as a CRC or similar mechanism to ensure end-to-end integrity of frames. The PW service-specific mechanisms must define whether the packet's checksum shall be preserved across the PW or be removed from PE-bound PDUs and then be recalculated for insertion in CE-bound data.

For protocols such as ATM and FR, the checksum is restricted to a single link (e.g., FR DLCI or ATM VPI/VCI).

For congestion consideration, if the traffic carried over the PW is known to be TCP friendly, packet discard in the PSN will trigger the necessary reduction in offered load, and no additional congestion avoidance action is necessary.

If the PW is operating over a PSN that provides enhanced delivery, PEs should monitor packet loss to ensure that the requested service is actually being delivered. If it is not, then the PE should assume that the PSN is providing a best-effort service and should use the best-effort service congestion avoidance measures described below.

4.4.4   Maintenance Reference Model

Signaling between CEs and PE is used to maintain the PW components as depicted in Figure 4.10.

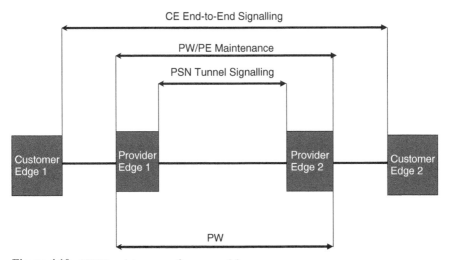

Figure 4.10   PWE3 maintenance reference model.

The end-to-end signaling is between the CEs. This signaling could be frame relay PVC status signaling, ATM SVC signaling, TDM CAS (channel-associated signaling) signaling, etc.

The PW/PE maintenance is used between the PEs (or NSPs) to set up, maintain, and tear down PWs, including any required coordination of parameters.

The PSN tunnel signaling controls the PW multiplexing and some elements of the underlying PSN. Examples are L2TP control protocol, MPLS LDP, and RSVP-TE.

4.5 CONTROL PLANE

PWE3 control-plane services include setting up and tearing down a PW, monitoring it, and dealing with various fault conditions. The control plane messages such as Ethernet Flow control and TDM tone signaling maybe exchanged in-band, while the messages such as the signaling VC of an ATM VP and TDM CCS signaling maybe exchanged out of band. The control-plane messages should be transported by either a higher priority or a reliable channel provided by the PW demultiplexer layer.

The control-plane services are as follows:

- *Setup or Teardown of PWs* —A PW must be set up before an emulated service can be established and must be torn down when an emulated service is no longer needed. Setup or teardown of a PW can be triggered by an operator command, from the management plane of a PE, by signaling setup or teardown of an AC (e.g., an ATM SVC), or by an autodiscovery mechanism.

- *Status Monitoring* —Some native services such as ATM have mechanisms for status monitoring. For these services, the corresponding emulated services must specify how to perform status monitoring. For status changes, including PW up and down, a notification should be sent to the management system. When the physical link (or subnetwork) between a CE and a PE fails, all the emulated services that go through that link (or subnetwork) will fail. Then, it is desirable that a single notification message be used to notify failure of the whole group of emulated services.

- *Misconnection and Payload Type Mismatch* —Misconnection can breach the integrity of the system, and the payload mismatch can disrupt the customer network. The tunneling mechanism and its associated control protocol can be used to deal with mismatch issues. For example, a PW-type identifier is exchanged during the PW setup, which is used to verify the compatibility of the ACs.

- *Packet Loss, Corruption, and Out-of-Order Delivery* —Packet loss, corruption, and out-of-order delivery on the PSN path between PEs may occur. For some payload types, these errors can be mapped either to a bit error burst or to loss of carrier on the PW. If a native service has some mechanism to deal with bit error, the corresponding PWE3 service should provide a similar mechanism.

- *Keep Alive*—If a native service has a keep-alive mechanism, the corresponding emulated service must provide a mechanism to propagate it across the PW.

In the following sections two examples are given.

4.5.1   PWE3 over an IP PSN

The PWE3 over an IP PSN protocol structure is depicted in Figure 4.11.

Timing and sequencing are provided by the RTP [3]. The encapsulation layer may also carry a sequence number. In that case, sequencing should be provided by either the PW encapsulation or the RTP.

The PW demultiplexing is provided by the PW label that can be an MPLS label, an L2TP session ID, or a UDP port number. When PWs are carried over IP, the PSN convergence layer will not be needed.

4.5.2   PWE3 over an MPLS PSN

The protocol layering for PWE3 over an MPLS PSN is given in Figure 4.12. An inner MPLS label is used to provide the PW demultiplexing function. A control word is used to carry most of the information needed by the PWE3 encapsulation layer and the PSN convergence layer in a compact format. The flags in the control word provide the necessary payload convergence. A sequence

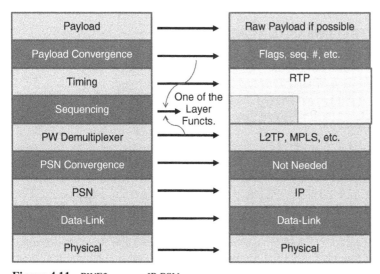

Figure 4.11   PWE3 over an IP PSN.

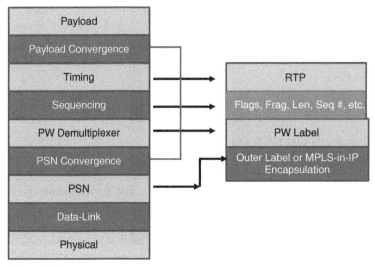

Figure 4.12   PWE3 over an MPLS PSN using a control word.

field provides support for both in-order payload delivery and a PSN fragmentation service within the PSN convergence layer (supported by a fragmentation control method).

Ethernet pads all frames to a minimum size of 64 bytes. The MPLS header does not include a length indicator. Therefore, to allow PWE3 to be carried in MPLS to pass correctly over an Ethernet data link, a length correction field is needed in the control word. As with an IP PSN, where appropriate, timing is provided by the RTP.

4.6 MULTISEGMENT ARCHITECTURE

PWs may span multiple domains of one or more networks. Figure 4.13 shows a multisegment case where Terminating PE1 (T-PE1) and Terminating PE3 (T-PE3) provide PWE3 service to CE1 and CE2. One PSN tunnel extends from T-PE1 to S-PE1 across PSN1, and a second PSN tunnel extends from S-PE1 to T-PE2 across PSN2 [4].

PWs are used to connect ACs attached to T-PE1 to the corresponding ACs attached to T-PE2. Each PW on PSN tunnel 1 is switched to a PW in the tunnel across PSN2 Sat S-PE1 to complete the multisegment pseudowire (MS-PW) between T-PE1 and T-PE2. S-PE1 is the PW switching point called as the *PW switching provider edge* (S-PE). PW1 and PW3 are segments of the same MS-PW, while PW2 and PW4 are segments of another PW. PW segments of the same MS-PW (e.g., PW1 and PW3) may be of the same PW type or different types, and PSN tunnels (e.g., PSN Tunnel 1 and PSN Tunnel 2) can be the same or different technology.

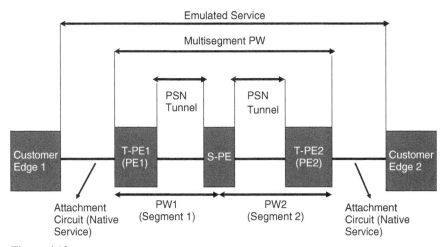

Figure 4.13 The PW switching reference model.

There are two methods for switching a PW between two PW domains. In the first method (Fig. 4.14), the two separate control-plane domains terminate on different PEs [5].

In Figure 4.15, PWs in two separate PSNs are stitched together using native service ACs. PE2 and PE3 only run the control plane for the PSN to which they are directly attached.

In Figure 4.15, SPE1 runs two separate control planes: one toward TPE1 and one toward TPE2. The PW switching point (S-PE) is configured to connect PW Segment 1 and PW Segment 2 together to complete the multisegment PW

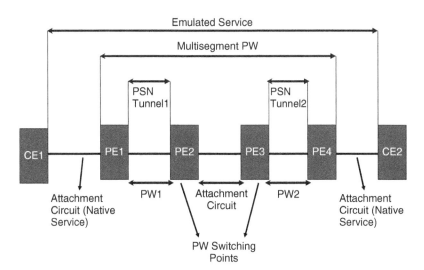

Figure 4.14 PW switching using AC reference model.

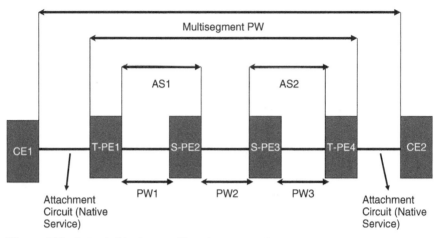

Figure 4.15 PW switching interprovider reference model.

between TPE1 and TPE2. PW Segment 1 and PW Segment 2 must be of the same PW type. However, PSN Tunnel 1 and PSN Tunnel 2 do not have to be the same technology since the PEs can adapt the PDU encapsulation between the different PSN technologies.

The PWs in each PSN are established independently, where each PSN is treated as a separate PW domain. For example, in Figure 4.15, PW1 can be set up between PE1 and PE2 using the LDP (label distribution protocol) targeted session, while PW2 is set up between PE3 and PE4 at the same time. On the other hand, the ACs are configured as the same PW type.

An S-PE switches an MS-PW from one segment to another based on the PW identifiers such as PW label of MPLS PWs. In Figure 4.15, the domains that PSN Tunnel 1 and PSN Tunnel 2 traverse could be IGP areas in the same IGP network.

A PW may transit more than one S-PEs along its path as shown in Figure 4.15, where S-PE2 at the border of one AS1 (Autonomous System 1) and S-PE3 at the border of AS2. The MS-PW between T-PE1 and T-PE4 is composed of three segments, PW1 segment in AS1, PW2 segment between two-border routers S-PE2, and S-PE3 acting as the switching PEs, and PWE3 segment in AS2. AS1 and AS2 could belong to the same provider or to two different providers.

PWE3 (Pseudowire Edge-to-Edge Emulation) defines the signaling and encapsulation techniques for establishing SS-PWs between a pair of terminating PEs (T-PEs), and in the vast majority of cases, this will be sufficient. MS-PWs may be useful in the following situations:

- A PW extends from a T-PE in one provider domain to a T-PE in another provider domain.

- It may not be feasible to establish a direct PW control channel between the T-PEs, residing in different provider networks, to set up and maintain PWs.
- PWE3 signaling protocols and PSN types may differ in different provider networks. The terminating PEs may be connected to networks employing different PW signaling and/or PSN protocols. In this case, it is not possible to use an SS-PW. An MS-PW with the appropriate signaling protocol interworking performed at the PW switching points can enable PW connectivity between the terminating PEs in this scenario.
- In deploying PWs edge to edge in large SP networks, the PWs will be tunneled over PSN TE tunnels with bandwidth constraints. A single-segment PW architecture would require that a full mesh of PSN TE tunnels be provisioned to allow PWs to be established between all PEs. In this environment, the network is either partitioned into a number of smaller PWE3 domains or consisting of a sparse mesh of PSN TE tunnels and PW signaling adjacencies
- SPs wish to extend PW technology to access and metro networks in order to reduce maintenance complexity and operational costs. For example, in hierarchical IP/MPLS networks, access networks connected to a backbone use PWs up to the edge of the backbone where they can be terminated or switched onto a PW segment crossing the backbone. The use of PWE3 switching between the access and backbone networks can potentially reduce the PWE3 control channels and routing information processed by the access network T-PEs.

4.7 MULTISEGMENT PSEUDOWIRE SETUP MECHANISMS

An MS-PW can traverse multiple SP administrative domains. Furthermore, it can traverse multiple autonomous systems within the same administrative domain or different administrative domains. As a result, PWs and PW control channels such as targeted LDP may cross AS (autonomous system) boundaries.

A multisegment PW can be configured statically, where the switching points (S-PEs) and T-PEs are manually provisioned for each segment. The static configuration of MPLS labels for MPLS-PW segments and the cross-connection of them are useful when an MS-PW crosses provider boundaries and two providers do not want to run any PW signaling protocol between them.

A multisegment PW can be also configured using signaling mechanisms via either predetermined routes or signaled dynamic routes. In predetermined route configuration, the PW is established along an administratively determined route using an end-to-end signaling protocol with automated stitching at the S-PEs. In signaled dynamic route configuration, the PW is established along a dynamically determined route using an end-to-end signaling protocol with automated stitching

at the S-PEs. The number of S-PEs traversed is only limited by the TTL field of
PW MPLS label, which is set by the originating PE. In establishing an MPLS-PW
via signaling, LDP with FEC 128 (i.e., PWid FEC Element) or LDP with FEC
129 (i.e., Generalized PWid FEC Element) can be used. In establishing layer 2
Tunneling Protocol version 3 (L2TPv3) PW, L2TPv3 is used.

When the MS-PW segments are dynamically signaled, the signaled MS-PW
segments can be on the path of a statically configured MS-PW, signaled/statically
routed MS-PW, or signaled/dynamically routed MS-PW. Segments are dynamically rerouted around failure points when segments are set up using the dynamic
setup method.

For the MPLS-PW setup, there would be the following four PW switching
alternatives:

- Switching between two static control planes
- Switching between a static and a dynamic LDP control plane
- Switching between two LDP control planes using the same FEC type
- Switching between LDP using FEC 128 and LDP using the generalized
  FEC 129.

For static control-plane switching, the S-PE is configured to direct the MPLS
packets from one PW into the other. There is no control protocol involved.
It is possible to have one of the control planes is a simple static PW configuration and the other control plane is either a dynamic LDP FEC 128 or
generalized PW FEC. In this case, the static control plane is considered similar
to an AC.

In switching between LDP using FEC 128 and LDP using the Generalized
FEC 129, the PE using the generalized FEC 129 can be active or passive. A
PE that assumes the active role will send the LDP PW setup message, while
a passive role PE will simply reply to an incoming LDP PW setup message.
The S-PE will always remain passive until a PWid FEC 128 LDP message is
received, which will cause the corresponding generalized PW FEC LDP message
to be formed and sent. If a generalized FEC PW LDP message is received, while
the switching point PE is in a passive role, the corresponding PW FEC 128 LDP
message will be formed and sent.

Control-plane switching between MPLS-PW and L2TPv3-PW can be static
or dynamic as well. The switching alternatives are as follows:

- Switching between static MPLS-PW and static L2TPv3 PW, where
  there is no control protocol involved. The S-PE maps MPLS-PW label
  to L2TPv3 Session ID as well MPLS tunnel label to PE destination IP
  address.
- Switching between a static MPLS-PW and a dynamic L2TPv3 PW, where
  the static control plane is considered identical to an AC.
- Switching between a static L2TPv3 PW and a dynamic LDP/MPLS-PW,
  where the static control plane is considered identical to an AC.

- Switching between a dynamic LDP/MPLS-PW and a dynamic L2TPv3 PW, where the switching point assumes an initial passive role and does not initiate an LDP/MPLS or L2TPv3 PW until it has received a label mapping or incoming-call request from one side of the node. MPLS PWs are made up of two unidirectional label switched paths (LSPs) bonded together by FEC identifiers. The L2TPv3 PWs are bidirectional in nature and set up via message exchanges.

In dynamic route selection for an MS-PW, S-PEs and T-PEs discover S-PEs on the path to a destination T-PE. After that, the S-PEs along the MS-PW are automatically selected.

4.7.1 LDP SP-PE TLV

The edge-to-edge PW might traverse several switching points, in separate administrative domains as described above. For management and troubleshooting reasons, it is useful to record information about the switching points at the S-PEs that the PW traverses. This is accomplished by a PW switching point PE TLV (SP-PE TLV) [5].

The SP-PE TLV may appear once for each switching point traversed and cannot be of length zero. The SP-PE TLV is appended to the PW FEC at each S-PE, and the order of the SP-PE TLVs in the LDP message must be preserved. The SP-PE TLV is necessary to support some of the virtual circuit connectivity verification (VCCV) functions for MS-PWs. The SP-PE TLV is encoded in Figure 4.16.

The SP-PE TLV format is shown in Figure 4.16. SP-PE TLV length field specifies the total length of all the following SP-PE TLV fields in octets. Sub-TLV type field encodes how the value field is to be interpreted. The SP-PE TLV contains sub-TLVs to describe various characteristics of the S-PE traversed. Length field specifies the length of the value field in octets. Value field is the octet string of length octets that encodes information to be interpreted as specified by the type field.

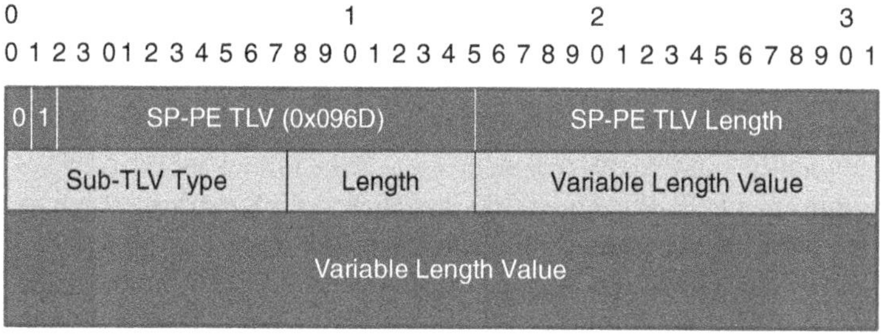

Figure 4.16 SP-PE TLV.

4.8 RESILIENCY

A PW segment, a contiguous set of PW segments, and the end-to-end path can be protected. The protection and primary paths for the protected segment(s) share the same respective segments endpoints. A protection path for a PW segment, sequence of segments, or end-to-end path is signaled.

Traffic is switched from a primary PW to secondary PW when an element on the path of a primary MS-PW fails. The primary and backup paths may be statically configured, statically specified for signaling, or dynamically selected via dynamic routing depending on the MS-PW establishment mechanism. Backup and primary paths should have the ability to traverse separate S-PEs. For example, a backup PW can be configured with a different T-PE from the primary path.

The protection mechanism can automatically revert to a primary PW from a backup PW, once the primary path is recovered from failures.

4.9 QUALITY OF SERVICE AND CONGESTION CONTROL

PWs are intended to support emulated services with strict packet/frame loss, delay, and jitter requirements satisfied by reserving sufficient network resources such as bandwidth and buffer and by providing appropriate scheduling priority and drop precedence throughout the network.

Path provisioning is frequently performed through QoS reservation protocols or network management protocols. QoS provisioning for MS-PWs, which may transmit across network domains under the control of multiple entities, is much more difficult than that for SS-PWs that remain within a single administrative domain.

When the T-PE attempts to signal an MS-PW, signaling identifies the CoS associated with an MS-PW, carries the traffic parameters used by the admission control for an MS-PW per CoS, and separates traffic parameter values to be specified for the forward and reverse directions of the PW.

The signaling protocol prioritizes the PW setup and support maintenance operation among PWs.

For SS-PWs, a traffic-engineered PSN tunnel (i.e., MPLS-TE) may be used to ensure that sufficient resources are reserved in the P-routers to provide QoS to PWs on the tunnel. In this case, T-PEs will provide admission control of PWs onto the PSN tunnel and accounting for reserved and available bandwidth on the tunnel.

For MS-PWs, each S-PE maps a PW segment to a PSN tunnel, where S-PEs and T-PEs automatically bind a PW segment to a PSN tunnel based on CoS and bandwidth requirements. S-PEs and T-PEs associate a CoS marking, such as EXP field value of MPLS PWs, with PW-PDUs to specify packet treatment.

Different administrative domains may use different CoS values to imply the same CoS treatment. S-PEs at administrative domain boundaries translate from one CoS value to another as a PW-PDU crosses from one domain to the next.

The CoS and bandwidth of the MS-PW are configurable at T-PEs and S-PEs. Each domain individually implements a method to control congestion. This can be by QoS reservation or other congestion control method.

Each PSN carrying the PW may be subject to congestion. Each PW segment will handle any congestion independently of the other MS-PW segments.

4.10   OPERATIONS AND MAINTENANCE (OAM)

The PE reports the status of the interface and tabulates statistics for PW that help monitoring the state of the network and measure service-level agreements (SLAs) for the PW. Typical counters are as follows:

- Counts of PW-PDUs sent and received, with and without errors,
- Counts of sequenced PW-PDUs lost,
- Counts of service PDUs sent and received over the PSN, with and without errors,
- Service-specific interface counts, and
- One-way delay and delay variation.

End-to-end connectivity and the exact functional path can be identified by connection verification and traceroute mechanisms available at PEs. Connection verification and other alarm mechanisms can alert the operator that a PW has lost its remote connection.

The OAM (operations and maintenance) mechanisms defined in ITU-T I.610 can be used for ACs to detect, localize, and diagnose defects in the network and communicates PW defect states on the PW AC.

Defect states for SS-PWs between AC and PWs are propagated across a PWE3 network following the failure and recovery from faults. For MS-PWs, a common PW OAM mechanism agreed by all PE routers along the MS-PW is supported end to end. Failure of a segment is notified to other segments of an MS-PW. At the S-PE, defects on an PSN tunnel is propagated to all PWs that utilize that particular PSN tunnel.

The S-PE can behave as a segment endpoint and pass T-PE to T-PE PW OAM messages transparently. Both MS-PWs and SS-PWs can measure round-trip delay, one-way delay, jitter, and packet loss ratio.

Single-segment PWs and multisegment PW capabilities are signaled using the VCCV parameter included in the interface parameter field of the PWid FEC TLV or the interface parameter sub-TLV of the generalized PWid FEC TLV. When a switching point exists between PE nodes, it is required to be able to continue operating VCCV end to end across a switching point and to provide the ability to trace the path of the MS-PW over any number of segments.

When an MS-PW includes SS-PWs that use the L2TPv3, the MPLS-PW OAM is terminated at the S-PE connecting the L2TPv3 and MPLS segments. Status information received in a particular PW segment can be used to generate the appropriate status messages on the following PW segment. In the case of L2TPV3, the status bits in the circuit status can be mapped directly to the PW status bits defined in Reference 2.

The VCCV messages are specific to the MPLS data plane and cannot be used for an L2TPv3 PW segment. The VCCV messages from L2TPv3 PW segments must be translated to those for MPLS-PW segments and vice versa.

As stated above, the S-PE performs a standard MPLS label swap operation on the MPLS PW label, where the PW label TTL is decreased at every S-PE. Once the PW label, TTL, reaches the value of 0, the packet is sent to the control plane to be processed. Hence, by controlling the PW TTL value of the PW label, it is possible to select exactly which S-PE will respond to the VCCV packet.

In the PW switching with ACs (Fig. 4.14), PW status messages indicating PW or AC faults is mapped to fault indications or OAM messages on the connecting AC.

In the PW control-plane switching (Fig. 4.17), the status of the PWs is forwarded unchanged from one PW to the other by the control-plane switching function.

Communication of fault status of one of the locally attached PW segments at an S-PE may be needed. For the LDP, this can be accomplished by sending an LDP notification message as shown in Figure 4.17.

This message is then relayed by each S-PE unchanged. The T-PE decodes the status message and the included SP-PE TLV to detect exactly where the fault occurred. At the T-PE, if there is no SP-PE TLV included in the LDP status

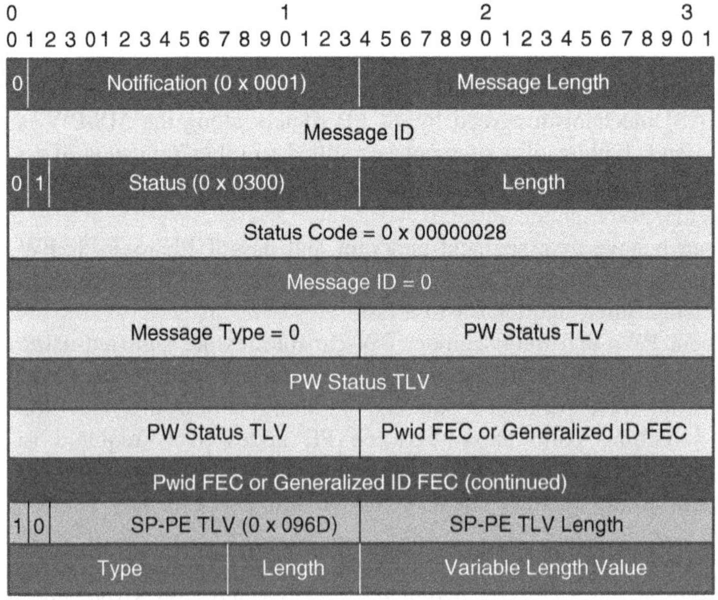

Figure 4.17 LDP notification message.

notification, then the status message can be assumed to have originated at the remote T-PE.

4.11 SECURITY

PWE3 provides no means of protecting the integrity, confidentiality, or delivery of the native data units. The relatively weak security mechanisms represent a greater vulnerability in an emulated Ethernet connected via a PW.

Controlling PSN access to the PW tunnel end point may protect against PW demultiplexer and PSN tunnel services disruption. By restricting PW tunnel end point access to legitimate remote PE sources of traffic, the PE may reject traffic that would interfere with the PW demultiplexing and PSN tunnel services. Security protocols such as IPSec may be used by the PW demultiplexer layer in order provide authentication and data integrity of the data between the PW demultiplexer end points.

Security needs to be provided for data plane and control plane [6]. For data-plane security, packets of an MS-PW traveling to a PE or an AC should be delivered to intended recipients. Packets from outside an MS-PW entering the MS-PW should be consistent with the policies of the MS-PW.

MS-PWs that cross SP domain boundaries may connect one T-PE in a SP domain to a T-PE in another provider domain. They may also transit other provider domains even if the two T-PEs are under the control of one SP.

When there is one or more PDUs that are falsely inserted into an MS-PW at any of the originating, terminating, or transit domains as a result of a malicious attack or fault in the S-PE, there should be a mechanism for the end-to-end authenticity of MS-PW PDUs.

For control-plane security, an MS-PW connects two ACs. It is important to make sure that PW connections are not arbitrarily accepted from anywhere, or else a local AC might get connected to an arbitrary remote AC.

Directly interconnecting the S-PEs using a physically secure link, and enabling signaling and routing authentication between the S-PEs, eliminates the possibility of receiving an MS-PW signaling message or packet from an untrusted peer.

S-PEs in different provider networks may reside at each end of a physically secure link, or be interconnected by a limited number of trusted PSN tunnels, each S-PE will have a trust relationship with a limited number of S-PEs in other ASs.

Static manual configuration of MS-PWs at S-PEs and T-PEs provides a greater degree of security. If an identification of both ends of an MS-PW is configured and carried in the signaling message, an S-PE can verify the signaling message against the configuration.

An incoming MS-PW request/reply is not accepted unless its IP source address is known to be the source of an "eligible" peer, which is an S-PE or a T-PE with which the originating S-PE or T-PE has a trust relationship.

The set of eligible peers could be preconfigured (either as a list of IP addresses or as a list of address/mask combinations) or automatically generated from the local PW configuration information.

The S-PE and T-PE drop the unaccepted signaling messages in the data path to avoid a denial-of-service (DoS) attack on the control plane. S-PEs that connect one provider domain to another provider domain usually rate limit signaling traffic in order to prevent DoS attacks on the control plane.

Even if a connection request appears to come from an eligible peer, its source address can be spoofed. Source address filtering at the border routers of that network could eliminate the possibility of source address spoofing.

A PW connects two ACs. In order to prevent a local AC getting connected to an arbitrary remote AC, the LDP connections are not arbitrarily accepted from anywhere. An incoming session request MUST NOT be accepted unless its IP source address is known to be the source of an "eligible" peer.

Even if a connection request appears to come from an eligible peer, its source address can be spoofed. A mean of preventing source address spoofing must be in place such as source address filtering at the border routers that could eliminate the possibility of source address spoofing.

The LDP MD5 authentication can be used to provide integrity and authentication for the LDP messages and protect against source address spoofing.

4.12 CONCLUSION

The PW3 is a mechanism to emulate telecommunications services across packet-switched networks. The emulated services are mostly T1/T3 leased lines, frame relay, Ethernet, and ATM to maximize return on existing assets and minimize operating costs of the SPs. Therefore, payloads can be packet, cell, bit stream, and structured bit stream.

A PW is a connection between two PE equipment connecting two ACs. It can span one or multiple networks. The PWE3 control-plane services provide capabilities for setting up and tearing down a PW, monitoring it and dealing with various fault conditions.

The PE reports the status of the interface and tabulates statistics for PW that help monitoring the state of the network and measure service-level agreements (SLAs) for the PW. The end-to-end connectivity and the exact functional path can be identified by connection verification and trace route mechanisms available at PEs.

The PWE3 provides no means of protecting the integrity, confidentiality, or delivery of the native data units. Security needs to be provided for data plane and control plane. For data-plane security, packets of an MS-PW traveling to a PE or an AC should be delivered to intended recipients. Packets from outside an MS-PW entering the MS-PW should be consistent with the policies of the MS-PW.

REFERENCES

1. RFC 3985, Pseudo Wire Emulation Edge-to-Edge (PWE3) Architecture, 2005.
2. RFC 3916, Requirements for Pseudo-Wire Emulation Edge-to-Edge (PWE3), Sept 2004.
3. RFC 3550, RTP: A Transport Protocol for Real-Time Applications, July 2003.
4. RFC 5254, Requirements for Multi-Segment Pseudowire Emulation Edge-to-Edge (PWE3), Oct 2008.
5. RFC 6073, Segmented Pseudowire, Jan 2011.
6. RFC 4447, Pseudowire Setup and Maintenance Using the Label Distribution Protocol (LDP), April 2006.

Chapter 5

Ethernet Protection

5.1 INTRODUCTION

Services such as broadcast video, voice over IP, and video on demand require five-nines availability, which means a high-level survivability against failures. When failures do occur, they are not supposed to be noticed by the subscriber. The main purpose of automatic protection switching (APS) is to guarantee the availability of resources in the event of failure and to ensure that switchover is achieved in less than 50 ms so that the network will not be affected, although 50 ms is currently being debated in the industry. In fact, MEF 32 [1] mandates 500-ms protection switching at User Network Interface (UNI) or External Network to Network Interface (ENNI). The smaller switching time increases the cost of the equipment. The author's view is that at higher speeds such as 1 G, 10 G, protection switching time should be even smaller than 50 ms in order to reduce frame loss so that the service can stay within FLR limits. The protection switching time is the *minimum possible time* for the sum of persistent (i.e., non-transient) failure detection, speed of light propagation, signaling protocol time, and regaining sync alignment.

To a degree, both the Spanning Tree Protocol (STP) and Rapid Spanning Tree Protocol (RSTP) have been proven effective for preventing loops and assuring backup paths are available. However, both protocols are slow to respond to network failures. They are slow on the order of 30 s or more. APS, Ethernet Protection Switched Rings (EPSRs), and Link Aggregation Groups (LAGs) are among the proposed solutions.

Maintenance operations can benefit from network flexibility and from the ability to transmit traffic over either the "working" entity or the "protection" entity. Therefore, the APS mechanism needs to support traffic in order to be manually switched over from one trunk to the other.

In a communication network, traffic protection can be implemented in many different ways for various network topologies such as star, loop, and mesh. Loop-based networks are attractive for enabling redundancy in a network consisting of a relatively large number of nodes. A redundant path for each node is provided by just one additional link that closes two adjacent branches to form a loop. However,

Networks and Services: Carrier Ethernet, PBT, MPLS-TP, and VPLS, First Edition. Mehmet Toy.
© 2012 John Wiley & Sons, Inc. Published 2012 by John Wiley & Sons, Inc.

loops are not allowed in Ethernet-based transport networks since Ethernet frames would circulate forever. So the loop has to be broken at some point so that Ethernet transport is enabled.

Ethernet automatic protection switching (EAPS) is an exemplary solution for layer-2 loop protection, which is comparable to solutions such as EPSRs and Ethernet ring protection (ERP) [2]. The terms loop and ring are synonyms here.

The EAPS ring consists of the master node and one or more transit nodes. The two ring ports of the master node are configured as primary port and secondary port. The master node blocks logically the secondary port except for a control VLAN. The master node sends periodic health-check packets from the primary port through the control VLAN toward the secondary port. When a fault occurs in the ring, the master detects this either by missing health-check packets or by special fault detection packets generated by one of the transit nodes.

EAPS and similar solutions are on/off-type mechanisms without adaptation to available link capacities in the ring. Therefore, they do not allow any load balancing. In the spanning tree, loops resulting from redundant paths are broken by the use of the STP algorithm. The STP breaks loops by disabling Ethernet switch ports so that the remaining active links build up a tree topology. In a failure case, when an active link breaks, STP calculates a new tree, taking then the appropriate so far disabled links into use. The RSTP converges faster. However, both STP and RSTP are on/off-type mechanisms without adaptation to available link capacities in the ring, therefore, not allowing any load balancing.

Resilient Packet Ring (RPR; IEEE 802.17), which is independent of the underlying physical layer, provides a fast protection switching (<50 ms. The RPR concept is based on two counterrotating rings designed to transport Ethernet frames efficiently. There are no dedicated protection resources. Both the rings transport traffic using the shortest paths RPR provides.

In this chapter, linear protection, ring-based protection, and LAGs are described.

5.2 AUTOMATIC PROTECTION SWITCHING (APS) ENTITIES

APS includes detecting *failures* (signal failure (SF) or signal degrade (SD)) on a *working channel*, switching traffic *transmission* to a *protection channel*, selecting traffic *reception* from the protection channel, and (optionally) reverting back to the working channel once failure is repaired. The entities are (Fig. 5.1)

- **working entity**, which is used when no failure exists;
- **protection entity**, which is used when a failure exists;
- **head end**, which is the entity transmitting data to working/protection channel;

Figure 5.1 APS entities.

- **tail end**, which is the entity receiving data from the working/protection channel;
- **bridge**, function at head end that connects traffic (including extra traffic) to the working and protection channels;
- **selector**, function at tail end that extracts traffic (perhaps extra traffic) from the working or protection channel;
- **APS signaling channel**, channel used to communicate between head end and tail end for APS purposes;
- **trail termination**, function responsible for failure detection including injection and extraction of OAM.

Protection switching is usually triggered by a *failure* although the operator may manually *force* a protection switch. A *failure* is declared when a fault persists long enough for the ability to perform the required function. Failures are SF or SD (of various types), and they may be detected by physical layer, indicated by signaling (e.g., AIS), and detected by OAM mechanisms. When there is no SF or SD, the state is called *no request* (*NR*). EAPS is a linear protection scheme designed to protect VLAN-based Ethernet networks.

With EAPS, a protected domain is configured with two paths, a working path and a protection path. Both the paths can be monitored using an OAM protocol such as Connectivity Fault Management (CFM). Normally, traffic is carried on the working path and the protection path is disabled. If the working path fails, APS switches the traffic to the protection path and the protection path becomes the active path.

The EAPS can be unidirectional and bidirectional. The unidirectional switching utilizes fully independent selectors at each end of the protected domain. Bidirectional switching attempts to configure the two end points with the same bridge and selector settings, even for a unidirectional failure. Unidirectional switching can protect two unidirectional failures in opposite directions on different entities.

Protection switching can be unidirectional and bidirectional. With unidirectional switching, the selectors at each end are fully independent. Therefore, the two ends must be coordinated (i.e., APS communications). Since there is no signaling, it could be faster than bidirectional switching and easy to implement. On the other hand, the bidirectional switching management can be easier since directions traverse the same network elements.

5.3 LINEAR PROTECTION

APS uses two modes of operation: the linear 1+1 protection switching archi-tecture and the linear 1 : 1 protection switching architecture. The former mode operates with either unidirectional or bidirectional switching, whereas the latter mode operates with bidirectional switching.

The linear protection architecture is defined in Reference 3, in which protec-tion switching occurs at the two distinct end points of a point-to-point Ethernet entity such as EVC. Between these end points, there will be both "working" and "protection" transport entities. In the linear 1+1 protection switching architec-ture, the normal traffic is copied and fed to both working and protection paths with a permanent bridge at the source of the protected domain. The traffic on the working and protection transport entities is transmitted simultaneously to the sink of the protected domain, where a selection between the working and protection transport entities is made.

In the linear 1 : 1 protection switching architecture, the normal traffic is trans-ported on either the working path or on the protection path using a selector bridge at the source of the protection domain. The selector at the sink of the protected domain selects the entity that carries the normal traffic.

Protection switching can be revertive or nonrevertive. In revertive operation, normal traffic signal is restored to the working unit after the condition(s) causing a switch has cleared. In the case of clearing a command, such as forced switch, this happens immediately. In the case of clearing of a defect, this generally happens after the expiry of a wait-to-restore (WTR) timer, to avoid chattering of selectors in the case of intermittent defects.

In the nonrevertive operation, normal traffic signal is allowed to remain on the protection transport entity even after a switch reason has cleared. This is generally accomplished by replacing the previous switch request with a "Do not Revert (DNR)" request.

Revertive mechanisms may be preferable when the working channel has better performance (free bandwidth, BER (bit error rate), delay), when there are frequent switches (easier to manage), and when there is extra traffic. However, the nonrevertive operation also has advantages of only one service disruption due to protection switching and may be simpler to implement.

Protection switching can be triggered by either an administrator or SFs. The details are discussed in the following section.

5.3.1 1+1 Protection Switching

1+1 protection is the simplest and fastest form of protection, but wastes the capacity of the protecting entity. Head-end bridge always sends data on both working and protecting entities, while the tail end selects the transport entity to use.

1+1 protection is often provisioned as nonrevertive to avoid a second inter-ruption to the normal traffic signal. However, there are reasons to be revertive as well. The primary path could be a favorable path for traffic in terms of delay, capacity, etc.

1+1 bidirectional linear protection switching architecture is shown in Figure 5.2 [3]. The protected Ethernet entity (i.e., EVC, OVC, Link) connection is permanently bridged to both the working transport entity and the protection transport entity. The Ethernet frames are received only from the working entity.

When a fault such as loss of signal occurs in the working entity, the traffic is switched from the working entity to the protecting entity (Fig. 5.3) [3]. Both directions are switched even when a unidirectional defect occurs. The APS coordination protocol is necessary to achieve coordination.

The 1+1 unidirectional linear protection switching architecture is shown in Figure 5.4 [3]. The protected Ethernet connection is permanently bridged to both the working transport entity and the protection transport entity. The traffic is received only from the working entity for both directions.

When a fault such as loss of signal occurs in the working entity in the west to east direction, each direction is switched independently. Selectors at the sink of the protected domain operate only based on the local information; therefore, the APS coordination protocol is not necessary (Fig. 5.5) [3]. Traffic that is not affected from this failure continues to be received via the working entity.

When there are failures in both working and protecting entities (Fig. 5.6) [3], both entities are switched. Unidirectional protection switching can protect this type of dual-defect scenarios, while bidirectional protection switching cannot.

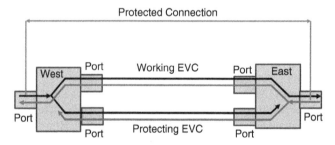

Figure 5.2   1+1 Bidirectional protection switching architecture.

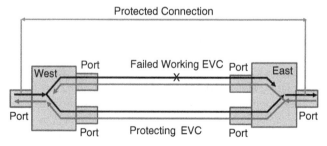

Figure 5.3   1+1 Bidirectional protection switching for failed EVC.

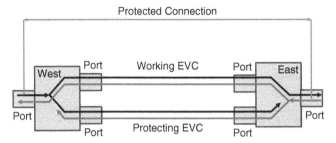

Figure 5.4 1+1 Unidirectional protection switching architecture.

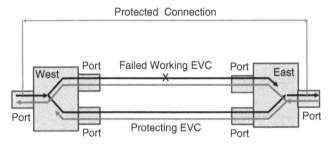

Figure 5.5 1+1 Unidirectional protection switching for failed EVC in the west to east direction.

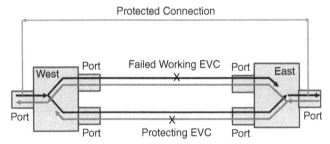

Figure 5.6 1+1 Unidirectional protection switching for failed EVC in both directions.

This is possible if each of the transport entities have a healthy path that does not use the faulty units to carry the traffic.

5.3.2 1 : 1 Protection Switching

In the 1 : 1 protection scheme, head-end bridge usually sends data on the working channel. When failure is detected, it starts sending frames over the protecting entity and the tail end selects the protecting entity. The protecting channel is used for traffic other than the one that is being protected, when it is not used

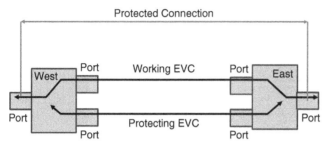

Figure 5.7 1:1 Bidirectional protection switching architecture.

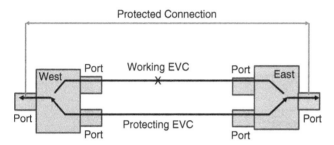

Figure 5.8 1:1 Bidirectional protection switching for failed EVC.

for protecting the working entity. This is the main difference between 1 : 1 and 1+1 protection schemes where the capacity of the protecting entity is not wasted. The failure is detected by the tail end via Continuity Check Messages (CCMs); therefore, the APS signaling is necessary.

1 : 1 protection is usually revertive, but it is possible to define the protocol to permit nonrevertive operation for 1 : 1 protection. In general, the choice of revertive or nonrevertive will be the same at both ends of the protection group. However, a mismatch of this parameter does not prevent interworking.

Figure 5.7 [3] illustrates the 1 : 1 linear protection switching architecture, where the normal traffic is being transmitted over the working entity and the tail-end protection switching process that determines the specific output for Ethernet connection A is transferred.

When there is a failure in the working entity (Fig. 5.8), at the head end, the traffic for connection A is forwarded to the protection transport entity. At the tail end, the traffic for connection A is received from the protection transport entity.

5.3.3 Protection Switching Triggers

Protection switching can be initiated by a network administrator or network failures. A network operator can trigger protection switching in the forced switch or the manual switch mode if it has a higher priority than any other local request

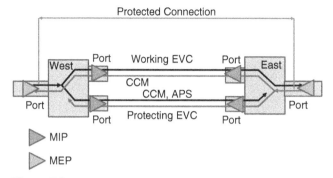

Figure 5.9 MEPs in 1+1 bidirectional EVC protection switching architecture.

	1								2								3								4						
8	7	6	5	4	3	2	1	8	7	6	5	4	3	2	1	8	7	6	5	4	3	2	1	8	7	6	5	4	3	2	1

MEL	Version (0)	OpCode (APS = 39)	Flags (0)	TLV Offset (4)

APS-Specific Information

End TLV (0)

Figure 5.10 APS PDU format.

or the tail-end request. Similarly, a declared SF or SD in the working entity can switch traffic to the protecting entity if the protecting entity is healthy and the detected SF condition has a higher priority than any other local request or the far-end request. In the bidirectional 1+1 and 1 : 1 architecture, the received APS protocol requests to switch have a higher priority than any other local request.

In order to monitor both the working and protecting entities, MEPs at head end and tail end must be active and must transmit CCMs. SF condition can be due to LOC (loss of connectivity) as a result of not receiving three CCMs within a configured time interval or other failure reasons discussed in Chapter 7.

The protection switching process also requires APS communication in order to coordinate its switching behavior with the tail end of the protected domain if the protection switching architecture is not 1+1 unidirectional protection switching. APS PDU is transmitted and received between the same MEP pair on the protection transport entity where CCM is transmitted for monitoring (Fig. 5.9) [3].

5.3.4 APS PDU Format

Protection switching control is achieved via Ethernet OAM PDU, as defined in Reference 4. Proactive OAM frames for the EAPS Function need to be counted. Four octets in the APS PDU are used to carry APS-specific information (Figs 5.10 and 5.11) [3]. The TLV Offset (type, length, and value) field is required to be set to 0x04.

1	2	3	4
8 7 6 5 4 3 2 1	8 7 6 5 4 3 2 1	8 7 6 5 4 3 2 1	8 7 6 5 4 3 2 1

Request/ State	Prot. Type A B D R	Requested Signal	Bridged Signal	Reserved

Figure 5.11   APS-specific information format.

Table 5.1 describes code points and values for APS-specific information.

Revertive and Nonrevertive Modes

In the revertive mode of operation, the flow reverts to the working facility as soon as the failure has been corrected. In conditions in which working traffic is being received via the protection entity, if local protection switching requests have been previously active and have then become inactive, a local WTR state is entered. Since this state now represents the highest priority local request, it is indicated on the transmitted Request/State (Table 5.1 [3]) information and maintains the switch. This state normally times out and becomes an NR state after the WTR timer has expired. The WTR timer is deactivated earlier if any local request of higher priority preempts this state.

In the nonrevertive mode of operation, the protection facility is treated as the working facility. In conditions in which working traffic is being transmitted via the protection entity, if local protection switching requests have been previously active and have then become inactive, a local *DNR state* (Table 5.1) is reached. Since this state now represents the highest priority local request, it is indicated on the transmitted Request/State information and maintains the switch, thus preventing reversion to the released bridge/selector position in nonrevertive mode under NR conditions.

Transmission and Acceptance of APS

Traffic units that carry APS PDU are called *APS frames*. The APS frames are transported only via the protection transport entity, inserted by the head end of the protected domain, and extracted by the tail end of the protected domain. A new APS frame must be transmitted immediately when a change in the transmitted status occurs.

The first three APS frames should be transmitted as fast as possible after the status change of the protection end point occurs so that fast protection switching is possible even if one or two APS frames are lost or corrupted. For the fast protection switching in 50 ms, the interval of the first three APS frames is desirable to be 3.3 ms, which is the same interval as CCM frames for fast defect detection. APS frames after the first three frames should be transmitted with an interval of 5 s.

Table 5.1   Code Points and Field Values for APS-Specific Information

Request/State		1111	Lockout of protection (LO)	Priority
		1110	Signal fail for protection (SF-P)	Highest
		1101	Forced switch (FS)	
		1011	Signal fail for working (SF)	
		1001	Signal degrade (SD) (Note 1)	
		0111	Manual switch (MS)	
		0101	Wait to restore (WTR)	
		0100	Exercise (EXER)	
		0010	Reverse request (RR) (Note 2)	
		0001	Do not revert (DNR)	
		0000	No request (NR)	Lowest
		Others	Reserved for future international standardization	
Protection Type	A	0	No APS channel	
		1	APS channel	
	B	0	1+1 (permanent bridge)	
		1	1:1 (no permanent bridge)	
	D	0	Unidirectional switching	
		1	Bidirectional switching	
	R	0	Nonrevertive operation	
		1	Revertive operation	
Requested Signal		0	Null signal	
		1	Normal traffic signal	
		2-255	(Reserved for future use)	
Bridged Signal		0	Null signal	
		1	Normal traffic signal	
		2-255	(Reserved for future use)	

Protection Types

The valid protection types are

000x	1+1 Unidirectional, no APS communication
100x	1+1 Unidirectional with APS communication
101x	1+1 Bidirectional with APS communication
111x	1 : 1 Bidirectional with APS communication

The default value, all zeros, matches the only type of protection that can operate without 1+1 unidirectional APS.

If the B bit mismatches, the selector is released since 1 : 1 and 1+1 are incompatible. This will result in a defect. If the A bit mismatches, the side expecting APS will fall back to 1+1 unidirectional switching without APS communication. If the D bit mismatches, the bidirectional side will fall back to unidirectional switching. If the R bit mismatches, one side will clear switches to the WTR state and the other will clear to the DNR state. The two sides will interwork and the traffic will be protected.

Requested signal indicates the signal that the near-end requests be carried over the protection path. For NR, this is the null signal when the far end is not bridging normal traffic signal to the protection entity. For LO, this is the null signal. For Exercise, this is the null signal when Exercise replaces NR or the normal traffic signal in the case where Exercise replaces DNR. For all other requests, this will be the normal traffic signal requested to be carried over the protection transport entity.

Bridged signal indicates the signal that is bridged onto the protection path. For 1+1 protection, this reflects the permanent bridge. For 1 : 1 protection, this indicates what is actually bridged to the protection entity, either the null signal or the normal traffic signal.

Timers

There are three types of timers: hold-off timer, WTR timer, and guard timer. The hold-off timer coordinates timing of protection switches at multiple layers or across cascaded protected domains, allowing either a server layer protection switch to have a chance to fix the problem before switching at a client layer or an upstream protected domain to switch before a downstream domain.

Each protection group should have a provisionable hold-off timer. The suggested range of the hold-off timer is 0–10 s in steps of 100 ms with an accuracy of ±5 ms.

When a new defect or more severe defect occurs, this event will not be reported immediately to protection switching, if the provisioned hold-off timer is set to a nonzero value: the hold-off timer will be started. When the hold-off timer expires, it will be checked whether a defect still exists on the trail that started the timer. If it does, that defect will be reported to protection switching.

WTR timer prevents frequent operation of the protection switch due to an intermittent defect; a failed working transport entity must become fault free. After the failed entity becomes fault free for the WTR period, a normal traffic can use it.

After the failed working transport entity meets this criterion, a fixed period shall elapse before a normal traffic signal uses it again. This period, called *WTR period*, may be configured by the operator in 1-min steps between 5 and 12 min; the default value is 5 min. An SF (or SD, if applicable) condition will override the WTR.

In the revertive mode of operation, when the failed working transport entity is no longer in SF (or SD, if applicable) condition, a local WTR state will be activated. The WTR timer deactivates earlier when any request of higher priority preempts this state.

Ring Automatic Protection Switching (R-APS) messages are continuously transmitted, copied, and forwarded at every ring node around the ring. This can result in a message corresponding to an old request, which is no longer relevant, being received by ring nodes. The reception of messages with outdated information could result in erroneous interpretation of the existing requests in the ring and lead to erroneous protection switching decisions. The guard timer is used to prevent ring nodes from receiving outdated R-APS messages. During the duration of the guard timer, all the received R-APS messages are ignored by the ring protection control process. This allows for the condition that old messages still circulating on the ring may be ignored; as a result, a node will be unaware of new or existing ring requests transmitted from other nodes.

The period of the guard timer may be configured by the operator in 10-ms steps between 10 ms and 2 s, with a default value of 500 ms. This duration should be greater than the maximum expected forwarding delay for which one R-APS message circles around the ring.

5.4   RING PROTECTION

EPSRs run over standard Ethernet interfaces and are similar to STP, providing a polling mechanism to detect ring-based faults and failover accordingly. But unlike STP, EPSRs use a fault detection scheme that alerts the ring that a break has occurred and indicates that it must take action instead of making a calculation. When a fault is detected, the ring automatically sends traffic over a protected reverse path and can converge in less than 50 ms.

EPSRs operate over standard Ethernet Ports in a domain and can be used with either the STP or the RSTP. The EPSR domain is a collection of VLANs of user frames, the control VLAN, and the associated switch ports. The controlling node

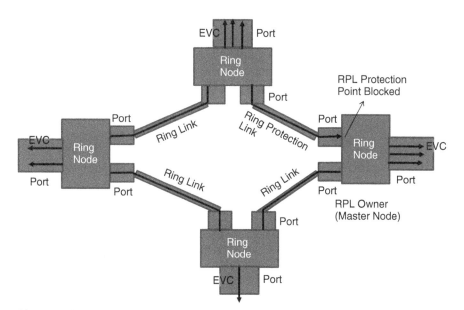

Figure 5.12   Ethernet ring protection switching architecture.

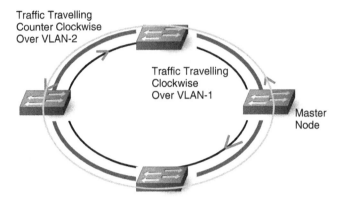

Traffic Travelling
Counter Clockwise
Over VLAN-2

Traffic Travelling
Clockwise
Over VLAN-1

Master
Node

Figure 5.13 Multiple ESR domains.

for an EPSR domain is called the *master node* (Fig. 5.12) [5], which is responsible for status polling, collecting error messages, and controlling the flow of traffic in an EPSR domain. All other nodes in an EPSR domain are transit nodes, which generate failure notices and receive control messages from the master node. An APS protocol is used to coordinate the protection actions over the ring.

The primary port of the master node determines the direction of traffic flow. This port is always operational. The secondary port of the master node remains active but blocks all protected VLANs from operating until ring failover. All control messages are sent and received over a Control VLAN. This is the only VLAN that is never blocked. Because traffic flow is based on the master node, its placement is critical. Most scenarios dictate that the least used node should be designated as the master node, thereby assuring optimum spatial reuse of the bandwidth.

Each ring node is connected to adjacent nodes participating in the same ring, using two independent links. The link between two adjacent nodes is called a *ring link*. A port on a ring link is called a *ring port*. The minimum number of nodes on a ring is two.

The ring protection switching architecture is based on the principle of loop avoidance and the utilization of learning, forwarding, and address table mechanisms, which is part of the Ethernet flow forwarding function. Loop avoidance in the ring is achieved by guaranteeing that, at any time, traffic may flow on all but one of the ring links. This particular link is called the *Ring Protection Link* (*RPL*), and under normal conditions, this link is blocked. One designated node, the RPL Owner, is responsible to block traffic over the RPL. Under a ring failure condition, the RPL Owner is responsible to unblock the RPL, allowing the RPL to be used for traffic. Under normal operation, the master node's secondary port is blocked for all protected VLANs; only the control VLAN remains unblocked on the secondary port. The master node periodically sends out health-check messages through its primary port at times specified by the user. These health-check messages are then received on the master node's secondary port.

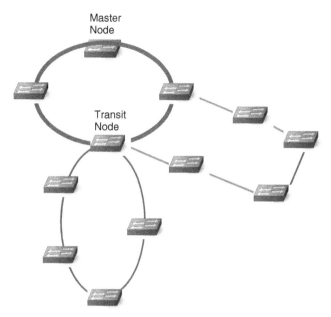

Figure 5.14 Conjoined Ethernet rings.

Multiple EPSR domains can operate over the same physical ring (Fig. 5.13) [5]. For example, while one EPSR domain operates with normal traffic flow through the westbound interface, a second EPSR domain could operate through the eastbound interface. This feature further enhances the spatial reuse capability of EPSR that allows the operator to accurately control the flow of traffic to maximize the available bandwidth of high-speed links in both directions.

The Ethernet rings could support a multiring/ladder network that consists of conjoined Ethernet rings (Fig. 5.14).

5.4.1 Protection Switching

The ring can enter into ring-fault operation if either the master node fails to receive two consecutive health-check messages on its secondary port or a transit node sends an EPSR LINK-DOWN control message to the master node. When the ring enters such a state, the master node unblocks its secondary port and flushes its forwarding database (FDB). It also sends a RING-DOWNFLUSH-FDB control message to all transit nodes instructing them to flush their FDBs as well. Then, by normal bridge learning, all paths and communication are restored (Fig. 5.15).

When the master node starts to receive its health-check messages on its secondary port or when the failed transit node sends a LINK_UP message, the master node will then restore the ring to its original topology by reblocking its secondary port, flushing its FDB, and sending a control message to all transit nodes instructing them to flush their FDBs. The transit node whose port has just

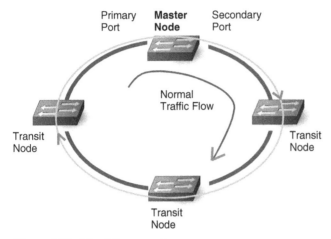

Figure 5.15   Ring failure operation.

come back up is responsible for preventing a network loop while the master node's secondary port is still in operation. It accomplishes this by blocking the protected VLANs on the restored port and placing them in a preforwarding state. When the ring restoration control message arrives from the master node, the transit node then flushes its FDB and unblocks the preforwarding port.

When a failure (i.e., defect condition such as SF) condition is detected, the nodes adjacent to the failed ring link initiate the protection switching mechanism.

In a revertive operation, after the clearing of an SF condition on a ring link with a defect, the position of the blocked port of the ring link is maintained until the WTR timer expires. An RPL owner will initiate reversion when the WTR timer expires, before any other higher priority event or command.

In an Ethernet ring, without congestion, with less than 1200 km of ring fiber circumference, and fewer than 16 nodes, the switching time for a failure on a ring link is expected to be less than 50 ms.

Protection switching is performed at all ring nodes as defined in G.8032 [6]. An APS protocol coordinates ring protection actions around the ring.

Protection switching is triggered when

- an SF is declared on one of the ring links, and the detected SF condition has a higher priority than any other local request or far-end request; or
- the received APS protocol requests to switch and it has a higher priority than any other local request.

However, manual switching is not defined in Reference 6.

In the revertive operation, the traffic channel is restored to the working transport entity, that is, blocked on the RPL, after the condition(s) causing a switch is cleared. In the case of clearing a defect, the traffic channel reverts after the expiry of the WTR timer, which is used to avoid toggling protection states in the case of intermittent defects.

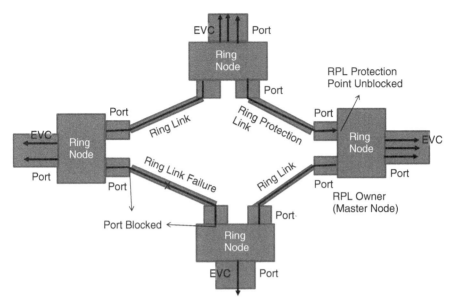

Figure 5.16   Ethernet ring protection switching architecture for a ring link failure.

In the nonrevertive operation, the traffic channel is allowed to use the RPL, if it is not failed, after a switch condition is cleared. Since in ERP the working transport entity resources may be more optimized, in some cases, it is desirable to revert to the normal path once all ring links are available. This is performed at the expense of an additional traffic interruption. In some cases, there may be no advantage to revert to the normal working transport entities immediately. In this case, a second traffic interruption is avoided by not reverting the protection switching.

Figure 5.16 illustrates a situation where a protection switch has occurred due to an SF condition on one ring link. In this case, traffic channel is blocked bidirectionally on the ports where the failure is detected and bidirectionally unblocked at the RPL connection point.

In the revertive operation, when the failure is recovered, the traffic channel will resume use of the recovered ring link only after the traffic channel has been blocked on the RPL.

The ring links of each node may be monitored by individually exchanging CCMs defined in Reference 5 on the maintenance entity group end points shown in Figure 5.17.

In Figure 5.17, MEPs on each ring port are used for monitoring the ring link.

If an MEP detects a defect, which contributes to an SF defect condition, this will inform the ERP control process that a failure condition has been detected. An ERP control function uses the ETH_CI_SSF information, forwarded from the ETHx/ETH-m_A_Sk, to assert the SF condition of the ring link.

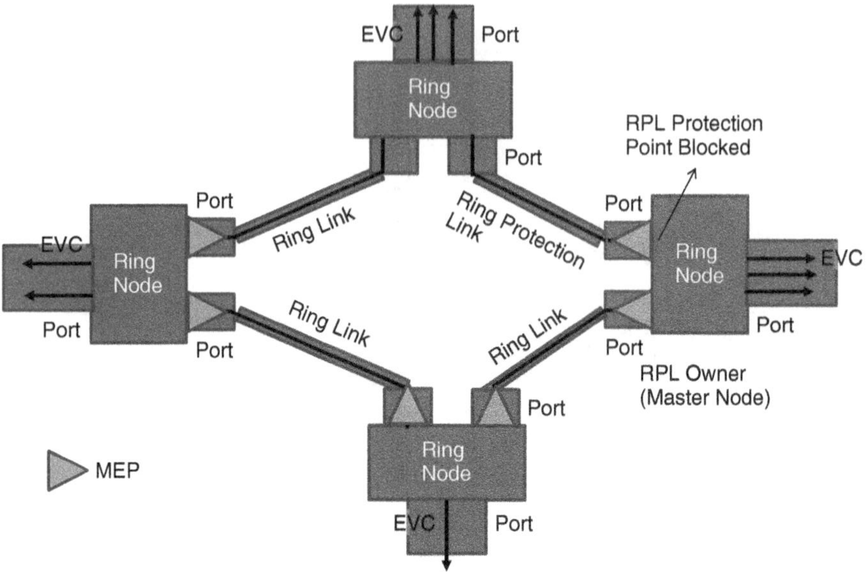

Figure 5.17   MEPs in Ethernet ring protection switching architecture.

The ring protection mechanism requires the APS protocol to coordinate the switching behavior among all ring nodes. The ring APS protocol communication is performed using R-APS PDUs. R-APS PDUs are transmitted and received at an ERP control process.

Blocking traffic from one or more VLANs is achieved by VLAN ID filtering on a port. This results in blocking the transmission and reception of traffic on one ring port.

R-APS channel VLAN traffic forwarding is always blocked at the same ports where traffic channel is blocked. This only prevents R-APS messages received at one ring port from being forwarded to the other ring port. This, however, does not prevent R-APS messages, locally generated at the ERP control process, from being transmitted over both ring ports and also allows R-APS messages received at each port to be delivered to the ERP control process.

A filtering database flush consists of removing the learned MAC (Media Access Control) addresses of the ring ports from the node's filtering database.

Ring protection is based on loop avoidance. This is achieved by guaranteeing that at any time traffic may flow on all but one of the ring links.

5.5 LINK AGGREGATION

IEEE 802.3ad-based link aggregation [6, 7] provides a method for aggregating two or more parallel physical links together to form LAGs such that an MAC

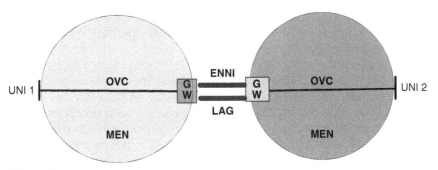

Figure 5.18 ENNI architecture with LAG.

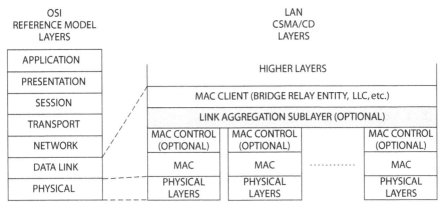

Figure 5.19 Link aggregation sublayer.

client can treat the LAG as if it were a single link. The link aggregation is able to increase the capacity and availability of the communications channel between devices using existing Fast Ethernet and Gigabit Ethernet technology. Two or more Gigabit Ethernet connections are combined in order to increase the bandwidth capability and to create resilient and redundant links. As a result, LAG is a mandatory capability at ENNI [8] when there are more than one link between two Metro Ethernet Networks (MENs) (Fig. 5.18).

Link aggregation also provides load balancing in which the processing and communications activity is distributed across several links in a trunk so that no single link is overwhelmed.

The link aggregation is an optional sublayer between an MAC client and the MAC, as depicted in Figure 5.19. It is dynamic and provides more functionality through the link aggregation control protocol (LACP) [9]. LACP dynamically detects whether links can be aggregated into an LAG, and it does so when links become available. IEEE 802.3ad was designed for point-to-point link aggregation only.

Once a port is a member of an LAG, it will always remain an LAG member even if there is just a single link active in the LAG. This has the benefit of improving resiliency.

LACP supports the automatic creation of LAGs by exchanging LACP packets between LAN ports. LACP frames (i.e., LACPDUs) are sent down to all links that have the protocol enabled. If they find a device on the other end of the link that is also LACP enabled, it will also independently send frames along the same links enabling the two units to detect multiple links between themselves and then combine them into a single logical link. LACP packets are exchanged only between ports in passive and active modes. In the active mode, they will always send frames along the configured links. In the passive mode, however, the port participates in the control if its partner is active.

In addition to automatic link aggregation, LACP maintains the LAG and therefore detects link layer failures. LACP packets are exchanged end-to-end; thus, if a link in the core were to fail and the local port(s) does not register the failure, LACP will time-out and remove the port from the LAG. The default LACP settings, with the long timers, will remove the port from the LAG in 90 s. If short timers were used, the port can be removed in 3 s.

5.5.1 LAG Objectives

The Standard 802.3ad [7] lists the following main goals for LAG.

- Increased bandwidth by combining multiple links into one logical link.
- Increased availability by not causing failure from the perspective of MAC client when a single link within LAG is failed or replaced.
- Linearly incremental bandwidth by increasing bandwidth in unit multiples (i.e., depending on physical layer options such as 100 Mbps or 1 GE) as opposed to the order of magnitude.
- Load sharing by distributing MAC client traffic across multiple links.
- Automatic configuration of LAGs and allocating individual links to those groups.
- Rapid configuration and reconfiguration in the event of changes in physical connectivity. Link aggregation quickly converges to a new configuration, typically on the order of 1 s or less.
- Deterministic behavior.
- Low risk of duplication or misordering of frames.
- Support of existing IEEE 802.3 MAC clients by not changing existing higher-layer protocols or application to use link aggregation.
- Backward compatibility with aggregation-unaware devices.
- Accommodation of differing capabilities and constraints by accommodating devices with differing hardware and software constraints on link aggregation.

- No change to the IEEE 802.3 frame format.
- Network Management Support–the standard specifies appropriate management objects for configuration, monitoring, and control of link aggregation.

Link aggregation does not support multipoint aggregations, dissimilar MACs, half-duplex operation, and operation across multiple data rates. Dissimilar MACs such as Gigabit Ethernet and FDDI are not supported. However, dissimilar PHYs such as copper and fiber are supported. All ports in an LAG must be operating in full-duplex mode. Also, all links in an LAG must operate at the same data rate.

5.5.2   Link Aggregation Operation

Link aggregation function consists of frame distribution, frame collection, aggregator parser/multiplexer, aggregation control, and control parser/multiplexer (Fig. 5.20).

The frame distribution function submits frames from MAC client to the appropriate port, based on a frame distribution algorithm. In reverse, frame collection block passes frames received from the various ports to the MAC client. The combination of distribution and collection is called the *aggregator*. The frame distribution may include an explicit marker protocol with marker generator/receiver that searches for a marker identifying the last frame of a conversation. On transmit, aggregator multiplexer passes frame transmission requests from the distributor, marker generator/responder to the appropriate port. On receive, aggregator parser distinguishes marker request, marker response, and MAC client PDUs and passes them to the appropriate marker responder, marker receiver, and collector, respectively. The aggregation control incorporates LACP. On transmit, the control parser/multiplexer passes frame transmission requests from the aggregator and control entities to the appropriate port. On receive, the control

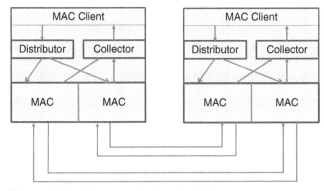

Figure 5.20   Link aggregation functional diagram.

parser/multiplexer distinguishes LACPDUs from other frames and passes LACP-DUs and other frames to the appropriate entities and ports.

Each network interface controller is assigned a unique MAC address. In most cases, this address is used as a source and destination address during the transmission of packets. Aggregated links appear as a single logical network interface with one virtual MAC address. The MAC address of one of the interfaces belonging to the aggregated link provides the "virtual" address of the logical link. All ports in an LAG must be in the same VLAN or VLANs. Ports in an LAG can be distributed over different modules.

LACP-enabled ports with the same key must have the same VLAN membership. In LACP-disabled ports with the same key, VLAN membership can be different. This usually happens when VLANs are added or deleted from these ports. But before LACP is re-enabled in these ports, VLAN membership must be the same for ports with the same key.

In order to change the VLANs' membership on an LAG, or the ports' membership within the VLANs, LACP needs to be disabled in the ports. Once the changes are complete, LACP can be enabled again on all appropriate port members.

In aggregated links, the link on which to transmit a given frame must be selected. Sending one long frame may take longer than sending several short ones; therefore, the short frames may be received earlier than one long frame. The order has to be restored at the receiver side. Thus, an agreement has been made: all frames belonging to one conversation (or EVC) must be transmitted through the same physical link. This guarantees correct ordering at the receiving end station. For this reason, no sequencing information may be added to the frames.

Traffic belonging to separate EVCs can be sent through various links in a random order. An EVC may need to be transferred to another link because the originally mapped link is out of service (failed or configured out of the aggregation) or a new link has become available relieving the existing ones. This can be realized either by means of a delay time that the distributor must determine somehow or through an explicit marker protocol that searches for a marker identifying the last frame of a conversation. The distributor inserts a marker message behind the last frame of a conversation. After the collector receives this marker message, it sends a response to the distributor, which then knows, that all frames of the conversation have been delivered. After that, the distributor can send frames of these types of conversations via a new link without delay.

5.5.3 LACP

The parameters used by LACP are listed in the following.

- *LACP priority*, which is configured at the system level and at the port level.

— *Port priority* can be configured automatically or through the command–line interfaceLine Interface (CLI). LACP uses the port priority to decide which ports should be put in standby mode when there is a hardware limitation that prevents all compatible ports from aggregating. LACP also uses port priority with the port number to form the port identifier.

— *System priority* is configured on each device running LACP. The system priority can be configured automatically or through the CLI. LACP uses system priority with the device MAC address to form the system ID and also during negotiation with other systems.

- *LACP keys* are used to determine which ports are eligible to be aggregated into an LAG where keys do not need to match between two LACP peers. LACP automatically configures an administrative key value on each port configured to use LACP. A port's ability to aggregate with other ports is determined by the LACP timers.

- *LACP timers determine* the failover times. The default timer settings are

  — time-out: 3
  — fast periodic time: 1000 (ms)
  — slow periodic time: 30,000 (ms)

    The user can choose to use either the fast or the slow timer that is set on the port level. By default, the long timer can be used. Hence, a link is determined ineligible to be aggregated if it does not receive an LACPDU for a period of time-out $\times$ slow periodic time $= 3 \times 30$ s $= 90$ s.

Link aggregation is compatible with the STP/RSTP/MSTP. LAGs must be in the same STP group(s). The operation of the LACP is only affected by the physical link state or its LACP peer status. When a link goes up and down, the LACP is notified. The STP forwarding state does not affect the operation of the LACP module. LACPDU can be sent even if the port is in the STP blocking state.

5.5.4 Limitations

802.3ad requires that all physical ports in the LAG must reside on the same logical device, which, in most scenarios, will leave a single point of failure when the physical switch to which both links are connected goes offline.

In most implementations, all the ports used in an aggregation consist of the same physical type, such as all copper ports (10/100/1000BASE-T), all multimode fiber ports, or all single-mode fiber ports. However, the IEEE standard requires that each link should be full duplex and all the links should have an identical speed.

Some equipment support static configuration of link aggregation. Link aggregation between similarly statically configured equipment will work, but will fail between a statically configured equipment and an equipment configured for LACP.

5.6 CONCLUSION

Protection switching is a key capability for the high availability of Metro Ethernet services. This chapter has described APS, ring protection, and LAG methods. The 50-ms protection switching time mentioned here has been challenged and 500 ms is recommended.

Currently, LAG is the mostly deployed protection scheme especially at ENNI and access. As applications requiring high availability are transported over Metro Ethernet services, more APS and ring protection deployments can be expected.

REFERENCES

1. MEF 32, Requirements for service protection across external interfaces. July 2011.
2. RFC 3619, Extreme networks' Ethernet automatic protection switching (EAPS), version 1. October 2003.
3. ITU-T G.8031, Ethernet linear protection switching. 2006.
4. ITU-T Y.1731, OAM functions and mechanisms for Ethernet based networks. 2008.
5. ITU-T G.8032, Ethernet ring protection switching. 2012.
6. IEEE Std 802.3ad, Amendment to carrier sense multiple access with collision detection (CSMA/CD) access method and physical layer specifications-aggregation of multiple link segments. 2000.
7. IEEE 802.1AX, Standard for local and metropolitan area networks—link aggregation. 2008.
8. MEF 26, External network network interface (ENNI) - phase 2. January 2012.
9. Link Aggregation Control Protocol (LACP) 802.3ad and VLACP for ES and ERS Technical Configuration Guide. Available at http://support.avaya.com/css/P8/documents/100093956.

Chapter 6

Carrier Ethernet Architectures and Services

6.1 INTRODUCTION

Ethernet is becoming a dominant technology in enterprise networks, allowing bandwidth partitioning, user segregation, and traffic prioritization, introduced by IEEE 802.1P and 1Q. In 1998, IEEE 802.1d combined VLAN and prioritization capabilities and introduced another service provider tag to scale the addressing structure. These capabilities formed the base for Carrier Ethernet. The Carrier Ethernet equipment business is estimated to be $30.7 billion in 2012 [1] by Vertical Systems.

In 2001, the MEF was formed to define Ethernet-based services. MEF defined Carrier Ethernet for business users as a ubiquitous, standardized, carrier-class Service and Network defined by five attributes (Standardized Services, QoS, Service Management, Reliability, Scalability) that distinguish it from familiar LAN-based Ethernet [2]. Similarly, it is defined for service providers as

- a set of certified network elements that connect to transport Carrier Ethernet services for all users, locally and worldwide
- Carrier Ethernet services that are carried over physical Ethernet networks and other legacy transport technologies.

Ethernet has been used as a LAN connectivity technology in Enterprise. With the introduction of Carrier Ethernet, Ethernet is being used as a service. It is an easy to use, widely available, and well-understood technology. Ethernet simplifies OAM&P (operations, administration, maintenance, and provisioning) and is cost-effective due to widespread usage of Ethernet interface. Bandwidth can be added in increments. Single user interface can connect to multiple services such as Internet, virtual private network (VPN), and voice and video services.

Ethernet was developed to connect computers on corporate networks. Since its inception, it has become the predominant technology used in corporate networks. Now carriers, which typically provide phone and Internet access services,

Networks and Services: Carrier Ethernet, PBT, MPLS-TP, and VPLS, First Edition. Mehmet Toy.
© 2012 John Wiley & Sons, Inc. Published 2012 by John Wiley & Sons, Inc.

are putting it to use in their metropolitan area networks. As a result, Ethernet is competing with wide area technologies such as Frame Relay, ATM (Asynchronous Transfer Mode), and dedicated leased line.

With Ethernet, the same protocol is being used for LAN and MAN. Therefore, there is no protocol conversion between LAN and WAN. People trained to provision and maintain WAN can be employed to provision and maintain LAN. This results in lower equipment and operational cost. Studies [3, 4] claim 30% reduction in total cost of ownership and delivery, 18–24% reduction in operational cost, 40–50% reduction in equipment cost compared to the traditional SONET/SDH (Synchronous Optical Network/Synchronous Digital Hierarchy) gear, and 70% reduction in equipment cost compared to ATM.

For Metro Ethernet to be successful, it has to display the same properties of current WAN technologies. The networks must scale to support the hundred thousands of customers to adequately address metropolitan and regional distances. The network availability needs to be 99.999%, with a range of protection mechanisms capable of sub-50-ms recovery. Service management systems must be available to provision new services, manage Service Level Agreements (SLAs), and troubleshoot the network under fault conditions. Standardized services must be offered to support existing applications and allow service providers to extend their geographic reach through interoperability agreements with competitors. QoS and traffic engineering schemes are needed to prioritize traffic streams and ensure that service-level parameters are agreed.

Ethernet is scalable from 10 Mbps to 100 Gbps with finer granularity as low as 56 kbps and 1 Mbps. Carrier Ethernet allows incremental bandwidth assignment. Most carriers provide services that start at 10 Mbps and can scale up to 1 Gbps or more, which are faster data rates than the rates of traditional Frame Relay and ATM services. With enhanced OAM, end-to-end monitoring and trouble isolation are possible. Link protection equipment [5] and ring protection mechanisms [6] have been developed to provide the necessary reliability for Ethernet networks.

With QoS and synchronization [7, 8] capabilities, applications with strict performance requirements can be supported.

Carrier requirements for access include fault detection and isolation, performance monitoring and statistics, and failover protection. A core network Ethernet mesh located in strategic Point of Presence (PoPs) will provide for the redundancy and performance required. The service is not required to be a fully protected service except at the core.

Ethernet connection is oriented by disabling unpredictable functions such as MAC learning, the Spanning Tree Protocol (STP), and "broadcast of unknown." Furthermore, this performance-grade Ethernet enhances the scalability and minimizes the cost and complexity of aggregation and transport. Connection-oriented Ethernet, combined with a transport-style intelligent control plane and service-level management, simplifies the operations of the network and allows Ethernet to be managed as easily as circuits, using the same procedures and systems.

6.2 STANDARDS

In addition to ITU and IEEE, MEF developed specifications for Carrier Ethernet architecture, service, management, and test and measurement. They are listed in Table 6.1.

Management specifications cover OAM requirements, models, and definitions. Architecture specs define an architectural reference model and a set of common linguistic tools for the technical teams. The architecture is layered. Each layer has been decomposed into the Adaptation, Connection, and Termination elements. Service specs define Carrier Ethernet service models, definitions, and service parameters and attributes. Test and Measurement define test methodologies and test suites that enable conformance to MEF services as defined in the MEF Specifications.

6.3 ARCHITECTURE

Ethernet transport network is a two-layer network [9, 27, 28] consisting of Ethernet MAC (ETH) layer network and Ethernet PHY (ETY) layer network. The ETY layer is the physical layer defined in IEEE 802.3. This layer can be called the section layer as well. The ETH layer is the pure packet layer, which is also called the path layer.

Ethernet transport network consists of a set of access groups of the same type, where the access groups demarcate the point of access into the ETH layer network. The information transferred between the access groups is the characteristic information. The association of two or more access points creates a connectionless transport entity called the Ethernet Connectionless Trail (ECT).

The ETH layer network depicted in Figure 6.1 is divided into ETH subnetworks that are also called ETH flow domain (EFD). An EFD is a set of all ETH (termination) flow points transferring information within a given administrative portion of the ETH layer network. EFDs may be partitioned recursively into sets of nonoverlapping EFDs that are interconnected by ETH links. An IEEE 802.1D bridge represents the smallest instance of an EFD.

A link is a topological component that describes a fixed topological relationship between subnetworks, along with the capacity supported by an underlying server Layer Network Domain (LND) trail (Fig. 6.3).

The termination of a link is called a flow point pool (FPP). The FPP describes the configuration information associated with an interface, such as a User Network Interface (UNI) or External Network to Network Interface (ENNI). The FPP is associated with the trail termination of the underlying server trail used to perform adaptation and transport of the characteristic information of the client LND.

A subset of flow points can form an FPP Link. The FPP link represents available capacity between a pair of flow domains, a flow domain and an access group, or a pair of access groups. EFD may be partitioned into smaller flow domains interconnected by FPP links (Fig. 6.2).

Table 6.1 List of MEF Standards

	MEF 3 Circuit Emulation Service Definitions, Framework and Requirements in Metro Ethernet Networks
	MEF 10.1 MEF 10.1 Ethernet Services Attributes Phase 2
	MEF 10.2 Ethernet Services Attributes Phase 2
	MEF 6.1 Metro Ethernet Services Definitions Phase 2
	MEF 8 Implementation Agreement for the Emulation of PDH Circuits over Metro Ethernet Networks
	MEF 10.1.1 Amendment to MEF 10.1 Ethernet Services Attributes Phase 2
	MEF 22 Mobile Backhaul Implementation Agreement (2/09)
	MEF 23 Class of Service Phase 1 Implementation Agreement (supersedes any file posted here before November 3, 2009)
	MEF 5 Traffic Management Phase 1 (superseded by MEF 10)
	MEF 10 Ethernet Services Attributes Phase 1
	MEF 22.1 Mobile Backhaul Phase 2 Implementation Agreement
	MEF 23.1 Class of Service Phase 2 Implementation Agreement
	MEF 26 ENNI and OVC Service Attributes Phase I
	MEF 26.0.1 Amendment to MEF 26. The Bandwidth Profile Algorithm.
	MEF 26.0.2 OVC Layer 2 Control Protocol Tunneling Amendment to MEF 26
	MEF 29 Ethernet Services Constructs
	MEF 33 Ethernet Access Services Definition
	MEF 32 Requirements for Service Protection Across External Interfaces
Architecture and Interfaces	MEF 2 Requirements and Framework for Ethernet Service Protection
	MEF 4 Metro Ethernet Network Architecture Framework Part 1: Generic Framework
	MEF 11 User Network Interface (UNI) Requirements and Framework
	MEF 12 Metro Ethernet Network Architecture Framework Part 2: Ethernet Services Layer
	MEF 13 User Network Interface (UNI) Type 1 Implementation Agreement
	MEF 20 UNI Type 2 Implementation Agreement
	MEF 26.1 External Network to Network Interface (ENNI)–Phase 2
	MEF 28 External Network to Network Interface (ENNI) Support for UNI Tunnel Access and Virtual UNI
	MEF 22.1 Mobile Backhaul Phase 2 Implementation Agreement
Management	MEF 7.1 Phase 2 EMS-NMS Information Model
	MEF 15 Requirements for Management of Metro Ethernet Phase 1 Network Elements
	MEF 16 Ethernet Local Management Interface

Table 6.1   (*Continued*)

	MEF 17 Service OAM Framework and Requirements
	MEF 30 Service OAM Fault Management Implementation Agreement
	MEF 31 OAM Fault Management Definition of Managed Objects
Test and	MEF 9 Abstract Test Suite for Ethernet Services at the UNI
Measurement	MEF 14 Abstract Test Suite for Traffic Management Phase 1
	MEF 18 Abstract Test Suite for Circuit Emulation Services
	MEF 19 Abstract Test Suite for UNI Type 1
	MEF 21 Abstract Test Suite for UNI Type 2 Part 1 Link OAM
	MEF 24 Abstract Test Suite for UNI Type 2 Part 2 E-LMI
	MEF 25 Abstract Test Suite for UNI Type 2 Part 3 Service OAM
	MEF 27 Abstract Test Suite for UNI Type 2 Part 5: Enhanced UNI Attributes & Part 6: L2CP Handling

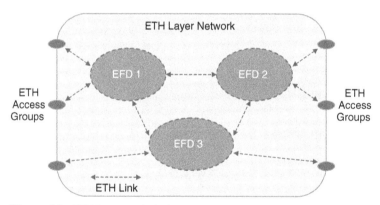

Figure 6.1   ETH layer topological components.

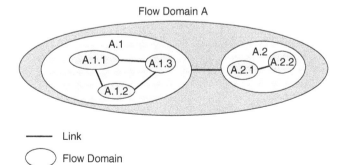

Figure 6.2   Flow domains connected via flow point pool links.

A group of collocated flow termination functions that are connected to the same flow domain or FPP link is called an access group.

The ETH layer is responsible for the instantiation of ETH-oriented connectivity services and the delivery of Ethernet PDUs presented across internal and external interfaces. All service-aware aspects associated with ETH flows, including OAM&P capabilities required to support Ethernet connectivity services, are functions of the ETH layer.

An LND represents an administration's view of the layer network responsible for transporting a specific type of characteristic information such as IP, MPLS, SONET. Layer networks may use transport resources in other layer networks. In Figure 6.3, the ETH LND uses the resources of the MPLS LND.

6.3.1 Protocol Stack

Protocol stack for a MEN with ETH layer can be represented as in Figure 6.4. The data plane, control plane, and management plane cross all these layers. Packets of application layer are represented in the form of Ethernet Service Layer PDUs at ETH Service. In turn, Ethernet Service PDUs are converted into transport

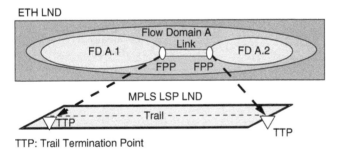

Figure 6.3 FPP, link and trail. TTP, trail termination point.

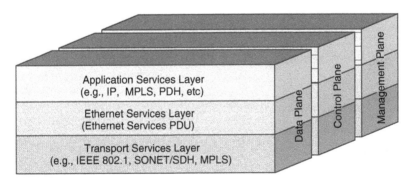

Figure 6.4 Metro Ethernet protocol layers.

layer frames or packets, depending on whether transport protocol is Ethernet, or SONET/SDH or MPLS, etc.

The data plane defines Ethernet frames, tagging, and traffic management. The control plane defines signaling and control. The management plane defines provisioning, device discovery, protection, and OAM.

Figure 6.5 depicts Ethernet services layer in a network encompassing multiple service providers.

The ETH layer network provides the transport of adapted information through an ETH connectionless trail between ETH access points [10]. An example of the ETH layer network and transport components is shown in Figure 6.6.

The transport entities provide transparent information transfer between layer network reference points. Two basic monitored entities of transport are flows and trails. Flows may be decomposed into network flows, flow domain flows, and link flows.

A flow is an aggregation of one or more traffic units with an element of common routing. It is unidirectional and can contain another flow. Flows can be multiplexed together in the same layer network.

A link flow represents the fixed relationship between the ends of the link. It is delimited by flow points and transfers information transparently across an FPP link. A link flow represents a pair of adaptation functions and a trail in the server layer network.

A grouping of traffic units that are transferred transparently across a flow domain is called a flow domain flow. It is delimited by the ports associated with flow points at the boundary of the flow domain. Flow domain flows are constructed from a concatenation of flow domain flows and link flows.

A network flow is a grouping of traffic units that are transferred transparently across a layer network. It is delimited by the termination flow points (TFPs). In general, network flows are constructed from a concatenation of flow domain flows and link flows. The TFP is formed by binding the port of a flow termination to either a flow domain port or a port on an FPP link.

A connectionless trail represents the transfer of monitored adapted characteristic information of the client layer network between access points. It is delimited by two access points, one at each end of the connectionless trail. It

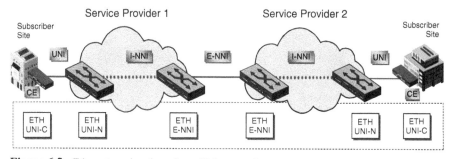

Figure 6.5 Ethernet services layer in multiple networks.

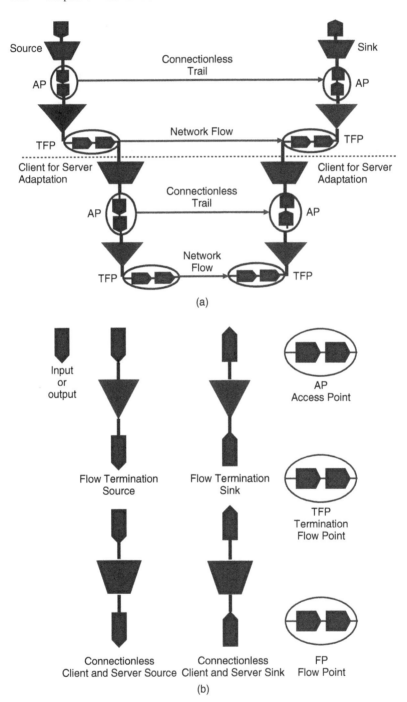

Figure 6.6   (a) Client–server relationship between a connectionless client layer network and connectionless server layer network. (b) Conventions.

represents the association between a source and destination on a per traffic unit or datagram basis.

Adaptation and flow termination functions are used in describing the architecture of connectionless layer networks. Adaptation source adapts the client layer network's characteristic information into a form that is suitable for transport over a trail in a connection-oriented server layer network or a connectionless trail in a connectionless server layer network in the server layer network. The adaptation source function input to output relation is a many-to-one or a one-to-many relationship.

Adaptation sink converts either the server layer network trail into a connection-oriented server layer network or the connectionless trail information in a connectionless server layer network information into the characteristic information of the client layer network. Labeling, scheduling, buffering, queuing, multiplexing, traffic dropping, segmentation, and reassembly are examples of an adaptation function.

Flow termination is accomplished by flow termination source and flow termination sink. Flow termination source accepts adapted characteristic information from a client layer network at its input, adds information to allow the connectionless trail to be monitored, and presents the characteristic information of the layer network at its output(s). A flow termination function adds or extracts information to monitor a connectionless trail and provides no overhead to monitor the connectionless trail. Flow termination sink accepts the characteristic information of the layer network at its input, removes the information related to connectionless trail monitoring, and presents the remaining information at its output.

Using these transport components, it is possible to completely describe the logical topology of a connectionless layer network. An example of the ETH layer network containing the transport processing functions, transport entities, topological components, and reference points is shown in Figure 6.7.

6.3.2   ETH Layer Characteristic Information (Ethernet Frame)

The Characteristic Information, which is the Ethernet frame, exchanged over ETH layer [11]. As discussed in Chapter 4, it consists of Preamble, Start Frame Delimiter (SFD), Destination MAC Address (DA), Source MAC Address (SA), optional 802.1QTag, Ethernet Length/Type (EtherType), user data, padding if required, frame check sequence (FCS), and extension field that is required only for 100 Mbps half-duplex operation (Fig. 6.8).

Seven bytes of Preamble allow physical layer to reach its steady-state synchronization with the received frame's timing.

The length of padding for n-byte-long user data is max [0, minFrameSize − $(8 \times n + 2 \times addressSize + 48)$] bits [11].

The maximum possible size of the data field is maxUntaggedFrameSize − (2addressSize + 48)/8 octets, where minFrameSize is 64 octets [11].

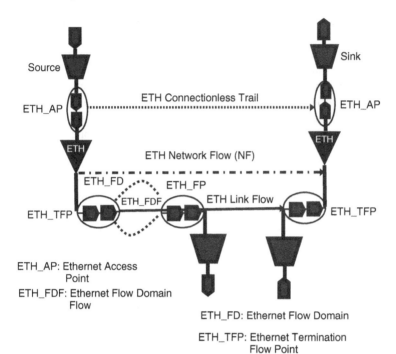

Figure 6.7   Ethernet layer network for unicast flow. NF, network flow; ETH_AP, Ethernet access point; ETH_FDF, Ethernet flow domain flow; ETH_FD, Ethernet flow domain; ETH_TFP, Ethernet termination flow point.

A cyclic redundancy check (CRC) is used by the transmit and receive algorithms to generate a CRC value for the FCS field. The FCS field contains a 4-octet (32-bit) CRC value. This value is computed as a function of the contents of the source address, destination address, length, data, and pad (i.e., all fields except the Preamble, SFD, FCS, and extension).

The optional 4-octet 802.1QTag is composed of a 2-octet 802.1QTagType and the tag control identifier (TCI). TCI contains 3 bits of CoS (Class of Service)/Priority information, the single-bit Canonical Format Indicator (CFI), and the 12-bit VLAN Identifier (VLAN ID). The VLAN ID and CoS/Priority are optional information elements of the frame. Figure 6.8 illustrates the ETH_CI for common IEEE 802.3-2002/2005 compliant frame formats.

CFI is always set to zero for Ethernet switches. CFI is used for compatibility reason between Ethernet type network and Token Ring type network. If a frame received at an Ethernet port with a CFI set to 1, then that frame should not be forwarded.

The user data conveys the APP layer PDU either in a raw format or as a logical link control (LLC)-encapsulated PDU where the LLC provides a link service access point (LSAP) for access to higher layers and additional LSAPs

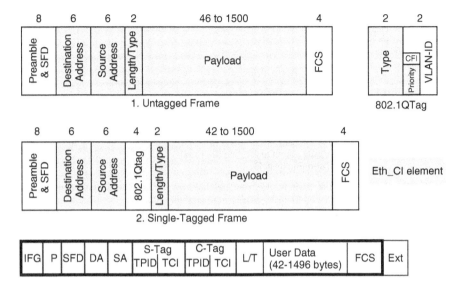

Figure 6.8   ETH layer characteristic information (ETH_CI).

are there for each implemented link layer protocol such as LLDP (Link Layer Discovery Protocol) and STP.

ETH layer PDUs are the data frames used to exchange the ETH_CI across standardized ETH layer interfaces and associated reference points (i.e., UNIs and ENNIs). The Ethernet Service Frame is the ETH layer PDU exchanged across the UNI.

With introduction of the 802.3z standard for gigabit Ethernet in 1998, an extension field was added to the end of the Ethernet frame to ensure it would be long enough for collisions to propagate to all stations in the network. The extension field is appended as needed to bring the minimum length of the transmission up to 512 bytes (as measured from the destination address field through the extension field). It is required only in half-duplex mode, as the collision protocol is not used in full-duplex mode. Nondata bits, referred to as extension bits, are transmitted in the extension field so the carrier is extended for the

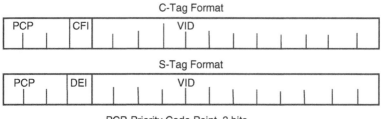

Figure 6.9 C-Tag and S-Tag. PCP, priority code point, 3 bits; CFI, canonical format indicator, 1 bit; VID, VLAN identifier, 12 bits (0–4094); DEI, drop eligibility bit.

minimum required time. Figure 6.9 illustrates a frame with an extension field appended.

CoS/Priority information associated with an ETH layer PDU can be conveyed explicitly in the 3-bit priority field.

The IEEE 802.1Q standard adds four additional bytes to the standard IEEE 802.3 Ethernet frame, which is referred to as the VLAN tag. Two bytes are used for the tag protocol identifier (TPID) and the other 2 bytes for tag control identifier (TCI). The TCI field is divided into Priority Code Point (PCP), CFI, and VID.

- TPID is a 16-bit field set to a value of 0x8100 in order to identify the frame as an IEEE 802.1Q-tagged frame. This field is located at the same position as the EtherType/Length field in untagged frames and is thus used to distinguish the frame from untagged frames.
- PCP of TCI is a 3-bit field that refers to the IEEE 802.1p priority. It indicates the frame priority level. Values are from 0 (best effort) to 7 (highest); 1 represents the lowest priority. These values can be used to prioritize different classes of traffic (voice, video, data, etc.).
- CFI of TCI is a 1-bit field. If the value of this field is 1, the MAC address is in noncanonical format. If the value is 0, the MAC address is in canonical format. It is always set to zero for Ethernet switches. CFI is used for compatibility between Ethernet and Token Ring networks. If a frame received at an Ethernet port has a CFI set to 1, then that frame should not be bridged to an untagged port.
- VLAN Identifier (VID) of TCI is a 12-bit field specifying the VLAN to which the frame belongs. The hexadecimal values of 0x000 and 0xFFF are reserved. All other values may be used as VLAN identifiers, allowing up to 4094 VLANs. The reserved value 0x000 indicates that the frame does not belong to any VLAN; in this case, the 802.1Q tag specifies only a priority and is referred to as a priority tag. On bridges, VLAN 1 (the default VLAN ID) is often reserved for a management VLAN; this is vendor specific.

IEEE 802.1ad introduces a service tag (S-Tag) next to SA. The S-Tag is used to identify the service. The subscriber's VLAN Tag (C-VLAN Tag) remains intact and is not altered by the service providers anywhere within the provider's network. TPID of 0x88a8 is defined for S-Tag.

Q-in-Q networks do not provide any separation between the provider's and subscribers' MAC addresses. Therefore, provider switches must learn all MAC addresses in the network, regardless of whether they belong to the service provider or to the subscriber.

Most Ethernet control protocols (i.e., Bridged Protocol Data Units or BPDUs) used by subscribers' networks must not interact with the provider's networking equipment. For example, STP instances in the subscriber network must not interact with STP instances used in the provider network. In the subscriber's STP, BPDUs need to be tunneled through the provider's network. BPDUs are identified by their destination MAC address.

The service frame is just a regular Ethernet frame beginning with the first bit of the destination address through the last bit of the FCS. A service frame that contains an IEEE 802.1Q tag can be up to 1522 bytes long, and a service frame that does not contain an IEEE 802.1Q tag can be up to 1518 bytes.

A service frame can be a unicast, multicast, broadcast, and L2CP frame. Unicast service frame is a service frame that has a unicast destination MAC address. Multicast service frame is a service frame that has a multicast destination MAC address. Broadcast service frame is a service frame with the broadcast destination MAC address. Layer 2 Control Protocol (L2CP) service frame is a frame whose destination MAC address is one of the addresses listed in Table 6.2. Some L2 Control Protocols share the same destination MAC address and are identified by additional fields such as the EtherType and a protocol identifier. Therefore, disposition of service frames carrying L2CPs may be different for different protocols that use the same destination MAC address. MEF 6.1 [12] contains some recommendations for the delivery of specific L2 Control Protocols. Additional addresses for identifying L2 Control Protocols may be used as well.

Service frame disposition can be discarding, tunneling, delivering unconditionally, and delivering conditionally where tunneling only applies to L2 Control Protocol service frames.

For example, all ingress service frames with an invalid FCS must be discarded by the MEN. For a point-to-point Ethernet Virtual Connection (EVC), service frames with correct FCS are delivered unconditionally across the egress UNI.

Table 6.2 MAC Addresses for L2 Control Protocols

MAC Addresses5	Description
01-80-C2-00-00-00 through 01-80-C2-00-00-0F	Bridge block of protocols
01-80-C2-00-00-20 through 01-80-C2-00-00-2F	Generic Attribute Registration Protocol (GARP) block of protocols
01-80-C2-00-00-10	All bridges protocol

6.3.3 ETH Layer Functions

The ETH layer functions may be categorized as

- ETH conditioning functions;
- ETH EVC Adaptation Function (EEAF);
- ETH EVC Termination Function (EETF);
- ETH Connection Function (ECF);
- APP to ETH Adaptation Function (EAF);
- ETH Flow Termination Function (EFTF);
- ETH to TRAN Adaptation Function (TAF).

Service frames are Ethernet frames that are exchanged between the Metro Ethernet network and the customer edge (CE) across the UNI. A service frame sent from the Metro Ethernet network to the customer edge at a UNI is called an egress service frame. A service frame sent from the CE to the Metro Ethernet network at a UNI is called an ingress service frame.

The ETH conditioning functions are frame classification, filtering, metering, marking, policing, and shaping that can be applied to flows. There are three types of ETH conditioning functions:

- The ETH Flow Conditioning Function (EFCF) is responsible for the conditioning of the subscriber flow into and out of a subscriber EFD. The ingress and egress process on flows to the MEN at the UNI-C is called EFCF.
- The ETH Subscriber Conditioning Function (ESCF) is for conditioning the subscriber flows toward and from the MEN. At the ingress toward the service provider EFD (MEN), per user contract with service provider (i.e., SLA), classification, filtering, CoS instance identification, policing, marking, and shaping are performed. At the egress, the same functions, except policing, are performed.
- The ETH Provider Conditioning Function (EPCF) is responsible for conditioning flow(s) between two MENs at the ENNI.

The EEAF is responsible for the adaptation of service frames into and out of the EVCs. The EEAF source

- maps conditioned ingress service frames into their corresponding EVC PDUs;
- adapts the subscriber CoS ID into the service provider CoS indication per CoS instance;
- multiplexes ingress service frames into their corresponding EVC;
- buffers and schedules ingress service frames according to a scheduling algorithm as per CoS instance.

On the other hand, the EEAF Sink demultiplexes egress service frames from various EVCs into their corresponding service flow instances and adapts the service provider CoS information into the subscriber CE-VLAN (Customer Edge VLAN)-CoS information, if applicable.

The EETF is responsible for the creation and termination of EVC trails. The EETF source multiplexes management (e.g., OAM), control, and data plane PDUs and relays adapted ETH layer PDU toward the service provider EFD. The sink receives the adapted EVC PDU from the service provider EFD and demultiplexes management (e.g., OAM), control, and data plane PDUs.

The ECF facilitates the creation of point-to-point or multipoint connections.

The ETH Adaptation Function (EAF) is responsible for the adaptation of the APP layer PDUs to the ETH layer. It forms LLC PDUs, allocates EtherType per client application and/or LLC type (if LLC is present), pads to minimum transmission unit size, and multiplexes adapted client PDUs toward EFTF, demultiplexes adapted client PDUs from EFTF, processes EtherType and performs decapsulation, and extracts LLC PDU (if LLC is present) and relays it to client process (as per EtherType).

The EFTF is responsible for the creation and termination of ETH network flows and supports the protocol interface between the APP layer and ETH layer. The EFTF source prepares Ethernet service frame, formats ETH layer PDU, and relays ETH layer PDU toward the target EFD. On the other hand, the EFTF Sink receives the ETH layer PDU from EFD, extracts user data, and relays adapted client PDU to EAF.

The ETH to TAF is responsible for the adaptation of the ETH layer PDUs to its serving the TRAN layer such as Ethernet, SONET/SDH, ATM, Frame Relay (FR), and MPLS. The TAF source buffers and schedules frames, allocates VLAN ID field value, pads payload to meet minimum transmission unit size, generates service frame FCS, and encapsulates/encodes (e.g., adaptation) ETH_CI according to the TRAN layer.

The TAF source also multiplexes EVC PDUs into ETH link, adapts rate into the TRANs layer, and inserts adapted ETH layer data stream into payload of the TRAN layer signal.

The sink TAF performs FCS verification of ETH frame and ETH frame filtering of subscriber frames not intended to be forwarded across the UNI, extracts adapted ETH_CI from payload of the TRAN layer signal, and demultiplexes encapsulated EVC PDUs.

6.3.4 ETH Links

There are two types of links, access links and trunk links. The link between the port in the subscriber CE supporting the UNI-C and the port in the service provider PE supporting the UNI-N is an example to an access link. On the other hand, the link that interconnects ports between the service provider NEs is an example of a trunk link.

For any access link at the UNI-C, there is a one-to-one relationship between the ETH access link and its underlying TRAN link.

Figure 6.10 illustrates the relationships between the UNI functional elements and the associated ETH access link, using the Ethernet layer functions defined in Section 5.2.

When any UNI-N is indirectly attached to an UNI-C via a transport multiplexing function (TMF), the ETH access link can be multiplexed by the TRAN layer. In this case there will be a many-to-one relationship between the ETH access link and its underlying TRAN link.

Figure 6.11 demonstrates the relationships between the UNI components and the associated ETH access link.

The UNI-C is always connected to the UNI reference point by an IEEE 802.3 PHY. The UNI-N implements the service provider side of the UNI functions. The reference point between the access network and the provider edge (PE) equipment is called service node interface (SNI) (Fig. 6.12).

In an ETH trunk link, multiple EVCs may ride on. An ETH trunk link is instantiated via TAF and underlying TRAN on an Internal-Network Network Interface (I-NNI) and an ETH Provider Conditioning and TAF along with an underlying TRAN trail on an ENNI.

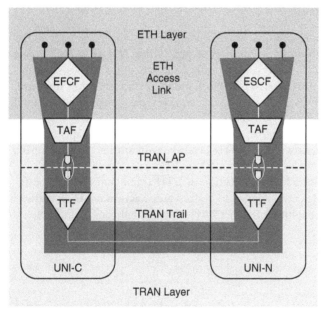

Figure 6.10 Access link for direct attachment of UNI-C to UNI-N. TRAN_AP, TRAN access point.

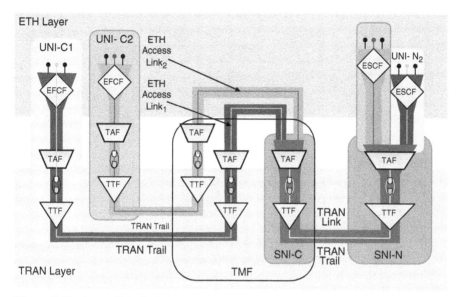

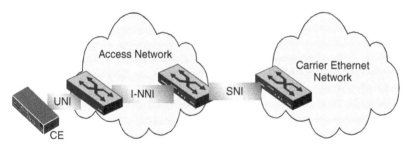

Figure 6.11   Access links for indirect attachment of a UNI-C to a UNI-N via a service node interface (SNI) from a TMF.

Figure 6.12   UNI and SNI reference points.

The ETH trunk link implemented via an I-NNI includes a TRAN-layer-specific Transport Adaptation Function (TAF) and an underlying TRAN trail.

Figure 6.13 illustrates the relationships between an I-NNI and the associated ETH trunk link. Figure 6.14 illustrates the relationships between the ENNI functional elements and the associated ETH trunk link.

6.4 INTERFACES

This section describes the interface between user and network, interface between networks, and connection types between users.

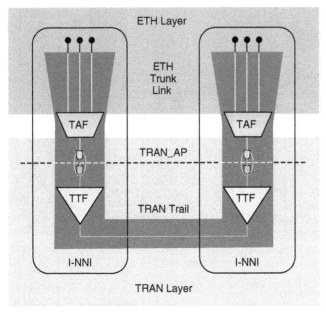

Figure 6.13   An ETH trunk link between two I-NNIs.

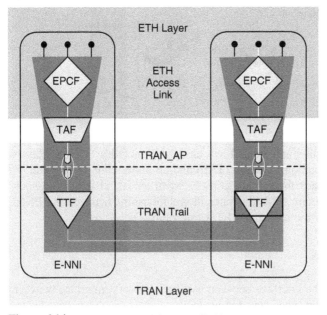

Figure 6.14   An ETH access link among ENNIs.

6.4.1 UNI

The standard user interface to a MEN is called the UNI [2]. UNI is dedicated physical demarcation point between the responsibility of the service provider and the responsibility of a subscriber.

UNI functions are distributed between Customer Premises Equipment (CPE) and MEN, as UNI-C and UN-N, respectively.

UNI-C executes the processes of the customer side, while UNI-N executes the processes of the network side. CPE may not support UNI-C functions and uses a Network Interface Device (NID) to support the demarcation point between the CPE and the MEN, as depicted in Figure 6.15, which is the most common implementation.

The UNI consists of a data plane, a control plane, and a management plane. The data plane defines Ethernet frames, tagging, and traffic management. The management plane defines provisioning, device discovery, protection, and OAM. The control plane is expected to define signaling and control, which is not addressed in the standards yet.

Ethernet Demarcation is analogous to a Channel Service Unit/Data Service Unit (CSU/DSU) in Frame Relay. It provides separation between the carrier WAN and enterprise LAN and enables testing and monitoring of both LAN and WAN.

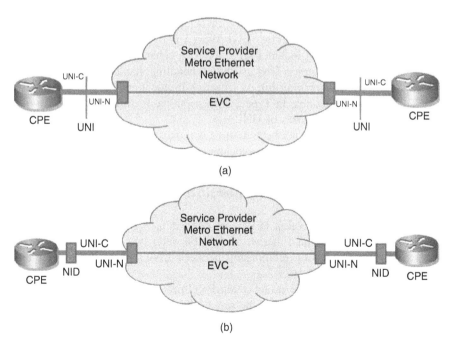

Figure 6.15 (a) CPE supporting UNI. (b) NID supporting UNI.

The physical medium for UNI can be copper, coax, or fiber. The operating speeds can be 1–10 Mbps, 100 Mbps, 1 Gbps, or 10 Gbps, where 10 Gbps rate is only supported over fiber.

Service frames are Ethernet frames that are exchanged between the MEN and the CE across the UNI. A service frame sent from the Metro Ethernet network to the CE at a UNI is called an egress service frame. A service frame sent from the CE to the Metro Ethernet network at a UNI is called an ingress service frame.

MEF defines UNI on a port or an EVC basis with CIR/EIR (Committed Information Rate/Excess Information Rate) and CBS/EBS (Committed Burst Size/Excess Burst Size).

Traffic is classified/prioritized based on Type of Service (TOS), Differentiated Services Code Point (DSCP), and 802.1P.

MEF 11 [13] introduced 3 types of UNI: Type 1, Type 2, and Type 3. In UNI Type 1, which is defined in MEF 13 [14], service provider and customer manually configure the UNI-N and UNI-C for services. MEF 13 divides UNI Type 1 into two categories:

- Type 1.1: Nonmultiplexed UNI for services such as Ethernet Private Line (EPL) and
- Type 1.2: Multiplexed UNI for Services such as Ethernet Virtual Private Line (EVPL).

UNI Type 2, which is defined in MEF 20 [15], supports an automated implementation model allowing UNI-C to retrieve EVC status and configuration information from UNI-N, enhanced UNI attributes, and additional fault management and protection functionalities. UNI Type 2 is divided into UNI Type 2.1 and Type 2.2. The Type 2.1 includes service OAM, enhanced UNI attributes such as bandwidth profile per egress UNI and L2CP handling mandatory features, and Link OAM, port protection via Link Aggregation Protocol and E-LMI (as defined in MEF 16) optional features. In Type 2.2, the optional features of Type 2.1 will become the mandatory features of UNI.

UNI Type 3, which has not been defined yet, may support UNI-C to request, signal, and negotiate EVCs and its associated service attributes to the UNI-N.

6.4.2   Ethernet Virtual Connection (EVC)

An Ethernet Virtual Connection (EVC) as defined in MEF 10.1 [26, 16] is a logical representation of an Ethernet service between two or more UNIs and establishes a communication relationship between UNIs. An UNI may contain one or more EVCs (Fig. 6.16).

While connecting two or more UNIs, EVC prevents data transfer between subscriber sites that are not part of the same UNI. The service frame cannot be delivered back to the originating UNI either.

The service frames cannot leak into or out of an EVC. This capability, which is similar to a Frame Relay or ATM Permanent Virtual Circuit (PVC), allows

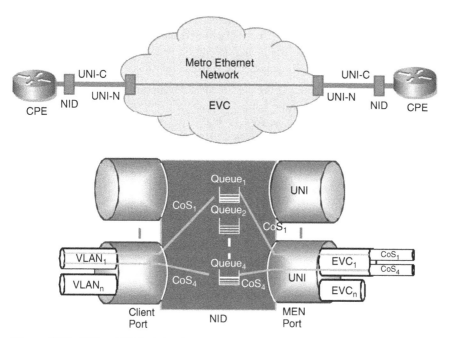

Figure 6.16 UNI and EVC relationship.

UNIs constructing their own Layer 2 Private Networks. At a UNI, multiple EVCs can be constructed to form multiple private line networks.

Point-to-point EVC (Fig. 6.15) supports communication between only two UNIs. All ingress service frames at one UNI, with the possible exception of L2CP messages, are typically delivered to the other UNI.

Multipoint-to-multipoint EVC supports any-to-any communication between two or more UNIs. This EVC creates a service that behaves like a switched Ethernet environment and is an essential component of E-LAN services. Additional UNIs can be added to the multipoint-to-multipoint EVC.

In a multipoint-to-multipoint EVC, a single broadcast or multicast ingress service frame, which is determined from the destination MAC address, at a given UNI is replicated in the MEN, and one copy would be delivered to each of the other UNIs in the EVC (Fig. 6.3).

Broadcast service frames defined by all 1's destination MAC addresses and multicast service frames with multicast destination MAC addresses are replicated and delivered to all other UNIs in the EVC.

Unicast service frames with unicast destination MAC addresses are managed in one of the two ways.

- They can be replicated and delivered to all other UNIs in the EVC. This makes the EVC behave like a shared-media Ethernet.
- The MEN can learn which MAC addresses belong to which UNIs by observing the source MAC addresses in service frames and can deliver a

service frame to only the appropriate UNI when it learns the destination MAC address. When it has not yet learned the destination MAC address, it replicates the service frame and delivers it to all other UNIs in the EVC. In this case, the MEN behaves like a MAC learning bridge.

Point-to-multipoint EVC, also called Rooted Multipoint EVC, supports communication between two or more UNIs but does not support any-to-any communication. UNIs are designated as root or leaf. Transmissions from the root are delivered to the leaves, and transmission from the leaves are delivered to the root(s). No communication can occur between the leaves or between the roots. This EVC can be used to create a hub-and-spoke communication arrangement without needing to configure multiple point-to-point EVCs. It is an essential component of a service type known as E-Tree (Figs 6.22 and 6.26).

In a given UNI, subscriber flows may be mapped into one or more EVCs. When multiple subscriber flows are mapped into a single EVC, there could be multiple bandwidth profiles associated with the EVC where there is a bandwidth profile per CoS Instance. The bandwidth profile defines bandwidth and performance parameters, such as delay, jitter, and availability per CoS instance.

Similarly, multiple CE-VLAN IDs can be mapped to one EVC where each CE-VLAN ID may belong to a different service (i.e., service multiplexing). This configuration is called a bundling map. When there is a bundling map, the EVC must have the CE-VLAN ID Preservation that allows the subscriber to use the same VLAN values at all sites.

6.4.3   Key UNI/EVC Attributes

Service Frame Disposition

An ingress service frame can be discarded, delivered conditionally, or delivered unconditionally. For example, frames with FCS error or containing a particular Layer 2 protocol can be discarded. In an E-Line service, all frames with no FCS error can be delivered unconditionally.

Service Frame Transparency

All fields of each egress service frame must be identical to the same fields of the corresponding ingress service frame. However, the egress service frame may have a CE-VLAN Tag (C-Tag)., whereas the corresponding ingress service frame does not and vice versa. In this case, the egress service frame must have a recalculated FCS. If both the egress service frame and corresponding ingress service frame have a CE-VLAN Tag, the contents of the Tag in the egress service frame may be different from the contents of the Tag in the corresponding ingress service frame. If the contents of the ingress and egress tags are different, the egress service frame must have a recalculated FCS.

The user Ethernet traffic enters the service provider network with either no VLAN tag or a preassigned VLAN tag. The service provider may choose to map the user tag to its own tag to uniquely identify each user within the network.

Table 6.3   PCP-CoS Mapping

EVC ID	Tagged or Untagged	PCP Values	CoS
10	Tagged	4,5,6,7	Gold
		0,2,3	Silver
		1	Discard
	Untagged	Traffic is treated as PCP = 0	Silver
20	Tagged	5	Best effort
		0,1,2,3,4,6,7	Gold
	Untagged	Traffic is treated as PCP = 0	Gold

CE-VLAN Tag (C-Tag) Preservation

C-Tag consists of CE-VLAN ID and PCPs that can be preserved for operational simplicity.

The ingress and egress service frames will have the same CoS.

Layer 2 Control Protocol Processing Service

An L2CP frame is discarded, tunneled, or peered for a given EVC at a given UNI. When user bridges attached to UNIs, BPDUs will only be processed after they exit the service provider network. When an L2CP is tunneled, the service frame at each egress UNI has to be identical to the corresponding ingress service frame. When an L2CP is peered, it is processed in the network.

Class of Service (CoS) Identifier

Service frame delivery performance is specified for all service frames transported within an EVC with a particular CoS identified by a CoS Identifier. The CoS Identifier can be based on EVC, PCP, or DSCP.

All frames with the same CoS Identifier such as Gold, Silver, or Best Effort can be mapped to the same EVC. Therefore, EVC ID will identify CoS as well. EVC ID-CoS mapping needs to be stored in operation systems.

PCP Field of C-Tag or S-Tag can be used to identify CoS. If the frame is untagged, it will have the same CoS Identifier as a frame with PCP field = 0. An example is given in Table 6.3.

Similarly, DSCP that is contained in TOS byte of IPv4 [29, 30] and Traffic Class Octet ion IPv6 can be used to identify CoS (Table 6.4).

EVC Performance

The EVC performance attributes specify the frame delivery performance for green colored frames of a particular CoS, arriving at the ingress UNI during a time

Table 6.4 DSCP-CoS Mapping

EVC ID	Tagged or Untagged	DSCP Values	CoS
10	Tagged	46	Gold
		26	Silver
		10	Discard
	Untagged	CoS per agreement	Silver
20	Tagged	10	Best effort
		46	Gold
	Untagged	CoS per agreement	Gold

interval T. These attributes are Frame Delay (FD), Interframe Delay Variation (IFDV), Mean Frame Delay (MFD), Frame Delay Range (FDR), Frame Loss Ratio (FLR), and Availability. The users choose to use FD and IFDV or MFD and FDR. These parameters are described later in detail.

EVC ID

The EVC is identified by an EVC ID. MEF defines it as an arbitrary string administered by the service provider to identify the EVC within the MEN for management and control purposes. The EVC ID is not carried in any field in the service frame but is stored in operation systems. ATIS [18] further defines EVC ID as 28 alphanumeric characters: <Prefix>. <Service Code/Modifier>.<Serial Number>.<Company ID>.<Segment Number>

- Prefix is a nonstandard code populated according to the special services circuit coding methodology of each carrier. It is one to two alphanumeric characters. This prefix may be used to identify a regional network of the carrier's nationwide network.
- Service Code is a standardized code that represents a tariff offering that requires special services circuit provisioning. It is two alphanumeric characters. Valid entries are defined in Reference 18.
- Service Code Modifier is a standardized code that designates the jurisdiction, networking application, and additional technical information of the service identified in the service code. It is two alphanumeric characters. Valid entries are defined in Reference 18 as listed in Table 6.5.

Table 6.5 EVC Types and Service Codes/Modifiers

EVC Types	Service Code/Modifier
Point-to-point	VLXP
Multipoint-to-multipoint	VLXM

- Serial Number is a serial number type code that uniquely identifies a special services circuit having the same prefix, service code, and service code modifier within a network. It is one to six numeric characters.

- Suffix is a serial number type code that relates a group of special services circuits having the same service code for the same customer and with similar termination equipment at each end. It is one to three numeric characters.

- Assigning Company ID is a standardized code that uniquely identifies the carrier assigning the circuit identification. Valid entries are outlined in Reference 19. It is two to three alphabets.

- Segment Number is a serial number type code that uniquely identifies each termination point of a special services circuit, when the circuit has more than two termination points, that is, multipoint circuit (one to three alphanumeric characters).

For example, 12.VLXP.123456..OB. represents carrier OB's point-to-point connection in region 12 where the suffix before the company code and the segment number after the company code are not utilized.

UNI ID

The UNI is identified by an UNI ID. MEF 10.2 [17] defines UNI ID as an arbitrary text string. The UNI ID must be unique among all UNIs for the MEN.

The ATIS standards define it as <Prefix>. <Service Code/Modifier>.<Serial Number>.<Company ID>.<Segment Number>, where the service code/modifier is given in Table 6.6.

For example, 15.KFGS.123457..BO. represents a 1-Gbps UNI in intrastate region 15 of the carrier BO.

6.4.4 ENNI (External Network to Network Interface)

Most of the domestic and international communications travel more than one service provider/operator network. This is expected to be true for Carrier Ethernet services as well. An EVC between two UNIs may have to travel multiple

Table 6.6   UNI Port Speeds and Associated Service Codes/Modifiers

UNI	Service Codes/Modifiers	
Port Speed	Inter	Intra
10 Mbps	KDGS	KDFS
100 Mbps	KEGS	KEFS
1 Gbps	KFGS	KFFS
10 Gbps	KGGS	KGFS

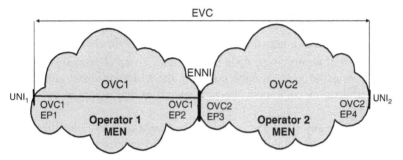

Figure 6.17 ENNI, EVC, and OVC.

Operator MENs. One operator/service provider may not be able to support all UNIs of a subscriber. As a result, MENs need to interface each other. This interface between two MENs is called the ENNI [ENNI Phase I], depicted in Figure 6.17.

An ENNI can be implemented with one or more physical links. However, when there is no protection mechanism such as LAG among multiple links connecting two Operator MENs, each link represents a distinct ENNI.

Similar to UNI, ENNI consists of a data plane, a control plane, and a management plane. The data plane defines Ethernet frames, tagging, and traffic management. The management plane defines provisioning, device discovery, protection, and OAM. The control plane is expected to define signaling and control, which is not addressed in the standards yet.

The physical medium for ENNI can be copper, coax, or fiber. However, the fiber interface between ENNI gateways is more common. The operating speeds can be 1Gbps or 10Gbps where 10Gbps is available.

Similar to UNI, ENNI has a bandwidth profile. The parameters of the bandwidth profile are CIR in bps, CBS in bytes that is greater than or equal to the largest MTU size allowed for the ENNI frames, EIR in bps, EBS in bytes that is greater than or equal to the largest MTU size allowed for the ENNI frames, Coupling Flag (CF), and Color Mode (CM). The CF has value 0 or 1, and CM has the value "color blind" or "color aware."

Operator Virtual Connection (OVC)

When the EVC travels multiple Operator MENs, the EVC is realized by concatenating EVC components in each operator network, which is called Operator Virtual Connections (OVCs). For example, in Figure 6.17, EVC is constructed by concatenating OVC 1 in MEN 1 with OVC 2 in MEN 2.

Each OVC End Point (OEP) is associated with either a UNI or an ENNI and at least one OEP is associated with an OEPs at an ENNI (Fig. 6.18).

Hairpin switching occurs when an OVC associates two or more OVC EPs at a given ENNI. An ingress S-Tagged ENNI frame at a given ENNI results in

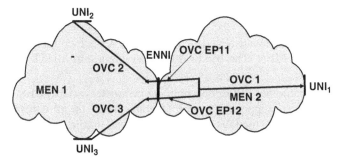

Figure 6.18   Example of supporting multiple OEPs (OEP-OVC End Point) for one OVC.

an egress S-Tagged ENNI frame with a different S-VLAN ID value at the given ENNI.

An OVC EP represents the OVC termination point at an external interface such as ENNI or UNI. The EP map specifies the relationship between S-Tagged ENNI Frame and OVC EP within an Operator MEN. An ingress S-Tagged ENNI frame that is not mapped to an existing EP will be discarded.

Table 6.7 illustrates an example of an EP map where S-VLAN IDs are mapped to EP Identifier, which is a string administered by the operator.

Multiple S-VLAN IDs can be bundled and mapped to an OVC EP. In that case, the OVC EP should be configured to preserve S-VLAN IDs. S-VLAN ID preservation cannot be supported in hairpin switching configuration.

When there is bundling, frames originated by more than one subscriber may be carried by the OVC, resulting in duplicate MAC addresses. To avoid possible problems due to this duplication, learning the MAC Address may be disabled on the OVC.

There are service attributes for each instance of an OVC End Point at a given ENNI that are summarized below.

- OVC EP Identifier: A string that is unique across the Operator MEN.
- CoS Identifiers: Identifies the CoS that is determined for an ENNI frame at each ENNI. The OVC associates the EP and the S-Tag PCP.
- Ingress Bandwidth Profile per OVC EP: Ingress policing parameters <CIR, CBS, EIR, EBS, CF, CM>, if supported, by the Operator MEN on all ingress ENNI frames are mapped to the OVC EP. The CM for the bandwidth profile algorithm MUST be color aware.

Table 6.7   End Point Map Example

S-VLAN ID Value	End Point Identifier	End Point Type
100	comcast-twc-enni_1_10	OVC end point
200	comcast-twc-enni_2_20	OVC end point

- Ingress Bandwidth Profile per ENNI CoS Identifier: Ingress policing parameters <CIR, CBS, EIR, EBS, CF, CM>, if supported, by the Operator MEN on all ingress ENNI frames with the CoS Identifier for the receiving Operator MEN. The bandwidth profile algorithm MUST be color aware.
- Egress Bandwidth Profile per EP: Egress policing parameters <CIR, CBS, EIR, EBS, CF, CM>, if supported, by the Operator MEN on all egress ENNI Frames are mapped to the OVC EP. The bandwidth profile algorithm MUST be color aware.
- Egress Bandwidth Profile per ENNI CoS Identifier: Egress policing parameters <CIR, CBS, EIR, EBS, CF, CM>, if supported, by the Operator MEN on all egress ENNI frames with the CoS Identifier for the receiving MEN. The bandwidth profile algorithm MUST be color aware.

6.4.5   VUNI/RUNI

Ordering and maintaining EVCs crossing an ENNI can be cumbersome due to coordinations between entities of two or more carriers. In order to simplify this process to an extent, the operator responsible for the end-to-end EVC may order a tunnel between its remote user that is connected to another operator's network and its ENNI gateway, which can accommodate multiple OVCs, instead of ordering the OVCs one by one. This access of the remote user to the service provider's network is called UNI tunnel access (UTA). The remote user end of the tunnel is called remote UNI (RUNI), while the service provider end of the tunnel is called virtual UNI (VUNI) (Fig. 6.19) [20].

The VUNI has service attributes similar to those of a UNI and is paired with an RUNI. Its main function is to specify the processing rules applicable to ENNI frames present in the VUNI Provider domain and associate them with a given UTA instance.

In Figure 6.20, the CE at UNI Y participates in EVCs 1, 2, and 3. These EVCs have the service-provider-agreed bandwidth profile attributes, as well

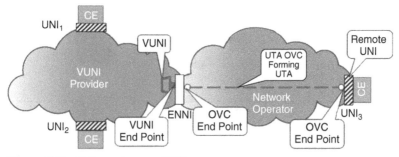

Figure 6.19   UNI tunnel access (UTA).

as CoS markings. At the RUNI, service frames of EVC 1, 2, and 3 are exchanged with the CE. Such frames may be C-tagged, priority tagged, or untagged. The RUNI is instantiated by the Operator as a UNI where the network operator maps all service frames to the single OVC EP supporting the UTA OVC.

A single bandwidth profile and CoS may be applied at this RUNI OVC EP. At the UTA OVC EP at the network operator's side of the ENNI, an S-VLAN ID is used to map ENNI frames to the OVC EP supporting the UTA and applies a UTA-specific single bandwidth profile and CoS.

Given there is a bandwidth profile per UTA OVC, it will become a challenge to satisfy bandwidth requirements per EVC basis. This issue is somewhat solvable when $CIR = 0$ and $EIR > 0$ and when $CIR > 0$ and $EIR = 0$. But it is very difficult to solve when $CIR > 0$ and $EIR > 0$.

In the VUNI provider's network, the relationship between the UTA OVC and the VUNI is realized by an S-VLAN ID present at the ENNI, whose value is negotiated between the VUNI provider and the network operator. At the ENNI, when receiving an ENNI frame, the VUNI provider maps (using the EP map) a single S-VLAN ID to a VUNI EP associated with a VUNI. The VUNI then maps frames based on their CE-VLAN ID to the appropriate OVC EP for OVCs A1, A2, and A3. In the reverse direction, the VUNI multiplexes frames from OVCs A1, A2, and A3 into a tunnel denoted by a unique S-VLAN ID, which is associated with the network operator's UTA OVC. CE-VLAN IDs of A1, A2, and A3 at VUNI do not overlap.

Multiple VUNIs can be associated with a single ENNI. Figure 6.21 shows how the UTA and a VUNI may be used to instantiate the services across ENNI.

At the ENNI between MEN B and MEN C, Figure 6.21 shows the mapping of frames with an S-VID of 100 to a VUNI EP representing VUNI A.

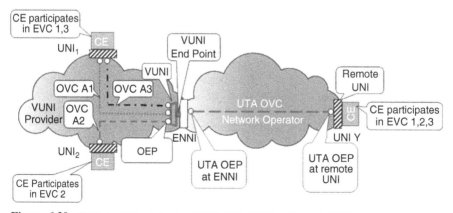

Figure 6.20 EVCs implemented using VUNI, UTA OVC, and remote UNI.

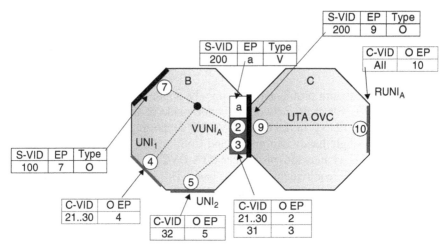

Figure 6.21 Example multiple EVCs supported by UTA and VUNI.

In summary, UTA is composed of a UTA OVC component in the Operator MEN and an associated VUNI in the Provider MEN and RUNI in the Operator MEN components. Requirements for each component are described in the following sections.

6.5 SERVICES

Carrier Ethernet services are defined in terms of what is seen by CE and independent from technology inside Metro Ethernet network with five attributes: QoS, standardized services, scalability, reliability, and service management. These attributes are reflected in the UNI and EVC parameters.

The physical demarcation point between a subscriber and a service provider is called UNI. A demarcation point between two service providers is called ENNI. If it is between service provider internal networks, it is called I-NNI.

The CE and MEN exchange service frames across the UNI where the frame is transmitted toward the service provider is called an ingress service frame and the frame transmitted toward the subscriber is called an egress service frame.

EVC is the logical representation of an Ethernet service, which is defined between two or more UNIs. EVCs can be point-to-point, point-to-multipoint (or rooted multipoint), or multipoint-to-multipoint, as depicted in Figure 6.22.

Carrier Ethernet services built over these EVCs are listed in Table 6.8. These services can support applications with strict performance requirements such as voice and best effort performance requirements such as Internet access (Table 6.9).

Ethernet service delivery is independent of the underlying technologies. At the access, Carrier Ethernet service may ride over the TDM, SONET, Ethernet,

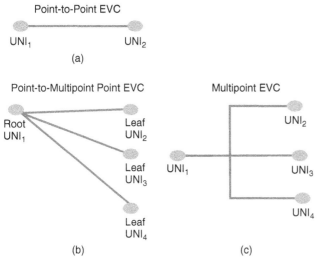

Figure 6.22 EVC types.

Table 6.8 Carrier Ethernet Services

Service Type	Port-Based (All to One Bundling)	VLAN-Based (Service Multiplexed)
E-Line over point-to-point EVC	Ethernet private line (EPL)	Ethernet virtual private line (EVPL)
E-LAN over multipoint-to-multipoint EVC	Ethernet private LAN (EP-LAN)	Ethernet virtual private LAN (EVP-LAN)
E-Tree over rooted multipoint EVC	Ethernet private tree (EP-Tree)	Ethernet virtual private tree (EVP-Tree)

Table 6.9 Possible Applications for Carrier Ethernet Services

EPL Applications	EVPL Applications	E-LAN Applications
Healthcare, financial, pharmaceuticals	Local/regional government, education, healthcare, financial	Local/regional government, primary and secondary education
Medical imaging, data center connectivity, business continuity, CAD engineering	VoIP, video, point-to-point intrametro connectivity, classes of service corporate networking, photo imaging, video conferencing	Collaboration, multisite connectivity, multipoint Ethernet, administration

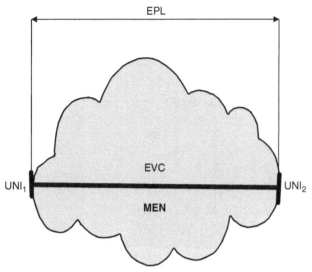

Figure 6.23 EPL example.

and PON technologies. At the backbone, Carrier Ethernet service may ride over the IP/MPLS network.

Table 6.10 compares Carrier-Ethernet-based VPN service and IP-based VPN services in access networks.

E-Line service can be divided into EPLs and virtual private lines. The EPL service is between two UNI ports where there is no multiplexing at UNI, and it provides transparency to the service frames (Fig. 6.16). On the other hand, EVPL may involve more than two UNI ports where there is multiplexing at UNI. E-Line service can support a specific performance assurance as well as no-performance assurance.

Table 6.10 Comparison of Ethernet and IP-based VPNs

Service Attribute	Ethernet Service	IP Service
Customer handoff	Ethernet UNI	Ethernet port (or PDH circuit)
Service identification	VLAN ID/EVC	IP address
CoS identification	PCP	DSCP/ToS
Packet/frame routing/forwarding	MAC address (E-LAN)	IP address
	VLAN ID (E-Line)	
Fault management	Link trace, continuity check (layer 2 ping), loopbacks	Traceroute, ICMP ping
Performance management	Frame Delay, Frame Delay Variation, Frame Loss Ratio, Service Availability	Packet delay, packet delay variation, packet loss

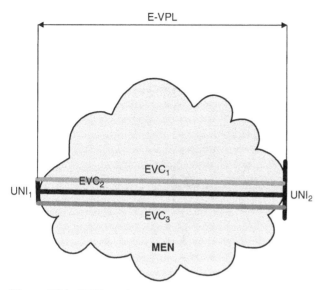

Figure 6.24 EVPL service.

EPL is used to replace TDM, FR, and ATM private lines (Fig. 6.23). There is a Single Ethernet Virtual Connection (EVC) per UNI. For cases where EVC speed is less than the UNI speed, the CE is expected to shape traffic.

An EVPL (Fig. 6.24) can be used to create services similar to the EPL. As a result of EVC multiplexing, an EVPL does not provide as much transparency of service frames as an EPL because some of the service frames may be sent to one EVC while other service frames may be sent to other EVCs.

E-LAN service is used to create multipoint L2 VPNs and transparent LAN service with a specific performance assurance as well as no-performance assurance. It is the foundation for IPTV and multicast networks.

In the EP-LAN service (Fig. 6.25), C-Tag can be preserved and L2CPs can be tunneled. This provides additional flexibility to users by reducing the coordination with service provider.

EVCs can be multiplexed on the same UNI to create EVP-LAN services (Fig. 6.26). Furthermore, E-LAN and E-Line services can be multiplexed in the same UNI, where some of the EVCs can be part of the E-LAN while the others can be part of E-Line services. Furthermore, UNIs can be added and removed from E-LAN service without disturbing the users on the E-LAN.

The E-Tree services depicted in Figures 6.27 and 6.28 are also a subset of E-LAN service where traffic from any "leaf" UNI can be sent/received to/from "Root" UNI(s) but never being forwarded to other "leaf" UNIs. The service is used for applications requiring point-to-multipoint topology such as video on

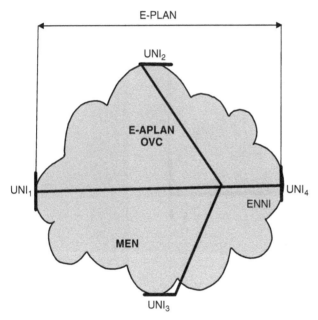

Figure 6.25   EP-LAN service.

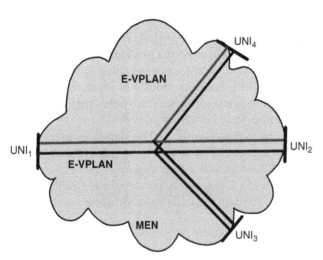

Figure 6.26   EVP-LAN service.

demand, Internet access, triple play backhaul, mobile cell site backhaul, and franchising applications. EP-Tree services have the same benefits as EP-LAN.

In the following sections, we describe attributes of each service.

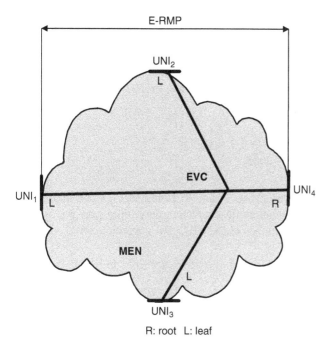

Figure 6.27   EP-Tree service.

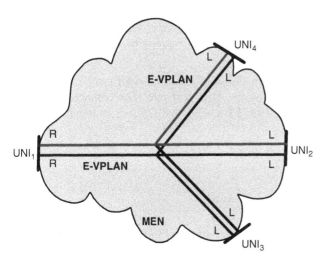

Figure 6.28   EVP-Tree service.

6.5.1   E-Line Service Attributes

The E-Line service attributes are UNI and EVC [12]. The UNI attributes include physical port parameters, bandwidth profile, L2CP treatment, and customer VLAN mapping.

- Port Name: If the device has a multislot chasse, then the name can be fe/6/1 (i.e., Fast Ethernet Slot 6 Port 1). If the device has a single line card such as an NID, it can be LAN-1, LAN-2, or WAN-1, etc.
- Speed: 10 Mbps, 100 Mbps, 10/100 Mbps autonegotiation, 10/100/1000 Mbps autonegotiation, 1 Gbps, 10 Gbps, or 100 Gbps, which is not being offered yet.
- Media Type: Fiber or copper.
- Mode: Full duplex.
- MAC Layer: IEEE 802.3–2005.
- MTU Size: ≥1522.
- UNI ID: Arbitrary text string. According to References 18, 19, and 21, it is possible to use 28 alphanumeric characters describing port rate, company name, company region or market ID, and a specific serial number (Section 6.4.3).
- Service Multiplexing: There is no multiplexing for EPL. In EVPL, multiple services can be multiplexed.
- All to One Bundling: It is possible for EPL, but not possible for EVPL.
- Bundling: It is possible for EVPL if CE-VLAN ID preservation is supported.
- Maximum Number of EVCs: It is 1 for EPL and ≥1 for EVPL.
- CE-VLAN ID for Untagged and Priority Tagged Service Frames: When there is no all to one bundling at the UNI for EPL, all frames need to have CE-VLAN IDs in the range of 1 to 4094. For EVPL, it is clear that we need CE-VLAN IDs for all frames.
- Ingress and Egress Bandwidth Profiles: The bandwidth profile for EPL is optional since it is defined by the UNI port rate. The bandwidth profile parameters are CIR, CBS, EIR, EBS, CM, and CF. For example,

  CIR = 10 Mbps
  CBS = 32 KB
  EIR = 2 Mbps
  EBS = 4 KB
  CM = "color blind" or "color aware"
  CF = 0 or 1

- L2CP Processing: L2 protocols can be treated in three ways:
  1. tunnel if only CPE equipment behind Ethernet devices process these protocols such as GARP/MRP (Multiple Registration Protocol) block;
  2. peer if Ethernet devices at both ends support these protocols such as E-LMI;
  3. discard if the device is not going to process the protocol, such as Link OAM.

L2CP treatment is described in Tables 6.11–6.19.

The EPL traffic is mapped to a single EVC. The EVC parameters include bandwidth profile per EVC or CoS and CE-VLAN ID to EVC mapping. The EVC parameters are listed in the following:

- EVC ID: A string formed by the concatenation of the UNI ID and the EVC ID (Section 6.4.3).
- CE-VLAN ID/EVC Map: All VLANs are mapped to one EVC.
- Ingress and Egress Bandwidth Profiles per EVC or CoS: The bandwidth profile parameters for an EVC or CoS are CIR, CBS, EIR, EBS, CM, and CF. It is possible to define just ingress bandwidth profile or both ingress and egress. If both egress and ingress parameters are defined, then both bandwidth profiles must be identical.

EVC parameters also include UNI list, bandwidth profile, L2CP treatment, and EVC performance.

- EVC ID
- EVC Type: Point-to-point
- UNI List
- Maximum Number of UNIs: 2
- EVC MTU Size: ≥ 1522
- CE-VLAN ID Preservation: Yes for EPL. Yes or No for EVPL.
- CE-VLAN CoS Preservation: Yes for EPL. Yes or No for EVPL
- Unicast Service Frame Delivery: EPL, deliver unconditionally. EVPL, deliver unconditionally or deliver conditionally. The condition must be specified. An example of such a condition is that the destination MAC address is known by the Metro Ethernet Network to be "at" the destination UNI.
- Multicast Service Frame Delivery: EPL, deliver unconditionally. EVPL, deliver unconditionally or deliver conditionally. The condition must be specified
- Broadcast Service Frame Delivery: EPL, deliver unconditionally. EVPL, deliver unconditionally or deliver conditionally. The condition must be specified.
- L2CP Processing that are Passed to EVC: L2CP treatment is listed in Tables 6.11–6.19.
- EVC Performance: At least one CoS is required. For each CoS, values for FD, FDV, and FLR may need to be specified if known.

6.5.2 EP-LAN and EVP-LAN Service Attributes

UNI parameters include physical port parameters, bandwidth profile, L2CP treatment, and customer VLAN mapping.

Table 6.11 L2CP Processing Requirements for EPL, EP-LAN, and EP-Tree Services[a]

Destination MAC Address	L2CP Action for EPL, EP-Tree, and EP-LAN
01-80-C2-00-00-00	MUST tunnel
01-80-C2-00-00-01 through	MUST NOT tunnel
01-80-C2-00-00-0A	
01-80-C2-00-00-0B	MUST tunnel
01-80-C2-00-00-0C	MUST tunnel
01-80-C2-00-00-0D	MUST tunnel
01-80-C2-00-00-0E	MUST NOT tunnel
01-80-C2-00-00-0F	MUST tunnel

[a]Ref. 22

Table 6.12 L2CP Action for EVPL, EVP-Tree, and EVP-LAN Services[a]

Destination MAC Address	L2CP Action for EVPL, EVP-Tree, and EVP-LAN
01-80-C2-00-00-00 through	MUST NOT tunnel
01-80-C2-00-00-0F	

[a]Ref. 22

Table 6.13 L2CP Processing Requirements for the EPL Option 1 Service

Protocol Type	L2CP Action
STP [6]/RSTP[6]/MSTP[7]	MUST Peer on all UNIs or Discard on all UNIs
PAUSE[8]	MUST Discard on all UNIs
LACP/LAMP[8]	MUST Peer or Discard per UNI
Link OAM[8]	MUST Peer or Discard per UNI
Port Authentication[10]	MUST Peer or Discard per UNI
E-LMI[12]	MUST Peer or Discard per UNI
LLDP[11]	MUST Peer or Discard per UNI
PTP Peer Delay	MUST Peer or Discard per UNI
ESMC	MUST Peer or Discard per UNI

Table 6.14 L2CP Processing Requirements for the EVPL Service

Protocol Type	L2CP Action
STP[6]/RSTP[6]/MSTP[7]	MUST Peer on all UNIs or Discard on all UNIs
PAUSE[8]	MUST Discard on all UNIs
LACP/LAMP[8]	MUST Peer or Discard per UNI
Link OAM[8]	MUST Peer or Discard per UNI
Port Authentication[10]	MUST Peer or Discard per UNI
E-LMI[12]	MUST Peer or Discard per UNI
LLDP[11]	MUST Discard on all UNIs
PTP Peer Delay	MUST Peer on all UNIs or Discard on all UNIs
ESMC	MUST Peer or Discard per UNI

Table 6.15 L2CP Processing Requirements for the EP-LAN Service

Protocol Type	L2CP Action
STP[6]/RSTP[6]/MSTP[7]	MUST Peer on all UNIs or Discard on all UNIs
PAUSE[8]	MUST Discard on all UNIs
LACP/LAMP[8]	MUST Peer or Discard per UNI
Link OAM[8]	MUST Peer or Discard per UNI
Port Authentication[10]	MUST Peer or Discard per UNI
E-LMI[12]	MUST Peer or Discard per UNI
LLDP[11]	MUST Discard on all UNIs
PTP Peer Delay	MUST Peer on all UNIs or Discard on all UNIs
ESMC	MUST Peer or Discard per UNI

Table 6.16 L2CP Processing Requirements for the EVP-LAN Service

Protocol Type	L2CP Action
STP[6]/RSTP[6]/MSTP[7]	MUST Peer on all UNIs or Discard on all UNIs
PAUSE[8]	MUST Discard on all UNIs
LACP/LAMP[8]	MUST Peer or Discard per UNI
Link OAM[8]	MUST Peer or Discard per UNI
Port Authentication[10]	MUST Peer or Discard per UNI
E-LMI[12]	MUST Peer or Discard per UNI
LLDP[11]	MUST Discard on all UNIs
PTP Peer Delay	MUST Peer on all UNIs or Discard on all UNIs
ESMC	MUST Peer or Discard per UNI

Table 6.17 L2CP Processing Requirements for the EP-Tree Service

Protocol Type	L2CP Action
STP[6]/RSTP[6]/MSTP[4]	MUST Peer on all UNIs or Discard on all UNIs
PAUSE[8]	MUST Discard on all UNIs
LACP/LAMP[8]	MUST Peer or Discard per UNI
Link OAM[8]	MUST Peer or Discard per UNI
Port Authentication[10]	MUST Peer or Discard per UNI
E-LMI[12]	MUST Peer or Discard per UNI
LLDP[11]	MUST Discard on all UNIs
PTP Peer Delay	MUST Peer on all UNIs or Discard on all UNIs
ESMC	MUST Peer or Discard per UNI

Table 6.18 L2CP Processing Requirements for the EVP-Tree Service

Protocol Type	L2CP Action
STP[6]/RSTP[6]/MSTP[7]	MUST Peer on all UNIs or Discard on all UNIs
PAUSE[8]	MUST Discard on all UNIs
LACP/LAMP[8]	MUST Peer or Discard per UNI
Link OAM[8]	MUST Peer or Discard per UNI
Port Authentication[10]	MUST Peer or Discard per UNI
E-LMI[12]	MUST Peer or Discard per UNI
LLDP[11]	MUST Discard on all UNIs
PTP Peer Delay	MUST Peer on all UNIs or Discard on all UNIs
ESMC	MUST Peer or Discard per UNI

Table 6.19 L2CP Processing Requirements for MRP Protocols

Service Type	L2CP Action
EPL, EP-LAN, EP-Tree	MUST tunnel all UNIs
EVPL, EVP-LAN, EVP-Tree	SHOULD peer or tunnel all UNIs

- Port Name: This parameter is not in MEF 6.1. If it is a multislot chasse, then the name can be fe/6/1. If the device is an NID, it can be LAN-1, LAN-2, or WAN-1, etc.
- Speed: 10 Mbps, 100 Mbps, 10/100 Mbps autonegotiation, 10/100/1000 Mbps autonegotiation, 1 Gbps, 10 Gbps, or 100 Gbps.
- Media Type: Fiber or copper.
- Mode: Full duplex.
- MAC Layer: IEEE 802.3–2005.
- MTU Size: Must be ≥ 1522.
- UNI ID: Arbitrary text string. According to References 18, 19, and 21, it is possible to use 28 alphanumeric characters describing port rate, company name, company region or market ID, and a specific serial number.
- Service Multiplexing: No for EP-LAN, and Yes or No for EVP-LAN. If Yes, CE-VLAN ID preservation must be supported.
- All to One Bundling: It is possible for EPL, but not for EVPL.
- Bundling: It is possible for EVPL if CE-VLAN ID preservation is supported.
- CE-VLAN ID for all Frames (Whether They are Tagged or Untagged): If VLAN is protected, there will be primary VLAN ID and secondary VLAN ID such as 10 and 20.
- Maximum Number of EVCs: It is 1 for EP-LAN and ≥ 1 for EVP-LAN.

- Ingress and Egress Bandwidth Profiles: Ingress BWP is optional for both EP-LAN and EVP-LAN. If supported, bandwidth profile parameters (CIR, CBS, EIR, EBS, CM, and CF) must be defined. For example,

  CIR = 10 Mbps
  CBS = 32 KB
  EIR = 2 Mbps
  EBS = 4 KB
  CM = "color blind" or "color aware"
  CF = 0 or 1 (if it is 0, unused bandwidth will not be shared).

Egress BWP is optional for both EP-LAN and EVP-LAN. For EVPL, egress values must be configured as the same as the ingress values.

EVC service parameters include UNI List, bandwidth profile, L2CP treatment, and EVC performance.

- EVC ID: An arbitrary string that is unique across the MEN.
- EVC Type: Multipoint-to-multipoint.
- UNI List: List of Root UNIs associated with this EVC.
- Maximum Number of UNIs: ≥ 2.
- EVC MTU Size: ≥ 1522.
- CE-VLAN ID Preservation: Yes for EP-LAN. Yes or No for EVP-LAN.
- CE-VLAN CoS Preservation: Yes for EP-LAN. Yes or No for EVP-LAN.
- Unicast Service Frame Delivery: For both EP-LAN and EVP-LAN: deliver unconditionally or deliver conditionally. The condition must be specified. An example of such a condition is that the destination MAC address is known by the Metro Ethernet Network to be "at" the destination UNI.
- Multicast Service Frame Delivery: For both EP-LAN and EVP-LAN: deliver unconditionally or deliver conditionally. The condition must be specified.
- Broadcast Service Frame Delivery: For both EP-LAN and EVP-LAN: deliver unconditionally or deliver conditionally. The condition must be specified.
- L2CP :
  - Tunnel if only CPE equipment behind Ethernet devices process these protocols such as GARP/MRP Block.
  - Peer if Ethernet devices at both ends support these protocols such as E-LMI.
  - Discard if the device is not going to process the protocol such as Link OAM.
- EVC Performance: At least one CoS is required. For each CoS, values for FD, FDV, and FLR may need to be specified if known.

6.5.3   EP-Tree and EVP-Tree Service Attributes

UNI Parameters include the following:

- Port Name: This parameter is not in MEF 6.1; therefore, there is no guideline for port naming. However, the port needs to be identified. If it is a multislot chasse, the name can be fe/6/1. If the device is an NID, it can be LAN-1, LAN-2, or WAN-1, etc.
- Speed: 10 Mbps, 100 Mbps, 10/100 Mbps autonegotiation, 10/100/ 1000 Mbps autonegotiation, 1 Gbps, 10 Gbps, or 100 Gbps.
- Media Type: Fiber or copper.
- Mode: Full duplex.
- MAC Layer: IEEE 802.3–2005.
- MTU Size: ≥ 1522.
- UNI ID: Arbitrary string.
- Bundling: No for EP-Tree, and Yes requiring CE-VLAN ID preservation or No for EVP-Tree.
- All to One Bundling: Yes for EP-Tree, and No for EVP-Tree.
- CE-VLAN ID for all Frames (Whether They are Tagged or Untagged): For EP-Tree, all frames must be mapped onto one EVC. If VLAN is protected, there will be primary VLAN ID and secondary VLAN ID such as 10 and 20.
- Maximum Number of EVCs: 1 for EP-Tree, and ≥ 1 for EVP-Tree.
- Ingress/Egress Bandwidth Profile: It is optional for both EP-Tree and EVP-Tree. If supported, bandwidth profile parameters (CIR, CBS, EIR, EBS, CM, and CF) must be defined. For example,

  CIR = 10 Mbps
  CBS = 32 KB
  EIR = 2 Mbps
  EBS = 4 KB
  CM = "color blind" or "color aware"
  CF = 0 or 1 (if it is 0, unused bandwidth will not be shared).
- L2CP Processing.

EVC attributes include the following.

- UNI EVC ID: A string formed by the concatenation of the UNI ID and the EVC ID.
- CE-VLAN ID/EVC Map: For EP-Tree, all VLANs will be mapped to one rooted multipoint EVC.

  VLAN ID EVC ID
  VLAN-1 bishopgate-UNI1-EVC1
  VLAN-2 bishopgate-UNI1-EVC1
  VLAN-3 bishopgate-UNI1-EVC1

For EVP-Tree, VLAN–EVC mapping could be such:

VLAN ID EVC ID

VLAN-1 bishopgate-UNI1-EVC1

VLAN-2 bishopgate-UNI1-EVC2

VLAN-3 bishopgate-UNI1-EVC2

- Ingress/Egress Bandwidth Profile per EVC: The bandwidth profile parameters (CIR, CBS, EIR, EBS, CM, and CF) can be specified for each CoS for both EP-Tree and EVP-Tree. For example,

  CIR = 10 Mbps

  CBS = 32 KB

  EIR = 2 Mbps

  EBS = 4 KB

  CM = "color blind" or "color aware"

  CF = 0 or 1 (if it is 0, unused bandwidth will not be shared among EVCs).

- Egress Bandwidth Profile per CoS ID: The bandwidth profile parameters (CIR, CBS, EIR, EBS, CM, and CF) can be specified for both EP-Tree and EVP-Tree. For example,

  CIR = 10 Mbps

  CBS = 32 KB

  EIR = 2 Mbps

  EBS = 4 KB

  CM = "color blind" or "color aware"

  CF = 0 or 1. if it is 0, unused bandwidth will not be shared among CoS flows.

6.5.4   EP-Tree EVC and EVP-Tree EVC Service

EVC Service attributes include the following:

- EVC ID
- UNI List
- Maximum Number of UNIs: ≥ 2
- EVC MTU Size: ≥ 1522
- CE-VLAN ID Preservation: Yes for EP-Tree. Yes or No for EVP-Tree.
- CE-VLAN CoS Preservation: Yes for EP-Tree. Yes or No for EVP-Tree.
- Unicast Service Frame Delivery: For both EP-Tree and EVP-Tree: deliver unconditionally or deliver conditionally. The condition must be specified.
- Multicast Service Frame Delivery and Broadcast Service Delivery: For both EP-Tree and EVP-Tree: deliver unconditionally or deliver conditionally. The condition must be specified.
- L2CP Processing that are passed to EVC.

- EVC Performance: At least one CoS is required. For each CoS, values for FD, FDV, and FLR may need to be specified if known.

6.5.5 ENNI Services

ENNI Phase I [23] covers only E-LINE services and has preliminary work on EP-LAN services as well. We expect all the Metro Ethernet services offered on a single network to be offered over multiple networks via the ENNI gateways (Figs 6.29–6.33). It is certainly somewhat easier to support EPL/EVPL services than to support EP-LAN and EVP-LAN services. We expect EPL/EVPL services to be widely implemented across ENNI.

In the following sections, we describe attributes for E-Line service parameters. EP-LAN, EVPL-LAN, EP-Tree, and EVP-Tree parameters defined in Sections 6.5.2 and 6.5.3 will stay the same as long as their segments crossing ENNI are EPL or EVPL.

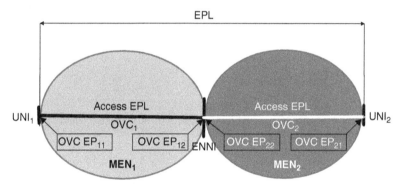

Figure 6.29   EPL across multiple networks.

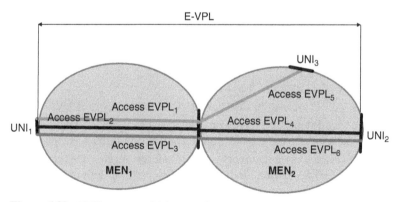

Figure 6.30   EVPL across multiple networks.

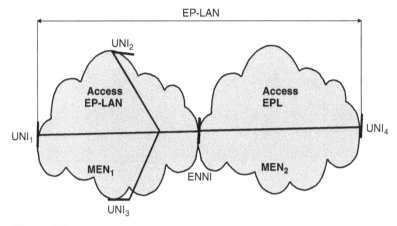

Figure 6.31   EP-LAN across multiple networks.

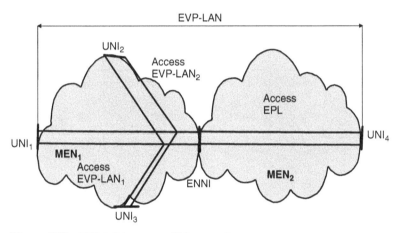

Figure 6.32   EVP-LAN across multiple networks.

6.5.6   Access EPL and Access EVPL Service Attributes

UNI attributes for EPL and EVPL in Section 6.5.1 are valid for access EPL and access EVPL, respectively. In addition, the UNI has an additional attribute, Maximum Number OVCs for the given UNI. ENNI attributes are listed in the following:

- ENNI ID: A string that is unique across the Operator MEN that does not contain more than 45 bytes [24]. Billing and Ordering Forum (ATIS) is also working on the ENNI ID. Our view is to use the format for UNI ID with an ENNI specific first two alphanumeric characters.

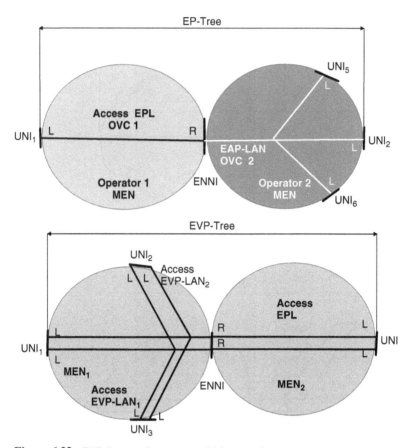

Figure 6.33   EVP-Tree service across multiple networks.

- Physical Layer: Possible values are 1000Base-SX, 1000Base-LX, 1000Base T, 10GBASE-SR, 10GBASE-LX4, 10GBASE-LR, 10GBASE-ER, 10GBASE-SW, 10GBASE-LW, and 10GBASE-EW (Chapter 2).
- Frame Format: As defined in IEEE 802.1ad-2005 [25], an ENNI frame can have zero or more VLAN tags. When there is a single tag, that tag is an S-Tag. When there are two tags, the outer tag is an S-Tag (TPID = 0x88A8) and the next tag is a C-Tag (TPID = 0x8100).
- Number of Links: It can be more than 1. If it is at least 2, LAG support is required.
- Protection Mechanism: Name of the protection protocol such as LAG between ENNI links.
- MTU Size: ≥ 1526 (i.e. additional 4 bytes due to S-Tag).
- EP Map: A table that maps each S-Tagged ENNI frame with an EP, with rows of the form <S-VLAN ID value, End Point Identifier, End Point Type>.

- Maximum Number of OVCs: An integer indicating the maximum number of OVCs that the operator can support at the ENNI.
- Maximum Number of OVC EPs per OVC: An integer indicating the maximum number of OVC EPs that the operator can support at the ENNI for an OVC.

On the other hand, OVC attributes are similar to EVC attributes.

- OVC Identifier: It is defined as "the string that is unique among all OVCs in the Operator MEN" [24]. Our view is to use the EVC ID format for OCs described in Section 6.4.3.
- OVC Type: Point-to-point.
- CoS ID: The OVC has a single CoS name.
- OVC EP List : A list of OVC EP Identifiers, exactly two, one OVC EP at the UNI and one at the ENNI.
- Maximum Number of UNI OVC EPs : One.
- Maximum Number of ENNI OVC EPs: One.
- OVC Maximum Transmission Unit (MTU) Size: $\geq$1526.
- CE-VLAN ID Preservation: Yes.
- CE-VLAN CoS Preservation: Yes.
- S-VLAN ID Preservation : This attribute can be used when there is more than one ENNI link. The value could be Yes or No.
- S-VLAN CoS Preservation: This attribute can be used when there is more than one ENNI link. The value could be Yes or No.
- Color Forwarding : Yes.
- Service Level Specification: One-way FD, one-way FDR, one-way MFD, IFDV, one-way FLR, one-way availability, one-way high loss intervals, and one-way consecutive high loss intervals.
- Unicast Service Frame Delivery: Deliver conditionally or unconditionally.
- Multicast Service Frame Delivery: Deliver conditionally or unconditionally.
- Broadcast Service Frame Delivery: Deliver conditionally or unconditionally.

In addition to the above attributes, for EVPL, OVC EP attributes need to be defined as below:

- OVC EP Identifier: A string that is unique across the Operator MEN [24]. Further work needs to be done to define this attribute.
- CoS Identifier for ENNI Frames: The OVC must have a single CoS.
- Ingress Bandwidth Profile per OVC EP: The bandwidth profile parameters are <CIR, CBS, EIR, EBS, CM, CF>. The ingress CIR for an OVC at the ENNI should be greater than the corresponding ingress CIR at the UNI

due to the presence of the added S-VLAN tag (4 bytes) at the ENNI. The following increments for CIR are recommended:

1–10 Mbps, increments of 1 Mbps;
10–100 Mbps, increments of 10 Mbps;
100–1000 Mbps, increments of 100 Mbps;
1–10 Gbps, increments of 1 Gbps.

It is also mandatory to support the bandwidth profile with EIR = 0, EBS = 0, CF = 0, CM = "color aware." The CBS values are ≥ 12176 bytes.

- Egress Bandwidth Profile per OVC EP: The egress bandwidth profile is needed since the egress parameters are equal to the ingress parameters.

When all CE-VLAN IDs map to the OVC at all the UNIs with an OVC EP associated by the OVC, at UNI, the ingress frames can be tagged or untagged, while at ENNI, egress frames toward receiving network will have an S-Tag. At UNI, the ingress frames with C-Tag will have an S-Tag and a C-Tag at ENNI as the egress frames. Their VLAN IDs in C-Tags will be equal to the C-Tags of the egress frames at ENNI. At ENNI, S-Tags will be preserved (i.e., S-Tag of the ingress frames is equal the S-Tag of egress frames).

When CE-VLAN CoS preservation is enabled, the PCP value of the ingress frames in C-Tag at UNI is equal to the PCP value of the egress frames in S-Tag and C-Tag at ENNI. Similarly, when the CoS preservation is enabled at ENNI, the PCP value of the ingress frames in C-Tag will be equal to the PCP value in the C-Tag and S-Tag of the egress frames at ENNI.

When Color Forwarding is enabled for the OVC, the OVC will mark yellow frame as yellow and cannot "promote" a frame from yellow to green. The color of the ingress frame will be the same as that of the egress frame.

6.6 CONCLUSION

Ethernet is becoming a dominant technology in enterprise networks, allowing bandwidth partitioning, user segregation, and traffic prioritization, introduced by IEEE 802.1d. It is scalable from 10 Mbps to 100 Gbps. Various services have been defined by MEF that can be used to support various user applications.

In this chapter, we have defined Metro Ethernet fundamentals in terms of architecture and services that can be used in service deployments on a single network as well as on multiple networks.

REFERENCES

1. Vertical Systems Group, "Worldwide Business Ethernet Services to Reach $31 Billion by 2012", October 2007 http://www.verticalsystems.com/prarticles/stat-flash-1007-wwethernetrev.html
2. MEF. Carrier Ethernet Services Overview; 2008.
3. Reardon M. , "Study: Ethernet in the city cuts carrier costs", January 2004 http://news.cnet.com/2100-1037-5146740.html

4. NokiaSiemens. Complete metro Ethernet solution brings immense ongoing savings, 2009. Available at http://www.nokiasiemensnetworks.com/sites/default/files/SWF38998Success_ Storymetro_Ethernet.pdf.
5. IEEE 802.3ad-2005. IEEE Standard for Carrier Sense Multiple Access With Collision Detection (CSMA/CD) Access Method and Physical Layer Specifications-Aggregation of Multiple Link Segments.
6. ITU-T G.8032. Ethernet ring protection switching; 2012.
7. IEEE 1588v2. IEEE standard for a precision clock synchronization protocol for networked measurement and control systems; 2008.
8. ITU-T G. 8261. Timing and synchronization aspects in packet networks; 2008.
9. ITU-T G.8010. Architecture of Ethernet layer networks; 2004.
10. ITU-T G.809. Functional architecture of connectionless layer networks; 2003.
11. IEEE 802.3–2005. Information technology—Telecommunications and information exchange between systems—Local and metropolitan area networks -Specific requirements—Part 3: Carrier sense multiple access with collision detection.
12. MEF 6.1. Ethernet Services Definitions - Phase 2; 2008 April.
13. MEF 11-User Network Interface (UNI). Requirements and framework; 2004 Nov.
14. MEF 13 -User Network Interface (UNI). Type 1 Implementation Agreement; 2005 Nov.
15. MEF 20 User Network Interface (UNI). Type 2 Implementation Agreement; 2008.
16. MEF 10.1. Ethernet Services Attributes Phase 2; 2006.
17. MEF 10.2. Ethernet Services Attributes Phase 2; 2009 Oct.
18. COMMON LANGUAGE® Special Service Circuit Codes (CLCI^{TM} S/S Codes)—Access. Telcordia Technologies Job Aid, JA-42, Issue 26; 2007 Aug (For BR-795-402-100, Issue 20, August 2007).
19. COMMON LANGUAGE® General Codes. Telecommunications Service Providers IAC Codes, Exchange Carrier Names, Company Codes, Telcordia Technologies System Documentation, BR-751-100-112, Issue 6; 2009 Sept.
20. MEF 28. External Network Network Interface (ENNI) Support for UNI Tunnel Access and Virtual UNI; 2010 Oct.
21. Ethernet Virtual Connection (EVC). Form Preparation Guide (Access Service Ordering Guidelines (ASOG) Industry Support Interface), Version 41, ATIS-0404016-0041.
22. L2 Protocol Handling Amendment to MEF 6.1; 2011 Oct 24.
23. MEF 33. Ethernet Access Services Definition; 2012 Jan.
24. MEF 26.External Network Network Interface; 2010 Jan.
25. IEEE 802.1ad. Local and Metropolitan Area Networks Virtual Bridged Local Area Networks; 2005.
26. MEF 10. Ethernet Services Attributes Phase 1; 2004.
27. MEF 12.1, Carrier Ethernet Network Architecture Framework Part 2: Ethernet Services Layer – Basic Elements, April 2010.
28. MEF 12.1.1, Carrier Ethernet Network Architecture Framework Part 2: Ethernet Services Layer – External Interface Extensions, October 2011.
29. RFC2474, "Definition of the Differentiated Services Field (DS Field) in the IPv4 and IPv6 Headers", December 1998.
30. RFC2475, "An Architecture for Differentiated Services", December 1998.

Chapter 7

Carrier Ethernet Traffic Management

7.1 INTRODUCTION

Data plane for Carrier Ethernet consists of Media Access Control; data parsing, which is parsing frame/packet headers for address or protocol information; classification, which is identifying frame against a criteria of a filtering/forwarding decision, quality of service (QoS), or accounting, etc.; data encapsulation, which is transformation of packet data between protocols; and traffic management, which is queuing, scheduling, and policing frames. Traffic management function is also called *packet conditioning*.

The classification function identifies packet flows using filters that have been statically configured in the node. It may consist of a series of filter rules specifying values of various fields in layer 2, layer 3 (IPv4 or IPv6), or layer 4 (TCP, UDP, or ICMP) headers such as

- MAC source and destination address,
- 802.1P priority,
- VLAN identifier,
- IPv4 protocol field,
- IPv4 source and destination addresses (plus associated masks),
- IPv4 DSCP (or IP-precedence field or TOS field),
- IPv6 source and destination addresses (plus associated masks),
- IPv6 class,
- IPv6 flow label,
- IPv6 next header,
- TCP/UDP source and destination port (range),
- TCP flags+mask, and
- ICMP type and code.

Networks and Services: Carrier Ethernet, PBT, MPLS-TP, and VPLS, First Edition. Mehmet Toy.
© 2012 John Wiley & Sons, Inc. Published 2012 by John Wiley & Sons, Inc.

In layer 2 systems, MAC DA, MAC SA, 802.1P, VLAN ID, Source Port Number/ID, Destination Port Number/ID are among the key parameters of the classification. These parameters can be fully specified, wild-carded, or where it is applicable, defined as a *prefix or value range*.

Packets are conditioned after classification. Once the classification processing has identified a particular flow to which the given packet belongs, the matching filter will specify a chain of actions to execute on the packet flow to yield appropriate conditioning. In layers 2 and 3 networks, the conditioning actions include:

- counting bytes and packets that pass and drop;
- policing packet flow according to predetermined burst/flow rate (includes both color-aware (CA) and color-unaware);
- setting color;
- setting 1P/DSCP/TOS priority value;
- remarking/remapping 1P/DSCP/TOS based on a predefined remarking table;
- setting next hop for policy-based routing (PBR);
- setting forwarding information base (FIB);
- shaping packets to conformance to predetermined flow rate; and
- sending packets to a particular queue.

In the upstream (i.e., customer to network) direction, the conditioning unit maps frames arriving on a UNI to one or more EVCs. The frames in EVCs are assigned to a flow, and a flow is assigned to a policer for optional rate limiting. A flow can consist of all the frames on a UNI or some subset of frames, based on VLAN and priority values. The resulting policed flows are then combined with QoS Queues, which are scheduled for transmission on the network EVCs. QoS queues can consist of one or more policed flows from a UNI.

In the downstream direction (i.e., network to customer), frames arrive from the network port and are classified into EVCs on the basis of VLAN tags. Each flow can consist of multiple classes of service (CoSs) that are mapped to schedulers for the transmission on a client UNI.

Ingress frames arriving on a UNI to one or more EVCs are mapped to a CoS flow for optional policing. A CoS flow may consist of the following.

- All frames belonging to the same UNI.
- A subset of frames belonging to one or more VLANs on the same UNI. Untagged frames can be assigned to a default port VLAN.
- A subset of frames belonging to a VLAN with one or more priority (802.1p or TOS or DSCP) levels. A single VLAN can have multiple policing flows in order to police the rate of different CoSs independently.

Once classified into a CoS flow, ingress frames are then policed according to the CoS bandwidth profile assigned to the flow. Following policing, a CoS flow is then assigned to a QoS queue for scheduling.

Frames in a CoS flow that are not discarded by the policer proceed to the next step where they are queued on a QoS flow queue associated with the destination EVC. The system determines whether to queue or discard the frame based on the color marking of the frame and the queue fill ratio. When queue reaches a certain level, let us say 70 or 80%, yellow frames will be discarded.

It is desirable, within the same EVC, to support multiple priorities to support applications such as VoIP to co-exist in the same flow with other data services.

Priority classification is enabled at the UNI (port) level by the user. Priority classifications can be 801.2P, TOS, and DSCP. Ingress frames may then be mapped to one of the CoS priorities in the system. In addition to assigning a CoS priority level, the table can be used to remark the priority level of the incoming frame's 802.1p field.

In the following sections, we describe the key components of Metro Ethernet traffic engineering including CoS and associated SLAs.

7.2 POLICING

CoS flows are policed according to CIR, CBS, EIR, EBS, CF, and CM. Policers employ the leaky bucket method to rate-limit traffic. Leaky bucket method is used to support single rate three color marking (SRTCM) [1] or two rate three color marking (TRTCM) [2].

Bandwidth profile parameters are **Committed Information Rate (CIR)**, **Excess Information Rate (EIR)**, **Committed Burst Size (CBS)**, **Excess Burst Size (EBS)**, **Coupling Flag (CF)**, **and Color Mode (CM)**.

CIR defines the average rate (bps) of ingress traffic up to which the device guarantees the delivery of the frames while meeting the performance objectives related to traffic loss, frame delay, and frame delay variations.

EIR defines the excess rate (bps) of ingress traffic up to which the device may deliver the frames without any performance objectives. EIR traffic may be oversubscribed on the WAN interface and may be dropped if the network cannot support it. Traffic offered above the EIR is dropped at the UNI.

CBS defines the maximum number of bytes available for a committed burst of ingress frames sent at the UNI line rate for which the device guarantees delivery.

EBS defines the maximum number of bytes available for a burst of excess ingress frames sent at the UNI line rate for which the device may deliver.

Policers employ the token bucket method to rate-limit traffic. The system can operate in CA mode or color-blind mode. The CF is used to share bandwidth between CIR and EIR settings on a flow.

Allocated tokens never exceed CBS + EBS. Token rates determine policing granularity. For example, *if* 1 token represents 64 bits and tokens are refreshed every millisecond then we can have a scheduler granularity of 64 kbps. An ingress frame is accepted if there is at least one token available (i.e., token count can become negative).

Most data traffic on typical networks exhibits a burstlike nature, consisting of short bursts of packets at line rate, interspersed with intervals of silence. The resulting average traffic rate is often much less than the line rate.

Traffic that exceeds the CIR setting, but is less than or equal to the EIR setting, is transported if bandwidth is available. The aggregate of CIR + EIR traffic on a device may exceed the line rate of the network interface.

Through combinations of CIR, EIR, CBS, and EBS settings, the system can be tuned to provide different QoS characteristics to match the bandwidth, traffic loss, latency, and jitter requirements for each traffic class.

The policing algorithm determines whether a customer frame is accepted from the UNI and if it has to be marked green or yellow. The algorithm is executed for each received frame.

CBS and EBS values depend on various parameters, including CIR, EIR, and loss and the delay to be introduced. In Reference 3, CBS is required to be greater than 12176 bytes.

7.3   QUEUING, SCHEDULING, AND FLOW CONTROL

Frames that are marked "green" by the policer are always queued, whereas frames that are marked "yellow" are queued only if the fill level of the queue is less than a threshold such as 70%. This ensures that "green" frames are always forwarded.

Within a QoS flow, unused tokens at priority level one cascade to lower priority levels. Committed tokens flow down to committed tokens and excess tokens to excess tokens.

Each CoS flow queue is assigned a set of buffers from a buffer pool. The size of the queue can be tuned to satisfy the performance characteristics of the associated flow. Buffers can be allocated on a per queue basis and are not shared between queues or services. Each EVC instance can be assigned multiple queues for ingress and each client UNI can be assigned multiple queues for egress to support prioritized traffic scheduling.

Flow control can be enabled on ports. As traffic arrives from client LAN ports, system buffers may begin to fill. At some point, the buffer usage may cross a Pause Threshold. If the service definition has enabled 802.3x [4] flow control, the system generates a PAUSE control frame to the sender. The sender, interpreting the PAUSE control frame, halts transmission for a brief period of time to allow the traffic in the bucket to flow on the network. This active flow control function permits the sending device (customer) to manage traffic priority, discard, and traffic shaping according to its own policies.

7.4   THREE CoS MODEL

The integration of real-time and non-real-time applications over data networks requires differentiating packets from different applications and provides differentiated performance according to the needs of each application. This differentiation is referred to as *CoS*.

MEF 23 [5] covers three CoSs, but does not exclude more classes. Applications such as e-mail, internet access, control traffic, VoIP, video, and pricing considerations due to distance differences such as New York–WDC versus New York–San Francisco or New York–London may require more than three classes. Three classes are defined as H, M, and L. Their priorities are indicated by PCP or DSCP or TOS bytes. Each CoS Label has its own performance parameters.

- MEF Class H is intended for real-time applications with tight delay/jitter constraints such as VoIP.
- MEF Class M is intended for time-critical data.
- MEF Class L is intended for non-time-critical data such as e-mail.

Tables 7.1 and 7.2 list the parameters of these three classes [6].

Table 7.1   Parameters of H, M, and L Classes

Performance Attribute	H	M	L
FD	≥99.9th percentile of FD values monitored over a month	≥99th percentile of FD values monitored over a month	≥95th percentile of FD values monitored over a month
MFD	Mean frame delay values calculated over a month	Mean frame delay values calculated over a month	Mean frame delay values calculated over a month
IFDV	≥99.9th percentile of FD values monitored over a month, with 1 s between frame pairs	≥99th percentile of FD values monitored over a month, with 1 s between frame pairs Or not supported	Not supported
FDR	≥99th percentile corresponding to minimum delay	≥99th percentile corresponding to minimum delay Or not supported	Not supported
FLR	Frame loss ratio values calculated over a month	Frame loss ratio values calculated over a month	Frame loss ratio values calculated over a month
Availability	To be defined by MEF	To be defined by MEF	To be defined by MEF
High loss interval	To be defined by MEF	To be defined by MEF	To be defined by MEF
Consecutive high loss interval	To be defined by MEF	To be defined by MEF	To be defined by MEF

Table 7.2 MEF Three CoS Bandwidth Constraints, and PCP and DSCP Mapping

CoS Label	Ingress EI Bandwidth Profile Constraints	CoS and Color Identifiers						CoS-only Identifiers			Example Applications
		C-Tag PCP		PHB (DSCP)		S-Tag PCP		C-Tag PCP	PHB (DSCP)	S-Tag PCP	
		Color Green	Color Yellow	Color Green	Color Yellow	Color Green CoS w/DEI	Color Yellow	PCP	PHB (DSCP)	PCP	
H	CIR > 0 EIR ≥ 0 CF = 0	5	Undefined	EF (46)	Undefined	5	Undefined	5	EF (46)	5	VoIP and Mobile Back-haul Control
M	CIR > 0 EIR ≥ 0	3	2	AF31 (26)	AF32 (28) or AF33 (30)	3	2	2–3	AF31-33 (26, 28, 30)	2–3	Near-Real-Time or Critical Data Apps
L	CIR ≥ 0 EIR ≥ 0	1	0	AF11 (10)	AF12 (12), AF13 (14) or Default (0)	1	0	0–1	AF11-13 (10, 12, 14) or Default (0)	0–1	Noncritical Data Apps

Performance parameters do not have to be uniform across CoS Labels, PTs, EVC/OVC Types. For example, the T associated with FLR may be different from the T associated with FD.

Delay and jitter values above are valid for point-to-point connections. Those constraints for multipoint connections have not been defined by MEF yet. We expect similar figures hold for multipoint connections as well.

Bandwidth Profile for CoS Label L allows CIR or EIR $= 0$. When CIR $= 0$, there will be no performance objectives, while the case of CIR > 0 will requires conformance with performance attribute objectives.

At ENNI, drop eligible indicator (DEI) is used to represent yellow, therefore, there is no need for DSCP/PCP values to represent color.

A CoS instance for a service frame is identified either by the EVC or by the combination of the EVC and the user priority field in tagged service frames (i.e., PCP) or DSCP values.

7.5   SLAs (SERVICE-LEVEL AGREEMENTS)

SLAs between a service provider and subscriber for a given EVC are defined in terms of delay, jitter, loss, and availability performances. In the following sections, we describe each of them.

7.5.1   Frame Delay Performance

Frame delay performance is a measure of the delays experienced by different service frames belonging to the same CoS instance.

The One-way frame delay for an egress service frame at a given UNI in the EVC is defined as the time elapsed from the reception at the ingress UNI of the first bit of the corresponding ingress service frame until the transmission of the last bit of the service frame at the given UNI. This delay definition as illustrated in Figure 7.1 is the one-way delay that includes transmission delays across the ingress and egress UNIs as well as that introduced by the MEN.

Frame delay performance is defined for a time interval T for a CoS instance on an EVC that carries a subset of ordered pairs of UNIs. Each frame delay performance metric **is** defined as follows:

$D_T^{\langle i,j \rangle}$, the set of one-way frame delay values for all qualified service frames at UNI j resulting from an ingress service frame at UNI$_I$, can be expressed as $D_T^{\langle i,j \rangle} = \{d_1^{\langle i,j \rangle}, d_2^{\langle i,j \rangle}, \ldots, d_{N_{\langle i,j \rangle}}^{\langle i,j \rangle}\}$, where $d_k^{\langle i,j \rangle}$ is the one-way frame delay of the kth service frame. Defining $\bar{d}_T^{\langle i,j \rangle}$ for $P>0$,

$$
\bar{d}_T^{\langle i,j \rangle} = \begin{cases} \min \left\{ d \,|\, P \le \frac{100}{N_{\langle i,j \rangle}} \sum_{k=1}^{N_{\langle i,j \rangle}} I \left(d, d_k^{\langle i,j \rangle} \right) \right\} & \text{if } N_{\langle i,j \rangle} \ge 1 \\ Undefined & \text{otherwise} \end{cases}
$$

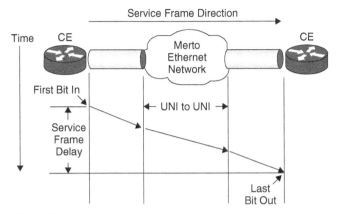

Figure 7.1 Frame delay for service frame.

where,

$$I(d, d_k) = \begin{cases} 1 \text{ if } d \geq d_k \\ 0 \text{ otherwise} \end{cases}$$

and $N_{(i,j)}$ is the number of frames involved in the measurement.

$\bar{d}_T^{\langle i,j \rangle}$ is the minimal delay during the time internal T that P percent of the frames do not exceed.

One-way frame delay Performance metric for an EVC can be expressed as

$$\bar{d}_{T,S} = \begin{cases} \max \left\{ \bar{d}_T^{\langle i,j \rangle} | \langle i,j \rangle \in S \text{ and where } N_{(i,j)} > 0 \right\} \\ Undefined \text{ when all } N_{(i,j)} = 0 | \langle i,j \rangle \in S \end{cases}$$

where S is a subset of ordered UNI pairs in the EVC.

7.5.2 Frame Delay Range Performance

In addition to FD and Interframe delay variation (IFDV), frame delay performance and frame delay range performance are defined in Reference 7. Frame delay range is the difference between the frame delay performance values corresponding to two different percentiles. Frame delay range performance is a measure of the extent of delay variability experienced by different service frames belonging to the same CoS instance.

The difference between the delay performance of two selected percentiles, P_x and P_y, can be expressed as

$$\bar{d}_{Tyx}^{\langle i,j \rangle} = \begin{cases} (\bar{d}_{Ty}^{\langle i,j \rangle} - \bar{d}_{Tx}^{\langle i,j \rangle}) & \text{if } N_{(i,j)} > 0 \\ Undefined & \text{if } N_{(i,j)} = 0 \end{cases}$$

Then a **one-way frame delay range performance metric** for an EVC can be expressed as

$$\bar{d}_{TyxS} = \begin{cases} \max\left\{\bar{d}_{Tyx}^{\langle i,j\rangle} | \langle i,j\rangle \in S \text{ and where } N_{\langle i,j\rangle} > 0\right\} \\ Undefined \text{ when all } N_{\langle i,j\rangle} = 0 | \langle i,j\rangle \in S. \end{cases}$$

7.5.3   Mean Frame Delay Performance

The **minimum one-way delay** is an element of $D_T^{\langle i,j\rangle}$, where $d_{\min}^{\langle i,j\rangle} \le d_k^{\langle i,j\rangle}$ (for all $k = 1, 2, \ldots, N_{\langle i,j\rangle}$) and is a possible selection as one of the percentiles. The minimum delay represents the $N_{\langle i,j\rangle}^{-1}$th percentile and all lower values of P as $P \to 0$.

Another One-way frame delay attribute is the arithmetic mean of $D_T^{\langle i,j\rangle}$, which can be expressed as,

$$\bar{\mu}_T^{\langle i,j\rangle} = \begin{cases} \dfrac{1}{N_{\langle i,j\rangle}} \displaystyle\sum_{k=1}^{N_{\langle i,j\rangle}} \left(d_k^{\langle i,j\rangle}\right) & \text{if } N_{\langle i,j\rangle} > 0 \\ Undefined & \text{if } N_{\langle i,j\rangle} = 0 \end{cases}$$

Then a **One-way mean frame delay (MFD) Performance metric** for an EVC can be expressed as

$$\bar{\mu}_{TS} = \begin{cases} \max\left\{\bar{\mu}_T^{\langle i,j\rangle} | \langle i,j\rangle \in S \text{ and } \bar{\mu}_T^{\langle i,j\rangle} \text{ where } N_{\langle i,j\rangle} > 0\right\} \\ Undefined \text{ when all } N_{\langle i,j\rangle} = 0 | \langle i,j\rangle \in S. \end{cases}$$

For a point-to-point EVC, S **may** include one or both the ordered pairs of UNIs in the EVC. For a multipoint-to-multipoint EVC, S **may** be any subset of the ordered pairs of UNIs in the EVC. For a rooted-multipoint EVC, S **must** be such that all ordered pairs in S contain at least one UNI that is designated as a root.

7.5.4   Frame Delay Variation

IFDV is the difference between the one-way delays of a pair of selected service frames. For a particular CoS identifier and an ordered pair of UNIs in the EVC, IFDV performance is applicable to Green color frames.

The IFDV Performance **is** defined as the P-percentile of the absolute values of the difference between the frame delays of all qualified service frame pairs given that the difference in the arrival times of the first bit of each service frame at the ingress UNI was exactly Δt.

The choice of the value for Δt can be related to the application timing information. As an example of voice applications where voice frames are generated at regular intervals, Δt may be chosen to be few multiples of the interframe time.

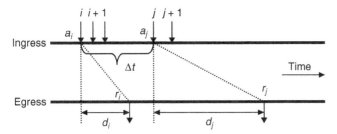

Figure 7.2   Interframe delay variation.

Let a_i be the time of the arrival of the first bit of the ith service frame at the ingress UNI, then the two frames i and j are selected according to the selection criterion:

$$\{a_j - a_i = \Delta t \quad \text{and} \quad j > i\}$$

Let r_i be the time frame and i be successfully received (last bit of the frame) at the egress UNI then the difference in the delays encountered by frame i and frame j is given by $d_i - d_j$, which is defined as

$$\Delta d_{ij} = |d_i - d_j| = |(r_i - a_i) - (r_j - a_j)| = |(a_j - a_i) - (r_j - r_i)|$$

If either or both frames are lost or not delivered because of, for example, FCS violation then the value Δd_{ij} is not defined and does not contribute to the evaluation of the IFDV.

Figure 7.2 shows a depiction of the different times that are related to IFDV performance.

7.5.5 Frame Loss Ratio

Frame loss ratio is a measure of the number of lost frames between the ingress UNI and the egress UNI. Frame loss ratio is expressed as a percentage.

One-way frame loss ratio performance is defined for a time interval T of a CoS instance on an EVC that carries a subset of ordered pairs of UNIs. One-way frame loss ratio performance metric **is** defined as follows:

- Let $I_T^{\langle i,j \rangle}$ denotes the number of ingress service frames at UNI i whose first bit arrived at UNI i during the time interval T, whose Ingress bandwidth profile compliance was green, and that should have been delivered to UNI j according to the service frame delivery service attributes.
- Let $E_T^{\langle i,j \rangle}$ denote the number of such service frames delivered to UNI j.

- FLR is defined as $\mathrm{FLR}_T^{\langle i,j \rangle} = \begin{cases} \left(\frac{I_T^{\langle i,j \rangle} - E_T^{\langle i,j \rangle}}{I_T^{i,j}} \right) \times 100 \ \text{if} \ I_T^{\langle i,j \rangle} \geq 1 \\ Undefined \qquad \text{otherwise} \end{cases}$

- Then the one-way frame loss ratio performance metric **is** defined as

$$\mathrm{FLR}_{T,S} = \begin{cases} \max \left\{ \mathrm{FLR}_T^{\langle i,j \rangle} | \langle i,j \rangle \in S \ \text{and where} \ I_T^{\langle i,j \rangle} \geq 1 \right\} \\ Undefined \ \text{when all} \ I_T^{\langle i,j \rangle} = 0 | \langle i,j \rangle \in S \end{cases}$$

Given T, S, and a one-way frame loss ratio performance objective, the one-way frame loss performance **is** defined as met over the time interval T for the subset S if and only if $\mathrm{FLR}_{T,S} \leq \hat{L}$.

Recall that if the one-way frame loss ratio performance is undefined for time interval T and ordered pair $\langle i,j \rangle$ then the performance for that ordered pair **is** excluded from calculations on the performance of pairs in S.

For a point-to-point EVC, S **may** include one or both of the ordered pairs of UNIs in the EVC.

For a multipoint-to-multipoint EVC, S **may** be any subset of the ordered pairs of UNIs in the EVC.

For a rooted-multipoint EVC, S **must** be such that all ordered pairs in S contain at least one UNI that is designated as a Root.

7.6 SLAs

MEF 23.1 [6] defines four performance tiers (PTs). Although PTs associate SLAs with distance which is an approximate network miles, the impact of distance on SLAs is not more than the propagation delay for the given distance. Users and service providers need to pick a PT based on their applications.

1. PT1 (Metro PT) for typical metro distances (< 250 km, 2 ms propagation delay),
2. PT2 (Regional PT) for typical regional distances (< 1200 km, 8 ms propagation delay),
3. PT3 (Continental PT) for typical national/continental distances (<7000 km, 44 ms propagation delay),
4. PT4 (Global PT) for typical global/intercontinental distances ($<27,500$ km, 172 ms propagation delay).

Tables 7.3–7.6 list SLA figures for these PTs.

In these tables, an user or carrier may choose to support FD or MFD for delay and IFDV or FDR for jitter.

7.7 APPLICATION-CoS-PRIORITY MAPPING

Defining the number of priorities to be supported has been a challenge for equipment vendors and service providers. More priorities may increase the cost of equipment and operation systems.

Table 7.3 SLAs for Point-to-Point EVCs of Three Service Categories (i.e., CoS), Spanning No More than 250 Network Kilometers with 2 ms Propagation Delay of Metro Distances

SLA Parameters Measured over a Month	H	M	L
FD (ms)	<8 for 99.9th percentile	<20 for 99th percentile	<37 for 95th percentile
MFD	—	—	—
IFDV (ms)	<2 for 99.9th percentile with 1 s pair interval	<8 for 99th percentile with 1 s pair interval	Not specified
FDR	—	—	—
FLR	$<10^{-5}$	$<10^{-5}$	$<10^{-3}$
Availability	Undefined	Undefined	Undefined

Table 7.4 SLAs for Point-to-Point EVCs of Three Service Categories (i.e., CoS), Spanning No More than 1200 km with 8 ms Propagation Delay of Regional Distances

SLA Parameters Measured over a Month	H	M	L
FD (ms)	$\leq$25 for 99.9th percentile	$\leq$75 for 99th percentile	$\leq$125 for 95th percentile
MFD	—	—	—
IFDV (ms)	$\leq$8 for 99.9th percentile with 1 s pair interval	$\leq$40 for 99th percentile1 s pair interval	Not specified
FDR	—	—	—
FLR	$\leq10^{-5}$	$\leq10^{-5}$	$\leq10^{-3}$
Availability	Undefined	Undefined	Undefined

MEF picked three priorities: H, M, and L. However, the number of priorities recommended by ITU, IETF, and 3GPP/3GPP2 are much higher. In this section, we summarize application prioritization and related performance parameters for these standard bodies.

An example by IETF recommended for CoSs-DSCP Code points-Applications mapping is given by RFC 2698 [2] is described in Table 7.7. This table suggests 10 CoS.

Similarly, ITU-T defines IP service classes, maps them onto applications, and identifies the performance parameters for each in Reference 9. In Table 7.8, six classes are defined.

Furthermore, ITU-T [9] adds two more classes and defines performance parameters for each as depicted in Table 7.9. Classes 6 and 7 are intended to

Table 7.5 SLAs for Point-to-Point EVCs of Three Service Categories (i.e., CoS), Spanning No More than 7000 km with 44 ms Propagation Delay of National/Continental Distances

SLA Parameters Measured over a Month	H	M	L
FD (ms)	$\leq$77 for 99.9th percentile	$\leq$115 for 99th percentile	$\leq$230 for 95th percentile
MFD	—	—	—
IFDV (ms)	$\leq$10 for 99.9th percentile with 1 s pair interval	$\leq$40 for 99th percentile with 1 s pair interval	Not specified
FDR	—	—	—
FLR	$\leq10^{-4}$	$\leq10^{-4}$	$\leq10^{-3}$
Availability	Undefined	Undefined	Undefined

Table 7.6 SLAs for Point-to-Point EVCs of Three Service Categories (i.e., CoS), Spanning No More than 27500 km with 44 ms Propagation Delay of Global/Intercontinental Distances

SLA Parameters Measured over a Month	H	M	L
FD (ms)	$\leq$230 for 99.9th percentile	$\leq$250 for 99th percentile	$\leq$390 for 95th percentile
MFD	—	—	—
IFDV (ms)	$\leq$32 for 99.9th percentile with 1 s pair interval	$\leq$40 for 99th percentile with 1 s pair interval	Not specified
FDR	—	—	—
FLR	$\leq10^{-4}$	$\leq10^{-4}$	$\leq10^{-3}$
Availability	Undefined	Undefined	Undefined

support performance requirements of applications with stringent loss/error constraints.

Similar effort has taken place in wireless standards. Table 7.10 describes CoS defined in 3GPP for 3G traffic [8].

ITU-T performance constraints in Reference 9 are also considered for mobile applications (Table 7.11). But the IPRR parameter of ITU-T is ignored.

7.7.1 CoS Identification

CoS identification is obtained by CoS label and color identifier. CoS label is a name of CoS identified by a "CoS identifier" at UNI and ENNI.

Table 7.7 An Example of IP Service Classes Configured with DiffServe and Mapping of Applications to These Classes

Service Class	DSCP Name	DSCP Binary (Decimal) Value	Application Examples
Network control	CS6	110000 (48)	Network routing
Telephony	EF	101110 (46)	IP telephony bearer
Signaling	CS5	101000 (40)	Video conferencing
Real-time interactive	CS4	100000 (32)	Interactive control (e.g., CAM), real-time e-learning, games, e-arts
Broadcast video	CS3	011000 (24)	Broadcast TV & live events
Multimedia streaming	AF31, AF32, AF33	011010 (26), 011100 (28), 011110 (30)	Streaming video and audio on demand
Low latency data	AF21, AF22, AF23	010010 (18), 010100 (20), 010110 (22)	Transactional applications, database access, interactive data applications
OAM	CS2	010000 (16)	OAM (e.g., SNMP, Ethernet CFM, proprietary NMS traffic)
High throughput data	AF11, AF12, AF13	001010 (10), 001100 (12), 001110 (14)	—
Standard	DF (CS0)+Other	000000 (0)	Undifferentiated applications

CoS identifier at UNI can be EVC, EVC+PCP (i.e., CE-VLAN service frame priority), EVC+DSCP, or EVC+L2CP. It is clear that the EVC ID is not carried by service frame, therefore, when CoS ID is an EVC ID, operator may use string within EVC ID indicating the CoS or tables in OSS (Operations Support System) grouping EVCs according to their CoS.

Untagged service frames without a CoS identifier will have the same CoS identifier as an ingress data service frame with PCP = 0. The L2CP frames whose CoS is not determined using the EVC may have a CoS label determined by their MAC DA and/or Ethertype (protocol type).

Table 7.8 IP Service Classes and Example Applications [3, 9]

QoS Class	Applications (Examples)	Node Mechanisms	Network Techniques
0 (Highest Priority)	Real-time, jitter-sensitive, high interaction (VoIP, VTC)	Separate queue with preferential servicing, traffic grooming	Constrained routing and distance
1	Real-time, jitter-sensitive, interactive (VoIP, Video Teleconference)		Less constrained routing and distances
2	Transaction data, highly interactive (Signalling)	Separate queue, drop priority	Constrained routing and distances
3	Transaction data, interactive		Less constrained routing and distances
4	Low loss only (short transactions, bulk data, video streaming)	Long queue, drop priority any route/path	Any route/path
5 (Lowest Priority)	Traditional applications of default IP networks	Separate queue (lowest priority)	Any route/path

Color of a service frame at UNI and ENNI is indicated by a color identifier when the EVC is in CA mode, as a part of the bandwidth profile. Color identification is accomplished via the PCP or Differentiated Services Code Point (DSCP) at the UNI and via the PCP or DEI at the ENNI. Thus the PCP or DSCP may convey both CoS and color. When frames are untagged at the UNI, DSCP can be used to indicate color.

In Service VLAN tag (S-Tag), the drop eligible parameter can be encoded in and decoded from the DEI.

In the CA ENNI, color is indicated using either the PCP field of the S-Tag or indicated separately using the S-Tag DEI field. The PCP field of the S-Tag **must** be used when the CoS identifier type is EVC+PCP. Use of DEI may free up additional values of the PCP.

PCP code point usage to indicate color and CoS is described in Reference 11 as depicted in Table 7.12. For example, in the first row, there are eight priorities, therefore there is no PCP bit representing discard eligibility (DE) (i.e., yellow-colored frames). In the second row, five PCPs representing priorities, thus three PCPs are available to represent DE.

Table 7.9 IP Service Classes and Example Applications

Panel (a)

Network Performance Parameters	Network Performance Objectives	Class 0	Class 1	Class 2	Class 3
Applications examples	—	Real-time, jitter-sensitive, high interaction (VoIP, VTC)	Real-time, jitter-sensitive, interactive (VoIP, VTC)	Transaction data, highly interactive (signalling)	Transaction data, interactive
IPTD	Upper bound on the mean IPTD	100 ms	400 ms	100 ms	400 ms
IPDV	Upper bound on the $1-10^{-3}$ ($1-10^{-5}$ for Classes 5 and 6) quantile of IPTD minus the minimum IPTD	50 ms	50 ms	undefined	Undefined
IPLR	Upper bound on the packet loss probability	10^{-3}	10^{-3}	10^{-3}	10^{-3}
IPER	Upper bound	10^{-4}	10^{-4}	10^{-4}	10^{-4}
IPRR	Upper bound	Undefined	Undefined	Undefined	Undefined

(continued)

Table 7.9 (*Continued*)

Panel (b)

Network Performance Parameter	Network Performance Objective	Class 4	Class 5	Class 6	Class 7
Applications examples	—	Low loss only (short transactions, bulk data, video streaming	Traditional applications of default IP networks	TV transport, high capacity TCP transfers, and TDM circuit emulation	TV transport, high capacity TCP transfers, and TDM circuit emulation
IPTD	Upper bound on the mean IPTD	1 s	Undefined	100 ms	400 ms
IPDV	Upper bound on the $1-10^{-3}$ ($1-10^{-5}$ for Classes 5 and 6) quantile of IPTD minus the minimum IPTD	Undefined	Undefined	50 ms	50 ms
IPLR	Upper bound on the packet loss probability	10^{-3}	Undefined	10^{-5}	10^{-5}
IPER	Upper bound	10^{-4}	Undefined	10^{-6}	10^{-6}
IPRR	Upper bound	Undefined	Undefined	10^{-6}	10^{-6}

IPTD, IP Packet Transfer Delay; IPDV, IP Packet Delay Variation; IPLR, IP Packet Loss Ratio; IPER, IP Packet Errored Ratio; IPRR, IP Packet Reordering Ratio

Table 7.10 Service Classes for 3G Traffic

Traffic Class	Conversational Class ConversationalT RT	Streaming Class Streaming RT	Interactive Class Interactive Best Effort	Background Class Background Best Effort
Fundamental characteristics	Preserve time relation (variation) between information entities of the stream	Preserve time relation (variation) between information entities of the stream	Request response pattern	Destination is not expecting the data within a certain time
—	Conversational pattern (stringent and low delay)	—	Preserve payload content	Preserve payload content
Example of the application	Voice	Streaming video	Web browsing	Background download of emails

At UNI, the service provider can choose to recognize service frame colors (i.e., CA mode) or not to recognize service frame colors (i.e., color-blind mode).

When the CM parameter of the ingress bandwidth profile of a CoS instance is set to CA at the UNI, the CM parameter of all egress bandwidth profiles at the UNIs and ingress/egress bandwidth profiles at the ENNIs associated with the CoS instance **must** also be set to CA, in order to support color awareness end-to-end.

When the CM parameter of the ingress bandwidth profile of a CoS instance is set to color blind at the UNI, the CM parameter for all egress bandwidth profiles at the UNIs and ingress/egress bandwidth profiles at the ENNIs associated with the CoS instance should be set to CA.

7.7.2 PCP and DSCP Mapping

For a single-CoS EVC, all ingress data service frames mapped to the EVC, regardless of PCP or DSCP value, **will** have the same CoS identifier. For a multi-CoS EVC where EVC+PCP is used for CoS identification, all possible PCP values (i.e., 0–7) need to be mapped to the CoS supported on a given EVC at the UNI. For a multi-CoS EVC where EVC+DSCP is used for CoS identification, all possible DSCP values (i.e., from 0 to 63) must be fully mapped to the CoS supported on a given EVC at the UNI.

Table 7.13 is an example of PCP mapping to MEF H, M, and L service categories. The mapping in this table ignores the color of user traffic.

Table 7.11 Service Classes Defined by ETSI for Mobile Applications that are Aligned with 3GPP/3GPP2

Network Performance Parameter	Nature of Network Performance Objective	Class 0	Class 1	Class 2	Class 3	Class 4	Class 5	Class 6	Class 7
—		Real-time, jitter-sensitive, high interaction (VoIP, VTC)	Real-time, jitter-sensitive, interactive (VoIP, VTC)	Transaction data, highly interactive (Signaling)	Transaction data, interactive	Low loss only (short transactions, bulk data, video streaming)	Traditional applications of default IP networks	Television transport, high capacity TCP transfers, and TDM circuit emulation	Television transport, high capacity TCP transfers, and TDM circuit emulation
IPTD	Upper bound on the mean IPTD	100 ms	400 ms	100 ms	400 ms	1 s	U	100 ms	400 ms
IPDV	Upper bound on the $1-10^{-3}$ quantile of IPTD minus the minimum IPTD	50 ms	50 ms	U	U	U	U	50 ms	50 ms
IPLR	Upper bound on the packet loss probability	1×10^{-3}	1×10^{-3}	1×10^{-3}	1×10^{-3}	1×10^{-3}	U	1×10^{-5}	1×10^{-5}
IPER	Upper bound	—	—	1×10^{-4}	—	—	U	U	U

Table 7.12   PCP Values for QoS and Color

PCP Allocation		PCP Values and Traffic Classes							
# PCP Priorities	# PCP Drop Eligible	PCP = 7	6	5	4	3	2	1	0
8	0	IEEE traffic class = 7	6	5	4		3 2		1 0
5	**3**	IEEE traffic class = 7	6	4	4 DE	2	2 DE	0	0 DE

Table 7.13   Example of PCP Mapping for MEF H, M, and L Service Categories in Color-Blind Mode

MEF CoS Combination Supported on EVC	PCP Mapping per Class of Service		
	H	M	L
H, M, L	5	2–4, 6, 7	0, 1
H, M	5	0–4, 6, 7	Not specified
H, L	5	Not specified	0–4, 6, 7
M, L	Not specified	2–7	0, 1

7.8 BANDWIDTH PROFILE

The bandwidth profile defines long-term average bandwidth limits on the amount of data burst by specifying the rate and buffering for the given data stream and how the Metro Ethernet network should treat a service frame depending on its level of compliance. It consists of CIR, CBS, EIR, EBS, CF that is either 0 or 1, and CM.

The compliance of the service frames is determined through two token bucket leaky algorithms (Fig. 7.3).

Each bucket holds up to CBS and EBS tokens, respectively, with one token representing one byte. Tokens are added to each bucket at the rate of CIR/8 and EIR/8, respectively. When the bucket becomes full, additional tokens overflow

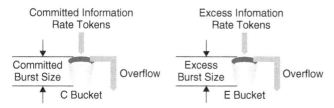

Figure 7.3   Two leaky buckets (i.e., C bucket is for CIR and E bucket EIR).

and are lost. When an ingress service frame is classified, its length is compared to the tokens in the C bucket. If the number of tokens is at least equal to the length of the service frame, the frame is declared "green," and tokens equal to the frame length are removed from the bucket. If there are no sufficient tokens in the C bucket, the service frame length is compared to the tokens in the E bucket and the process is repeated. If there are sufficient tokens in the E bucket, the frame is declared "yellow," otherwise it is declared "red."

Figure 7.4 shows the precise algorithm defined in Reference 7. Here, the service frames have lengths $\{l_j\}$ and arrival times $\{t_j\}$ for $j = 0, 1$, etc. The number of tokens in each token bucket is denoted by $B_c(t_j)$ and $B_e(t_j)$, respectively where $B_c(t_0) = \text{CBS}$ and $B_e(t_0) = \text{EBS}$.

The CF is set to either 0 or 1. When CF is set to 0, the long-term average bit rate of service frames that are declared yellow and admitted to the network are bound by EIR. When CF = 0, the two token buckets operate independently.

When CF is set to 1, the long-term average bit rate of service frames that are declared yellow and admitted to the network and those that are declared green and admitted to the network are bound by CIR + EIR.

In both the cases mentioned above, the burst size of the service frames that are declared yellow and admitted to the network is bounded by EBS. When CF = 1, it is possible that unused tokens from the C bucket can be put into the E bucket. This could happen if EIR is smaller than CIR and a long burst of

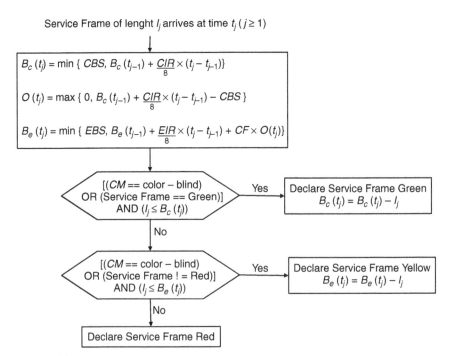

Figure 7.4   The bandwidth profile algorithm [7].

service frames is followed by a period of no traffic. At the end of the burst, both token buckets would be essentially empty. During the quiet period, the C bucket could completely fill before the E bucket, resulting in the "overflow" tokens (represented by $O(t_j)$ in Figure 7.4) being placed in the E bucket.

Green frames are delivered according to the service-level specification performance objectives, while *Yellow* frames are not. *Red* frames are discarded.

A bandwidth profile can be applied to all ingress service frames at UNI, EVC, and CoS, as depicted in Figures 7.5–7.7. In all the cases, there will be one profile applied to the service frames.

The use of bandwidth constraints of a bandwidth profile can lead to excessive frame loss and poor performance of an application using the service. For example, a host can send a TCP window of packets back to back at a UNI. When the corresponding frames cause the token bucket to exhaust, service frames are discarded. The resulting lost packets are detected by TCP in the host and retransmitted. However, TCP also slows down when loss is detected, and the net result is throughput degradation. This behavior can be suppressed by making CBS large or by implementing shaping in the customer edge at the UNI. Shaping service frames allow the frames that would be declared yellow or red to be declared green.

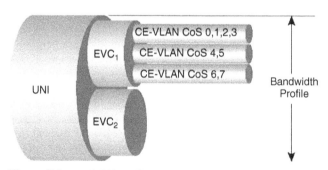

Figure 7.5   Bandwidth profile per UNI.

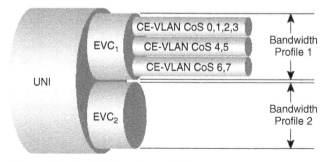

Figure 7.6   Bandwidth profile per EVC.

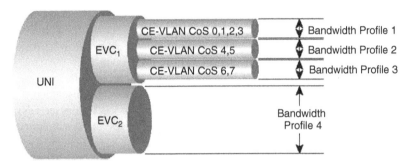

Figure 7.7 Bandwidth profile per CoS.

7.9 CONCLUSION

As users and carriers become familiar with Metro Ethernet technology, we expect applications with strict performance requirements and best effort applications to be mixed in the same port for economical reasons. As a result, we would need to prioritize and segregate one flow from another. Each flow will have its own bandwidth parameters and associated SLAs (i.e., delay, jitter, loss, and availability). In this chapter, we have summarized the guidelines to support applications requiring multi-CoS.

REFERENCES

1. RFC 2697. A Single Rate Three Color Marker; 1999 Sept.
2. RFC 2698. A Two Rate Three Color Marker; 1999 Sept.
3. MEF 33. Ethernet Access Services Definition; 2012 Jan.
4. IEEE 802.3x. IEEE Standards for Local and Metropolitan Area Networks: Supplements to Carrier Sense Multiple Access With Collision Detection (CSMA/CD) Access Method and Physical Layer Specifications—Specification for 802.3 Full Duplex Operation and Physical Layer Specification for 100Mb/s Operation on Two Pairs of Category 3 Or Better Balanced Twisted Pair Cable (100BASE-T2); 1997.
5. MEF 23. Carrier Ethernet Class of Service—Phase 1, Implementation Agreement; 2009 Jun.
6. MEF 23.1, Carrier Ethernet Class of Service—Phase 2, Implementation Agreement, January 2012.
7. MEF 10.2. Ethernet Services Attributes Phase 2; 2009 Oct.
8. RFC 4594. Configuration Guidelines for DiffServe Service Classes; 2006 Aug.
9. ITU-T Y.1541. Network Performance Objectives for IP-Based Services; 2011.
10. 3GPP TS 23.107 v3.0. QoS Concept and Architecture; 1999–10.
11. IEEE 802.1ad. Local and Metropolitan Area Networks Virtual Bridged Local Area Networks; 2005.

Chapter 8

Carrier Ethernet OAM&P (Operations, Administration, Management, and Performance)

8.1 INTRODUCTION

The advent of Ethernet as a service rather than for transport in metropolitan and wide-area networks has accelerated the need for a new set of in-band OAM (operations, administration, and maintenance) protocols. End-to-end Ethernet service, namely, EVC, often involves one or more operators in addition to a service provider.

An EVC is initiated with a service order from user. The order consists of various EVC information including UNI locations, bandwidth, CoS, SLAs. After provisioning the EVC, the service provider and operator conduct turnup testing to ensure that EVC is operational and support the user contract.

While the EVC is in use, all the parties involved in EVC (i.e., subscriber, operators, and service provider) want to monitor the same EVC to ensure Service Level Agreements (SLAs) such as delay, jitter, loss, throughput, and availability. Service provider and operators need to identify service anomalies in advance through measurements and TCAs (threshold crossing alerts) and take appropriate actions. When there is a problem identified through various alarms, service provider and operators need to isolate the problem and identify faulty components to repair them within availability constraints.

IEEE, ITU, and MEF standards provide the tools to support the goals above. Ethernet OAM consists of fault management, performance management, testing, and service monitoring capabilities to support Ethernet services described in Chapter 5. It is mainly addressed in References 1–3. Measurements, events/ alarms, and provisioning attributes for interfaces (i.e., UNI, VUNI, ENNI), EVC, and OVC are defined.

Networks and Services: Carrier Ethernet, PBT, MPLS-TP, and VPLS, First Edition. Mehmet Toy.
© 2012 John Wiley & Sons, Inc. Published 2012 by John Wiley & Sons, Inc.

Ethernet OAM can be categorized as

- link layer OAM defined by IEEE 802.3ah OAM [3];
- service layer OAM defined by IEEE 802.1ag [4], ITU Y.1731 [5], and MEF implementation agreements [1, 6, 7];
- Ethernet Local Management Interface (E-LMI) defined by MEF 16 [8].

Figure 8.1 depicts how these three protocols complement each other in fault monitoring of Ethernet networks. Figure 8.2 depicts operating boundaries of these protocols for an EVC.

IEEE 802.1ag-2007 [4] and ITU-T Recommendation Y.1731 [5] allow service providers to manage each customer service instance, EVC, individually and enable them to know if an EVC has failed, and if so, provide the tools to rapidly isolate the failure.

IEEE Std 802.3ah [3] enables service providers to monitor and troubleshoot a single Ethernet link. Remote end MAC address discovery, alarms for link failures, and loopback are supported.

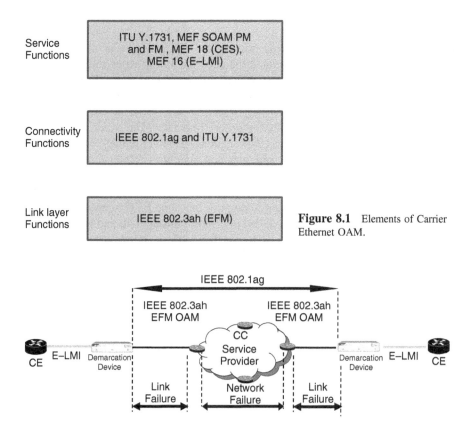

Service Functions	ITU Y.1731, MEF SOAM PM and FM , MEF 18 (CES), MEF 16 (E–LMI)

Connectivity Functions	IEEE 802.1ag and ITU Y.1731

Link layer Functions	IEEE 802.3ah (EFM)

Figure 8.1 Elements of Carrier Ethernet OAM.

Figure 8.2 OAM process operating boundaries.

Services and Performance (*ITU Y.1731/MEF*)	Basic Connectivity (*IEEE 802.1ag, ITU*)	Transport/Link (*802.3ah EFM*)
Discovery	Discovery	Discovery
Continuity check (keep alive)	Continuity Check	Remote failure indication: Dying gasp, link fault & critical event
Loopback (non-intrusive and intrusive	Loopback	Remote, local loopback
AIS/RDI/test		Fault isolation
Link Trace	Link Trace	Performance monitoring with threshold alarms
Performance management		Status monitoring

Figure 8.3   Summary of the features of the Ethernet OAM standards.

E-LMI protocol defined in Reference 8 operates between the customer edge (CE) device and network edge (NE) of the service provider It enables the service provider to automatically configure the CE device to match the subscribed service.

This automatic provisioning of the CE device reduces not only the effort to set up the service but also the amount of coordination required between the service provider and the enterprise customer. Furthermore, the enterprise customer does not have to learn how to configure the CE device, thus reducing barriers to adoption and greatly decreasing the risk of human error.

In addition to automatic provisioning of the CE device, E-LMI can provide EVC status information to the CE device. If an EVC fault is detected, the service provider edge device can notify the CE device of the failure so that traffic can be quickly rerouted to a different path. However, these days, E-LMI is not widely deployed in the industry to take advantage of its capabilities.

Figure 8.3 summarizes OAM functions provided by these standards. In addition to in-band procedures defined by these protocols, network management through EMS (Element Management System), NMS, and OSS (Operations Support System) via in-band and out-of-band communication channels is also needed for timely management of Carrier Ethernet networks.

In the following sections, we describe link and EVC level OAM functions in details.

8.2 LINK OAM

Link OAM as defined in IEEE 802.3ah provides mechanisms to monitor link operation and health, and improves fault isolation. It can be implemented on any full-duplex point-to-point or emulated point-to-point Ethernet link. OAM protocol data units (OAM PDUs) cannot propagate beyond a single hop within an Ethernet network and are not forwarded by bridges. OAM data is conveyed in untagged 802.3 slow protocol frames; therefore, OAM frame transmission rate is limited to a maximum of 10 frames per second. The major features covered by this protocol are:

- discovery of MAC address of the next hop,
- remote failure indication for link failures,
- power failure reporting via Dying Gasp, and
- remote loopback.

OAM PDUs are identified by MAC address and Ethernet Length/Type/sub-type field. This protocol uses a protocol sublayer between physical and data link layers. Ethernet in the First Mile (EFM) OAM is enabled/disabled on a per port basis.

Discovery is the first phase of link layer OAM. It identifies the devices at each end of the link along with their OAM capabilities. Discovery allows provider edge switch to determine the OAM capability of the remote demarcation device (Fig. 8.4). If both ends support OAM, then the two ends exchange state and configuration information:

- mode: active versus passive,
- OAM PDU size,
- identity, and
- loopback support.

Link monitoring OAM serves for detecting and indicating link faults under a variety of conditions. It provides statistics on the number of frame errors (or percentage of frames that have errors) as well as the number of coding symbol errors.

Failure conditions of an OAM entity are conveyed to its peer via OAM PDUs. For example, the failure conditions can be communicated via a loss of signal in one direction on the link (Fig. 8.5).

Ethernet AIS/RDI is used to suppress downstream alarms and eliminate alarm storms from a single failure (Fig. 8.6). This is similar to AIS/RDI in legacy services such as SONET, TDM, and frame relay.

Demarcation devices can be configured to propagate link failures downstream and vice versa. In other words, WAN port failures can be propagated to LAN ports and vice versa (Fig. 8.7). The fault propagation may be used for link-level protection schemes such as Hot Standby Redundant Routing Protocols (HSRPs) allowing transparent failover [9].

Dying Gasp message is generated for AC or DC power failures and sent immediately and continuously as depicted in Figure 8.8.

An OAM entity can put its remote peer that is at the other side of a link into loopback mode using the loopback control OAM PDU. In loopback mode, every frame received is transmitted back unchanged on the same port (except for

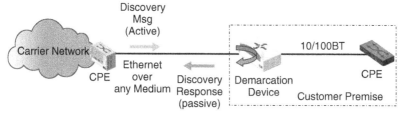

Figure 8.4   OAM discovery.

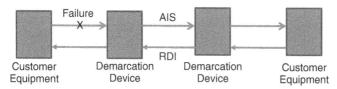

Figure 8.5   AIS/RDI.

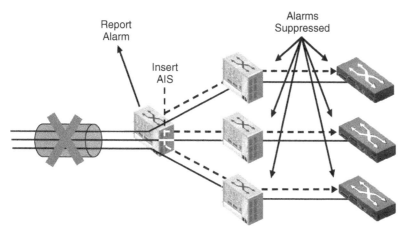

Figure 8.6   Alarm propagation to downstream devices.

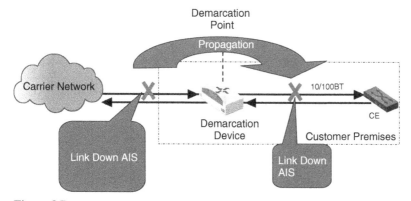

Figure 8.7   Alarm propagation from network to customer access.

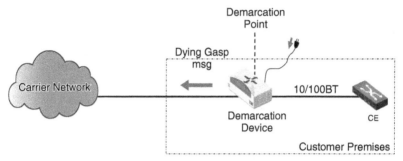

Figure 8.8 Dying Gasp message propagation.

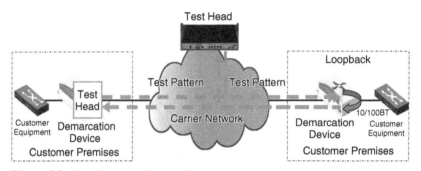

Figure 8.9 Loopback test using a test head.

OAM PDUs, which are needed to maintain the OAM session). Port-level testing is used in service turnup and in troubleshooting.

This feature can be configured such that the service provider device can put the customer device into loopback mode, but not conversely.

In the loopback mode, a central test head or testing software embedded in the device can perform RFC 2544 tests to measure throughput, delay, jitter, and dropped packets (Fig. 8.9).

Loopbacks can be facility or terminal, where terminal loopback signal travels through the device. Applying both the loopbacks will identify whether a problem is with the facility or the device (Fig. 8.10).

In addition to IEEE 802.3ah [3], some vendors support Time Domain Reflectometer (TDR) for cable integrity testing to identify opens, shorts, and impedance problems with CAT-5 cable on customer premise (Fig. 8.11). This testing method helps to identify patch-panel and a large percentage of customer-induced issues.

8.3 SERVICE OAM

The service OAM (SOAM) is addressed by IEEE 802.1ag (Connectivity Fault Management (CFM)) [4], ITU Y.1731 [5], RFC 2544 [10], Y.1564 [11],

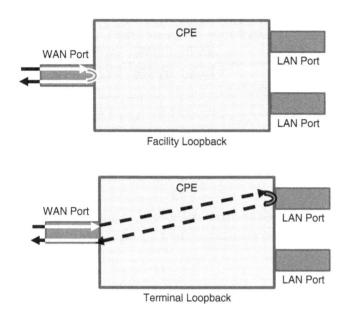

Figure 8.10   (a) Facility and (b) terminal loopbacks.

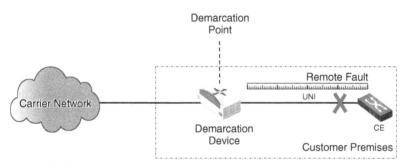

Figure 8.11   Cable integrity testing using TDR.

and MEF specifications [7, 12, 21]. Continuity check, loopback, link trace, delay/jitter measurements, loss measurements, and in-service and out-of-service testing protocols are defined in these standards to monitor SLAs (i.e., Service Level Specifications (SLSs)) and to identify service-level issues and debug them.

CFM defined in Reference 4 creates a maintenance hierarchy by defining maintenance domain (MD) and MD levels where Maintenance End Points (MEPs) determine domain boundaries. Maintenance points (MPs) that are between these two boundary points, MEPs, are called Maintenance Intermediate Points (MIPs). Maintenance entities defined for CFM are depicted in Figure 8.12.

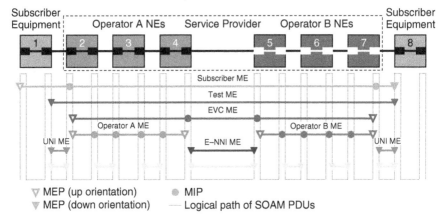

Figure 8.12 Maintenance entities and MA-MD level mapping, also described in Table 8.1.

An EVC service between two subscribers may involve multiple operators, in addition to a service provider. Customers purchase Ethernet service from service providers. Service providers may use their own networks or the networks of other operators to provide connectivity for the requested service. Customers themselves may be service providers, for example, a customer may be an Internet service provider that sells Internet connectivity or a cloud service broker orchestrating Ethernet Cloud Carrier access.

In all scenarios, the service provider and operators want to monitor and maintain the part of service that they are responsible for and the subscriber wants to monitor the service he/she pays for. In order to have all parties to operate without interfering each other, MDs and MD levels are defined. A MD is an administrative domain for the purpose of managing and administering a network. There are eight domain levels. Each domain is assigned a unique maintenance level by the administrator. MDs may nest or touch, but cannot intersect. If two domains nest, the outer domain must have a higher maintenance level than the one inside.

In this chapter, we describe OAM, service monitoring, and testing capabilities.

8.4 MAINTENANCE ENTITIES

An entity that requires management is called a management entity in ITU-T and MEF. A set of MEs that satisfy the following conditions form a Maintenance Entity Group (MEG) [5, 13]:

- MEs in an MEG exist in the same administrative domain and have the same MEG level.

- MEs in an MEG belong to the same service provider VLAN (S-VLAN).

For a point-to-point Ethernet connection, an MEG contains a single ME. For a multipoint Ethernet connection, an MEG contains $n \times (n-1)/2$ MEs, where n is the number of Ethernet connection end points.

An MEP is a maintenance functional entity that is implemented at the ends of an ME. It generates and receives OAM frames. An ME represents a relationship between two MEPs.

An MEP can be Up MEP or Down MEP. Up MEP is an MEP residing in a bridge that transmits CFM PDUs toward, and receives them from, the direction of the Bridge Relay Entity. Down MEP is an MEP residing in a bridge that receives CFM PDUs from, and transmits them toward, the direction of the LAN. The MEP that notifies the ETH layer MEPs on failure detection is called a server MEP.

An MIP is located at intermediate points along the end-to-end path where Ethernet frames are bridged to a set of transport links. It reacts and responds to OAM frames.

In case MEGs are nested, the OAM flow of each MEG has to be clearly identifiable and separable from the OAM flows of the other MEGs. This is accomplished by MEG level/maintenance domain and maintenance domain levels.

Eight MEG levels can be shared among the customer, provider, and operator to distinguish between OAM frames belonging to nested MEGs of customers, providers, and operators. The following lists the default MEG level assignment among the customer, provider, and operator roles:

- Customer role is assigned three MEG levels: 7, 6, and 5.
- Provider role is assigned two MEG levels: 4 and 3.
- Operator role is assigned three MEG levels: 2, 1, and 0.

MEF FM IA [12] recommends the default values in Table 8.1.

Table 8.1   MEF Recommendation for MA-MD Level Mapping and MEP Types

MA (Maintenance Entity in Y.1731)	Default MD Level (i.e., Maintenance Entity Group Level in Y.1731)	MEP Type
Subscriber-MA	6	Up
Test-MA	5	Down
EVC-MA	4	Up
Operator-MA	2	Up
UNI-MA	1	Down
ENNI-MA	1	Down

The service network is partitioned into the customer, provider, and operator maintenance levels. Providers have the end-to-end service responsibility. Operators provide service transport across a subnetwork.

The OAM architecture is designed such that an MEP at a particular MEG level transparently passes SOAM traffic at a higher MEG level, terminates traffic at its own MEG level, and discards SOAM traffic at a lower MEG level. This results in a nesting requirement where an MEG with a lower MEG level cannot exceed the boundary of an MEG with a higher MEG level [7].

The domain hierarchy provides a mechanism for protecting an MP (MEP or MIP) from other MPs with which the MP is designed to communicate. However, it is possible for an MP to flood one or more of its peer MPs with SOAM PDUs. This can result in a denial of service by forcing the receiving MPs to use computing resources for processing the SOAM PDUs from the flooding MP. Therefore, an NE supporting MPs needs to limit the number of SOAM PDUs per second that are processed at the local node.

8.5 MAINTENANCE POINTS

Any port of a bridge is referred to as an MP. MPs are unique for a given MD level. An MP may be classified as a maintenance end point or an MIP. According to Reference 4, MEPs and MIPs are the short form of MA (maintenance association) End Points and Intermediate Points (Table 8.3). They are functionally equivalent to ITU-T MEPs and MIPs.

MEG End Point (MEP) marks the end point of an ETH MEG that is capable of initiating and terminating OAM frames for fault management and PM. The OAM frames are distinct from the transit ETH flows and subject to the same forwarding treatment as the transit ETH flows being monitored.

An MEP does not terminate the transit ETH flows, even though it can observe these flows, for example, count the frames. It can generate and receive CFM PDUs and track any responses. It is identified by the maintenance association end point identifier (MEP ID), which is a small integer, unique over a given MA.

A server MEP represents both the server layer termination function and server/ETH adaptation function, which is used to notify the ETH layer MEPs on failure detection by the server layer termination function or server/ETH adaptation function.

On fault detection, AIS is generated in the upstream direction at the maintenance level of the client layer. The transport layer can be a MPLS pseudowire path, a SONET path, or an Ethernet link. At this layer, it is necessary to implement an Ethernet OAM AIS generation function for the Ethernet client layer at the appropriate MA level.

A server MEP needs to support ETH-AIS function, where the server/ETH adaptation function is required to issue frames with ETH-AIS information on detection of defect at the server layer by the server layer termination and/or adaptation function.

Table 8.2 Maintenance Point Functions

Functions	MEP	MIP	Transparent Point
Initiate CFM messages related to continuity check, loopback, and link trace.	Yes	No	No
Respond to loopback and link trace messages	Yes	Yes	No
Catalog continuity check information received	Yes	Yes	No
Forward CFM messages	No	Yes	Yes

The MEG Intermediate Point (MIP) is an intermediate point in an MEG, which is capable of reacting to some OAM frames. An MIP does not initiate OAM frames. An MIP takes no action on the transit ETH flows. MP functions are listed in Table 8.2.

An MIP will forward CFM packets unless it is a loopback or link trace destined for that intermediate point, whereas an MEPs do not forward CFM frames because they must keep them within the domain.

OAM frames belonging to an administrative domain originate and terminate in MEPs present at the boundary of that administrative domain. The MEP allows OAM frames from outside administrative domains belonging to higher level MEs to pass transparently but blocks those belonging to same or lower level MEs. The MEP is at an edge of a domain filter by maintenance level; therefore, OAM messages at their own level and at higher levels of the MEPs are filtered. With this, OAM frames can be prevented from leaking.

An MEP must be provisioned with information about its peer MEPs. This information can be potentially discovered. MEPs can proactively discover other MEPs by continuity check messages (CCMs). A multicast loopback can be used to discover other MEPs on an on-demand basis as well. MIPs can be discovered using link trace.

The principal operational issue for Ethernet OAM is scalability. A CCM can be sent as fast as every 3.3 ms. There can be 4094 VLANs per port and up to eight maintenance levels. This yields a worst-case CCM transmission rate of 9.8 million CCMs per second.

The main operational issue for link trace is the Ethernet MAC address learning and aging. When there is a network fault, the MAC address of a target node can age out in several minutes (e.g., typically 5 min). Solutions are to launch link trace within the age-out time or to maintain a separate target MEP database at intermediate MIPs. However, this requires an MIP CCM database.

Fault notification and alarm suppression is accomplished by Simple Network Management Protocol (SNMP) notifications and AIS/RDI. AIS can provide both

Table 8.3 Terminology Mapping

Y.1731 Term	802.1ag Term
MEG	MA
MEG ID	MAID (Domain Name + Short MA Name)
MEG level	MA level

alarm suppression and upstream notification. RDI provides downstream notification. The main issues with AIS are multipoint service instances and the potential interaction with Ethernet Spanning Tree Protocol (STP) loop prevention and recovery. An STP-based network reconfiguration may result in AIS interruption or redirection. The issues with RDI are multipoint service instances and bidirectional faults, which would block RDI downstream transmission.

The customer and service provider can independently use all the eight MEG levels, as well as mutually agree how to share the MEG levels.

The customer must send OAM frames as VLAN-tagged or priority-tagged frames to independently utilize all eight MEG levels. However, if the customer uses untagged OAM frames, the MEG levels may not be independent anymore and the customer and provider MEG levels need to be mutually agreed between the customer and the service provider.

Figure 8.1 illustrates the MDs and maintenance entities (MEs) by the MEF where pairs of MEPs (thus MEs) are communicating across various MDs that are hierarchically related to each other.

The scope of an MD is restricted to its associated VLAN, which has implications when VLAN identifiers are stacked. For example, if an ingress subscriber frame with a C-tag is stacked with an S-tag by the service provider, then the subscriber and test MDs are in a VLAN different from that of the EVC and operator MDs. Therefore, MD levels may not need to be coordinated between the subscriber and service provider.

An MA is identified by the Maintenance Association Identifier (MAID), which is unique over the domain. The MAID has two parts: the MD Name and the Short MA Name (Tables 8.4 and 8.5). The MD Name has the null format, and the Short MA Name has the text format. Using the text format for the Short MA Name allows for a maximum length of 45 octets for the Short MA Name (Table 8.3).

The MAID is carried in CFM PDUs to identify inadvertent connections among MEPs. A small integer, the MEP ID, uniquely identifies each MEP among those configured on a single MA.

A set of MEPs, each configured with the same MAID and MD level, established to verify the integrity of a single service instance is called an MA. An MA can also be thought of as a full mesh of maintenance entities among a set of MEPs so configured.

An MEP is associated with exactly one MA and identified for management purposes by MAID identifying the MA and MEP ID, where the MEP ID is an

Table 8.4 Maintenance Domain Name Format

Maintenance Domain Name Format Field	Value
Reserved for IEEE 802.1	0
No Maintenance Domain Name present	1
Domain Name based string denoting the identity of a particular Maintenance Domain	2
MAC address + 2-octet integer	3
Character string which is an IETF RFC 2579 DisplayString, with the exception that character codes 0–31 are not used	4
Reserved for IEEE 802.1	5–31
Defined by ITU-T Y.1731	32–63
Reserved for IEEE 802.1	64–255

Table 8.5 Short MA Name Format

Short MA Name Format Field	Value
Reserved for IEEE 802.1	0
Primary VID	1
Character string which an IETF RFC 2579 DisplayString, with the exception that character codes 0–31 are not used	2
2-octet integer	3
RFC 2685 VPN ID	4
Reserved for IEEE 802.1	5–31
Defined by ITU-T Y.1731	32–63
Reserved for IEEE 802.1	64–255

integer in the range of 1 to 8191. It provides each MEP with the unique identity required for the MEPs to verify the correct connectivity of the MA. The MEP is identified for data forwarding purposes by a set of VIDs, including a primary VID, inherited from the MA. From the MA, the MEP inherits an MD, and from this it inherits an MD level, inherited from its MD, and a primary VID, inherited from its MA (Tables 8.4 and 8.5).

8.6 OAM ADDRESSING AND FRAME FORMAT

OAM frames are identified by a unique EtherType value, that is, 0x8902. OAM frames processing and filtering at an MEP is based on the OAM EtherType and MEG level fields for both unicast and multicast DA (destination address).

IEEE 802.1ag supports two addressing modes, the bridge port model and the master port model. In the bridge port model, MEPs and MIPs assume the same MAC address as the bridge port. For the master port model, MEPs are implemented in a logical bridge master port such as a CPU. All MEPs use the same master port MAC address. These master port MEPs have an ambiguity in the identification of an MEP to which a loopback reply (LBR) is destined. This model is for legacy devices that cannot support port-based SOAM.

The source MAC address for all OAM frames is always a unicast MAC address. The destination MAC address may be either a unicast or a multicast address. Two types of multicast addresses are required depending on the type of OAM function:

- Multicast DA Class 1: OAM frames that are addressed to all MEPs in a MEG (e.g., CCM, multicast LBM (loopback message), AIS)
- Multicast DA Class 2: OAM frames that are addressed to all MIPs and MEPs associated with an MEG (e.g., LTM (link trace message)).

A multicast DA could also implicitly carry the MEG level, which will result in eight distinct addresses for each of the multicast DA classes 1 and 2 for the eight different MEG levels.

Multicast addresses for classes 1 and 2 are 01-80-C2-00-00-3x and 01-80-C2-00-00-3y, respectively, where x represents MEG level with x being a value in the range of 0 to 7 and y represents MEG Level with y being a value in the range of 8 to F.

The following lists the usage of unicast DA and multicast DA by each OAM messages per Y.1731.

- CCM frames can be generated with a specific multicast class 1 DA or unicast DA. When a multicast DA is used, CCM frames allow discovery of MAC addresses associated with MEPs and detection of misconnections among domains.
- LBM frames can be generated with unicast or multicast class 1 DAs.
- LBR frames are always generated with unicast DAs.
- LTM frame is generated with a multicast class 2 DA. A multicast DA is used instead of a unicast DA for LTM frames, because in current bridges, ports do not look at the EtherType before looking at the DA; therefore, the MIPs would not be able to intercept a frame with a unicast DA that was not their own address and would forward it.
- LTR frames are always generated with unicast DAs.
- AIS frame can be generated with a multicast class 1 DA in a multipoint EVC and with unicast DA for point-to-point EVCs if unicast DA of the downstream MEP is configured on the MEP transmitting AIS.
- LCK frames can be generated with a multicast class 1 DA in a multipoint EVC and with unicast DA for point-to-point EVCs if unicast DA of the downstream MEP is configured on the MEP transmitting LCK.

- TST frames are generated with unicast DAs. TST frames may be generated with multicast class 1 DA if multipoint diagnostics are desired.
- APS (Automatic Protection Switching defined G.8031) frames can be generated with a specific multicast class 1 DA or unicast DA.
- MCC (Maintenance Communication Channel) frames are generated with unicast DAs.
- LMM (Loss Measurement Message) frames are generated with unicast DAs. LMM frames may be generated with multicast class 1 DA if multipoint measurements are desired.
- LMR (Loss Measurement Reply) frames are always generated with unicast DAs.
- One-way delay measurement (1DM) frames are generated with unicast DAs. 1DM frames may be generated with multicast class 1 DA if multipoint measurements are desired.
- DMM (Delay Measurement Message) frames are generated with unicast DAs. DMM frames may be generated with multicast class 1 DA if multipoint measurements are desired.
- DMR (Delay Measurement Reply) frames are always generated with unicast DAs.

Figure 8.13 illustrates the common Ethernet SOAM frame format. Each specific OAM message type will add additional fields to the common PDU format.

As with all Ethernet frames, the destination and source MAC address (DA/SA) are preceded by an 8-octet preamble and a 1-octet start of frame delimiter. The frame may or may not include a customer VLAN tag. On the other hand, it can include an S-Tag.

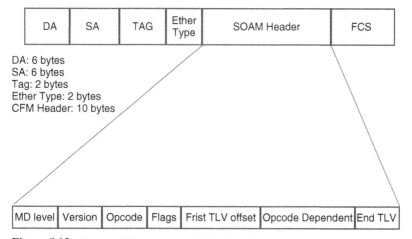

Figure 8.13   Common Ethernet service OAM frame format.

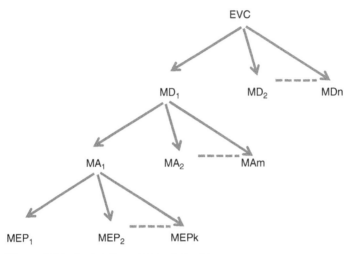

Figure 8.14   EVC, MD, and MA relationship.

IEEE 802.1ag also supports the LLC/SNAP (Subnetwork Access Protocol)-encoded frame format, which includes an LLC header in addition to the OAM EtherType. The following information elements are common across the OAM PDUs.

- MEG Level (i.e., MD Level) is a 3-bit field. It contains an integer value that identifies the MEG level of OAM PDU. Value ranges from 0 to 7. The MD level corresponds to the administrative domains shown in Figure 8.12. MD levels 5 through 7 are reserved for customer domains, MD levels 3 and 4 are reserved for service provider domains, and MA levels 0 through 2 are reserved for operator domains. Relationships among EVCs, MDs, MAs, and MEPs are depicted in Figure 8.14.
- Version is a 5-bit field containing an integer value that identifies the OAM protocol version. In Y.1731, the version is always 0.
- OpCode is a 1-octet field containing an OpCode that identifies an OAM PDU type (i.e., CCM, LTM, LBM). OpCode is used to identify the remaining content of an OAM PDU. Continuity check, loopback, and linktrace use OpCodes 0–31, whereas Y.1731 performance management functions use OpCodes 32–63 (Table 8.6).
- Flags is an 8-bit field. Its usage is dependent on the OAM PDU type.
- TLV (Type, Length, and Value) Offset is a 1-octet field. It contains the offset to the first TLV in an OAM PDU relative to the TLV Offset field. The value of this field is associated with an OAM PDU type. When the TLV Offset is 0, it points to the first octet following the TLV Offset field.

The following is a set of TLVs defined by IEEE 802.1ag for various CFM PDU types. Each TLV can be identified by the unique value assigned to its type field. Some type field values are reserved.

- TLVs Applicable for CCM: end TLV, sender ID TLV, port status TLV, interface status TLV, organization-specific TLV.
- TLVs Applicable for LBM: end TLV, sender ID TLV, data TLV, organization-specific TLV.
- TLVs Applicable for LBR: end TLV, sender ID TLV, data TLV, organization-specific TLV.
- TLVs Applicable for LTM: end TLV, LTM egress identifier TLV, sender ID TLV, organization-specific TLV.
- TLVs Applicable for LTR: end TLV, LTR egress identifier TLV, reply ingress TLV, reply egress TLV, sender ID TLV, organization-specific TLV.

Table 8.6 OpCodes for OAM PDUs [5]

OpCode Value	OAM PDU Type	OpCode Relevance for MEPs/MIPs
1	CCM	MEPs
3	LBM	MEPs and MIPs (connectivity verification)
2	LBR	MEPs and MIPs (connectivity verification)
5	LTM	MEPs and MIPs
4	LTR	MEPs and MIPs
33	AIS	MEPs
35	LCK	MEPs
37	TST	MEPs
39	Linear APS	MEPs
40	Ring APS	MEPs
41	MCC	MEPs
43	LMM	MEPs
42	LMR	MEPs
45	1DM	MEPs
47	DMM	MEPs
46	DMR	MEPs
49	EXM	
48	EXR	
51	VSM	
50	VSR	

The general format of TLVs is shown in Figure 8.15. Type values are specified in Table 8.7.

In an End TLV, both length and value fields are not used.

Priority and drop eligibility of a specific OAM frame are not present in OAM PDUs but conveyed in frames carrying OAM PDUs.

Type (1 byte)
Length (2-3 bytes)
Value which is optional (4 bytes)

Figure 8.15   Generic TLV format [2].

Table 8.7   Type Value [5]

Type Value	TLV Name
0	End TLV
3	Data TLV
5	Reply Ingress TLV
6	Reply Egress TLV
7	LTM Egress Identifier TLV
8	LTR Egress Identifier TLV
32	Test TLV
33–63	Reserved

8.7 CONTINUITY CHECK MESSAGE (CCM)

The OAM PDU used for Ethernet continuity check (ETH-CC) information is CCM. Frames carrying the CCM PDU are called CCM frames.

CCMs are "heartbeat" messages issued periodically by MEPs (Fig. 8.16). They are used for proactive OAM to detect loss of continuity (LOC) among MEPs and for the discovery of each other in the same domain. MIPs will discover MEPs that are in the same domain using CCMs as well. In addition, CCM can be used for loss measurements and for triggering protection switching.

CCM also allows detection of unintended connectivity between two MDs (or MEGs), namely, Mismerge; unintended connectivity within the MEG with

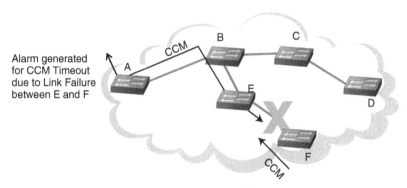

Figure 8.16   CCM.

an unexpected MEP (UnexpectedMEP),; and other defective conditions such as unexpected MEG level and unexpected period. CCM is applicable for fault management, PM, or protection switching applications.

When CCM transmission is enabled, the MEP periodically transmits frames to all other MEPs in the same domain. The CCM transmission period must be the same for all MEPs in the same domain. When an MEP is enabled to generate frames with CCM information, it also expects to receive frames with ETH-CC information from its peer MEPs in the MEG.

A CCM has the following attributes (Fig. 8.16 and Table 8.8) to be configured:

- An MEG ID (i.e., MAID) is 48 octets and identifies the MEG to which the MEP belongs. The MAID used by CCM must be unique across all MAs at the same MD level. The default MA values of CCM are untagged UNI-MA, untagged ENNI-MA, and EVC-MA.
- An MEP ID is a 2-octet field whose 13 least significant bits identify the MEP within the MEG.
- List of peer MEP IDs in the MEG. For a point-to-point MEG with a single ME, the list would consist of a single MEP ID for the peer.
- MEG level at which the MEP exists.
- CCM transmission period is encoded in the three least significant bits of the Flags field. The default transmission period for fault management, PM, and protection switching are 1 s, 100 ms, and 3.33 ms, respectively. Transmission periods of 10 s, 1 min, and 10 min are also used. If the port associated with an MEP experiences a fault condition, the MEP will encode RDI in the Flags field.
- Priority of frame with ETH-CC information. By default, the frame with ETH-CC information is transmitted with the highest priority available to the data traffic.
- Drop eligibility of the frame with ETH-CC information that is always marked as drop ineligible.

On the other hand, an MIP is transparent to the CCM information and does not require any configuration information to support CCM.

When CCM is used to support dual-ended loss measurement, the PDU includes the following (Figs 8.17–8.19):

- TxFCf is a 4-octet field that carries the value of the counter of in-profile data frames transmitted by the MEP toward its peer MEP, at the time of CCM frame transmission.
- RxFCb is a 4-octet field that carries the value of the counter of data frames received by the MEP from its peer MEP, at the time of receiving the last CCM frame from that peer MEP.
- TxFCb is a 4-octet field that carries the value of the TxFCf field in the last CCM frame received by the MEP from its peer MEP.

Table 8.8 Continuity Check Message Format Octet

Field	Length/Comment
Common CFM header	1–4
Sequence number	5–8
Maintenance Association End Point Identifier	9–10
Maintenance Association Identifier (MAID)	11–58
Defined by ITU-T Y.1731	59–74
Reserved for definition in future versions of the protocol	Zero length
Optional CCM TLVs	First TLV Offset + 5 (octet 75 for transmitted CCMs)
End TLV (0)	First TLV Offset + 5 if no optional CCM TLVs are present

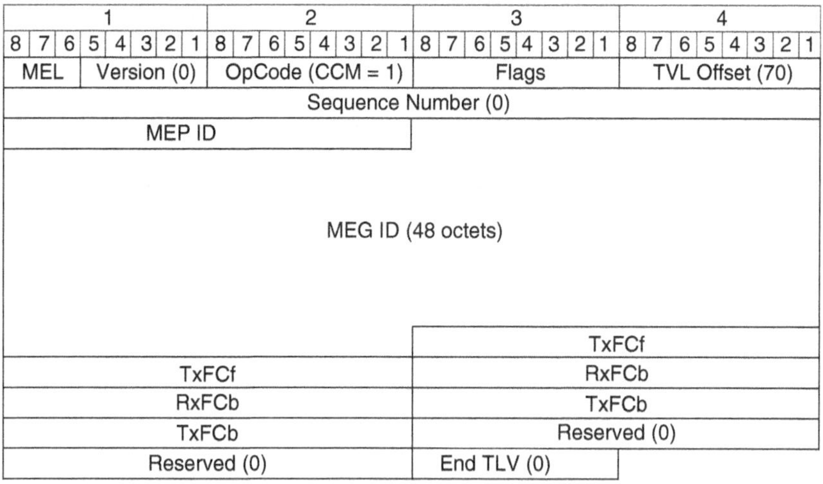

1								2								3								4							
8	7	6	5	4	3	2	1	8	7	6	5	4	3	2	1	8	7	6	5	4	3	2	1	8	7	6	5	4	3	2	1
MEL			Version (0)					OpCode (CCM = 1)								Flags								TVL Offset (70)							
Sequence Number (0)																															
MEP ID																															
MEG ID (48 octets)																															
																TxFCf															
TxFCf																RxFCb															
RxFCb																TxFCb															
TxFCb																Reserved (0)															
Reserved (0)																End TLV (0)															

Figure 8.17 CCM PDU.

A sender ID TLV, if included, indicates the management address of the source of the CCM frame. This can be used to identify the management address of a remote device rather than just its MAC address, making the identification of the device possible in a large network where MAC addresses are not well known.

The Port Status and Interface Status TLVs indicate the status of various parameters on the sender of the CCM. The Interface Status TLV can indicate to the far end that a problem has occurred on the near end. When applied to the EVC-MA, the Interface Status TLV can be used to indicate to the far end that the local UNI is down, even though the CCMs on the EVC are successful.

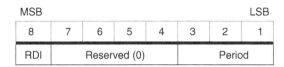

Figure 8.18 Format of CCM PDU flags.

MSB							LSB

8	7	6	5	4	3	2	1
RDI	Reserved (0)				Period		

MSB															LSB
Octet 9								Octet 10							
8	7	6	5	4	3	2	1	8	7	6	5	4	3	2	1
0	0	0	MEP ID												

Figure 8.19 MEP ID format in CCM PDU.

MEPs are configured (MEG ID and MEP ID) with a list of all the other MEPs in their maintenance level. Every active MEP maintains a CCM database. As an MEP receives CCMs, it catalogs them in the database indexed by the MEP ID. Exchange of CCMs between MEPs in an MEG allows for the detection of the following defects.

8.7.1 Loss of Continuity (LOC) Condition

If no CCM frames from a peer MEP are received within the interval equal to 3.5 times the receiving MEP's CCM transmission period, LOC with peer MEP is detected as a result of hard failures (e.g., link failure, device failure) or soft failures (e.g., memory corruption, misconfigurations). LOC state is exited during an interval equal to 3.5 times the CCM transmission period configured at the MEP, if the MEP receives at least three CCM frames from that peer MEP.

There is a direct correlation between the CCM frame transmission periods supported and the level of resiliency a network element can offer to a specific EVC. Three CCM messages must be lost before a failure is detected across a specific MA. A failure must be detected before any protection switching mechanisms can enable a new path through the network. For example, to enact a protection switching mechanism that claims a maximum correction time of 50 ms and uses CCMs to detect the failure, the CCM frame transmission period must be 10 ms or less. Otherwise, just detecting the failure would take more than 50 ms.

8.7.2 Unexpected MEG Condition

If a CCM frame with a MEG level lower than the receiving MEP's MEG level is received, the unexpected MEG level is detected.

8.7.3 Mismerge Condition

If a CCM frame with the same MEG level but with an MEG ID different than that of the receiving MEP is received, Mismerge is detected. Such a defect is most likely caused by misconfiguration but could also be caused by a hardware/software failure in the network. Mismerge state is exited when the MEP does not receive CCM frames with incorrect MEG ID during an interval equal to 3.5 times the CCM transmission period configured at the MEP.

8.7.4 Unexpected MEP (UnexpectedMEP) Condition

An MEP detects UnexpectedMEP when it receives a CCM frame from a remote MEP, which is not in the remote MEP list of this local MEP mostly due to misconfiguration, but has a correct MEG level with an unexpected MEP ID. Determination of unexpected MEPID is possible when the MEP maintains a list of its peer MEPIDs. UnexpectedMEP state is exited when the MEP does not receive CCM frames with an unexpected MEPID during a specified interval such as that equal to 3.5 times the CCM transmission period configured at the MEP.

8.7.5 Unexpected Period (UnexpectedPeriod) Condition

If a CCM frame is received with a correct MEG level, MEG ID, and MEPID, but with a period field value different from the receiving MEP's own CCM transmission period, UnexpectedPeriod is detected. Such a defect is most likely caused by misconfiguration. UnexpectedPeriod state is exited when the MEP does not receive CCM frames with incorrect period field value during an interval equal to 3.5 times the CCM transmission period configured at the MEP.

8.7.6 Signal Fail (SignalFail) Condition

An MEP declares signal fail condition on detection of defective conditions including LOC, Mismerge, UnexpectedMEP, and unexpected MEG level. In addition, signal fail condition may also be declared by the server layer termination function to notify the server/ETH adaptation function about the defective condition in the server layer.

A receiving MEP notifies the equipment fault management process when it detects the above defects.

8.8 LOOPBACK AND REPLY MESSAGES (LBM AND LBR)

Ethernet loopback (ETH-LB) function is an on-demand OAM function that is used to verify connectivity of an MEP with a peer MP(s). Loopback is transmitted by an MEP on the request of the administrator to verify connectivity to a particular MP. Loopback indicates whether the destination is reachable or not; it does not allow hop-by-hop discovery of the path. It is similar in concept to the ICMP Echo (Ping).

An ETH-LB session is defined as a sequence that begins with management initiating the transmission of (n) periodic LBM frames from an MEP to a peer MP and ends when the last LBR frame is received or incurs a time-out.

The number of LBM transmissions in an ETH-LB session is configurable in the range of 0 through 3600. Two to three LBM messages are common. The period for LBM transmission is configurable in the range of 0 through 60; 1 s is more common.

LBM and LBR message sizes can be from 64 bytes to maximum MTU sizes of the EVC.

There are two ETH-LB types, unicast ETH-LB and multicast ETH-LB. In order to dynamically discover MAC address of the remote MEP(s) on an MA, in addition to supporting a configurable destination to any unicast MAC address, an MEP may also support multicast MAC addresses corresponding to the reserved CFM multicast addresses for CCM.

Unicast ETH-LB can be used to verify bidirectional connectivity of an MEP with an MIP or a peer MEP and to perform a bidirectional in-service or out-of-service diagnostics test between a pair of peer MEPs. This includes verifying bandwidth, detecting bit errors, etc.

When unicast ETH-LB is used to verify bidirectional connectivity, an MEP sends a unicast frame with ETH-LB request information and expects to receive a unicast frame with ETH-LB reply information from a peer MP within a specified period of time. The peer MP is identified by its MAC address. This MAC address is encoded in the DA of the unicast request frame. If the MEP does not receive the unicast frame with ETH-LB reply information within the specified period of time, loss of connectivity with the peer MP can be declared.

For performing bidirectional diagnostic tests, an MEP sends unicast frames with ETH-LB request information to a peer MEP. This ETH-LB request information includes test patterns. For out-of-service diagnostic tests, MEPs are configured to send frames with ETH-LCK information.

For bidirectional connectivity verification, an MEP transmits a unicast LBM frame addressed to the remote peer MP with a specific transaction ID inserted in the Transaction ID/Sequence Number field. After unicast LBM frame transmission, an MEP expects to receive a unicast LBR frame within 5 s. The transmitted transaction ID is therefore retained by the MEP for at least 5 s after the unicast LBM frame is transmitted. A different transaction ID must be used for every unicast LBM frame, and no transaction ID from the same MEP may be repeated within 1 min. The maximum rate for frames with unicast ETH-LB should be at a value that does not adversely impact the user traffic. MEP attributes to be configured to support unicast ETH-LB are the following:

- MEG level at which the MEP exists.
- Unicast MAC address of remote peer MP.
- A test pattern and an optional checksum, which is optional. The test pattern can be a pseudorandom bit sequence (PRBS) $(2^{31}-1)$, an all "0" pattern,

frames of various sizes, etc. For bidirectional diagnostic test application, a test signal generator and a test signal detector associated with the MEP need to be configured.

- Priority of frames with unicast ETH-LB information.
- Drop eligibility of frames with unicast ETH-LB information to be discarded when congestive conditions are encountered.

A remote MP, on receiving the unicast frame with ETH-LB request responds with a unicast frame with ETH-LB reply information.

Unicast frames carrying the LBM PDU are called Unicast LBM frames. Unicast frames carrying the LBR PDU are called Unicast LBR frames. Unicast LBM frames are transmitted by an MEP on an on-demand basis.

Whenever a valid unicast LBM frame is received by an MIP or MEP, an LBR frame is generated and transmitted to the requesting MEP. A unicast LBM frame with a valid MEG level and a destination MAC address equal to the MAC address of receiving MIP or MEP is considered to be a valid unicast LBM frame. Every field in the unicast LBM frame is copied to the LBR frame with the following exceptions:

- the source and destination MAC addresses are swapped, and
- the OpCode field is changed from LBM to LBR.

An LBR is valid if the MEP receives the LBR frame within 5 s after transmitting the unicast LBM frame and the LBR has the same MEG level. Otherwise, the LBR frame addressed to it is invalid and is discarded.

When an MEP configured for a diagnostic test receives an LBR frame addressed to it with an MEG Level same as its own MEG Level, the LBR frame is valid. The test signal receiver associated with MEP may also validate the received sequence number against expected sequence numbers. If an MIP receives an LBR frame addressed to it, such an LBR frame is invalid and the MIP should discard it.

For diagnostic tests, an MEP transmits a unicast LBM frame addressed to the remote peer MEP with a test TLV. The test TLV is used to carry the test pattern generated by a test signal generator associated with the MEP. When the MEP is configured for an out-of-service diagnostic test, the MEP also generates LCK frames at the client MEG level in a direction opposite to the direction where LBM frames are issued.

Multicast ETH-LB is an on-demand OAM function that is used to verify bidirectional connectivity of an MEP with its peer MEPs. Multicast frames carrying the LBM PDU are called as multicast LBM frames. When a multicast ETH-LB function is invoked on an MEP, the MEP returns to the initiator of the multicast ETH-LB a list of its peer MEPs with which the bidirectional connectivity is detected.

When multicast ETH-LB is invoked on an MEP, a multicast frame with ETH-LB request information is sent from an MEP to other peer MEPs in the same MEG. The MEP expects to receive unicast frame with ETH-LB reply information from its peer MEPs within a specified period of time. On reception of a multicast frame with ETH-LB request information, the receiving MEPs validate the multicast frame with ETH-LB request information and transmit a unicast frame with ETH-LB reply information after a randomized delay in the range of 0 to 1 s.

In order for an MEP to support multicast ETH-LB, the MEG level at which the MEP exists and the priority of multicast frames with ETH-LB request information need to be configured.

Multicast frames with ETH-LB request information are always marked as drop ineligible; therefore, they are not configured.

On the other hand, an MIP is transparent to the multicast frames with ETH-LB request information and therefore does not require any information to support multicast ETH-LB.

After transmitting the multicast LBM frame with a specific transaction ID, the MEP expects to receive LBR frames within 5 s. The transmitted transaction ID is therefore retained for at least 5 s after the multicast LBM frame is transmitted. A different transaction ID must be used for every multicast LBM frame, and no transaction ID from the same MEP may be repeated within 1 min.

Whenever a valid multicast LBM frame is received by an MEP, an LBR frame is generated and transmitted to the requesting MEP after a randomized delay in the range of 0 to 1 s. The validity of the multicast LBM frame is determined based on the correct MEG level. Every field in the multicast LBM frame is copied to the LBR frame with the following exceptions:

- The source MAC address in the LBR frame is the unicast MAC address of the replying MEP. The destination MAC address in the LBR frame is copied from the source MAC address of the multicast LBM frame, which should be a unicast address.
- The OpCode field is changed from LBM to LBR.

When an LBR frame is received by an MEP with an expected transaction ID and within 5 s of transmitting the multicast LBM frame, the LBR frame is valid. If an MEP receives an LBR frame with a transaction ID that is not in the list of transmitted transaction IDs maintained by the MEP, the LBR frame is invalid and is discarded. If an MIP receives an LBR frame addressed to it, such an LBR frame is invalid and the MIP should discard it.

For the connectivity test, LBMs can be transmitted with either a unicast or a multicast DA. This address can be learned from the CCM database. The unicast DA can address either an MEP or MIP, while the multicast DA is only used to address MEPs. The LBM includes a transaction ID/sequence number, which is retained by the transmitting MEP for at least 5 s. After a unicast LBM frame transmission, an MEP expects to receive a unicast LBR frame, with the same transaction ID/sequence number within 5 s.

When an LBM is received by a remote MEP/MIP, which matches its address, an LBR will be generated. Every field in the LBM is copied to the LBR with the exception that (i) the source and destination MAC addresses are swapped and (ii) the OpCode field is changed from LBM to LBR. The Transaction ID/Sequence Number and Data TLV fields are returned to the originating MEP unchanged. These fields are verified by the originating MEP. For multipoint loopback, each MEP returns an LBR after a randomized delay. A loopback diagnostic text is only for point-to-point applications between MEPs and uses unicast destination MAC addresses. The LBM includes a test pattern, and the LBR returns the same test pattern.

As with CCMs, the sender ID TLV can be used to identify the management address of the MEP to its peers, thus providing IP correlation to the MAC address across a large network. However, for security reasons, the management domain and address can be empty in the sender ID TLV.

For an LB session, the number of LBMs transmitted, the number of LBRs received, and the percentage of responses lost (timed out) are reported by the initiating MEP. For an LB session, the round-trip time (RTT) min/max/average statistics MUST be supported by the initiating MEP.

LBM and LBR PDUs, as depicted in Figures 8.20–8.24, contain the mandatory Transaction ID/Sequence Number field and the optional Data/Test Pattern TLV field. Transaction ID/Sequence Number is a 4-octet field that contains the transaction ID/sequence number for the LBM. The length and contents of the data/test pattern TLV are determined by the transmitting MEP. In LBM PDU, Version is 0, OpCode is 3, and Flags are all ZEROes.

Let us consider an example where a service provider is using the networks of two operators to provide service (Fig. 8.24). The customer level allows the customer to test connectivity (using connectivity checks) and isolate issues (using loopback and link trace).

The customer could use CFM loopback or link trace to isolate a fault between the MEP on the CPE and the MIP on CE. By definition, the link between the CPE and CE is a single hop, and therefore, the customer would know which link has the fault. However, if the fault is between the two MIPs, the customer will need to rely on the service provider to determine between which MEPs or MIPs the fault has occurred. Even then, the service provider may simply isolate the fault to a specific operator's network and will in turn rely on the operator to isolate the fault to a specific link in its network.

Therefore, all the parties have the ability to isolate the fault within their domain boundaries, without the service provider having to share its network information with the customer or the operator having to share its network information with the service provider.

8.9 LINK TRACE AND REPLY MESSAGES (LTM AND LTR)

Ethernet link trace (ETH-LT) function is an on-demand OAM function initiated in an MEP on the request of the administrator to track the path to a destination

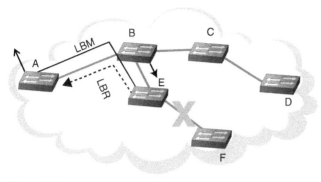

Figure 8.20 LBM/LBR.

	1	2	3	4
	8 7 6 5 4 3 2 1	8 7 6 5 4 3 2 1	8 7 6 5 4 3 2 1	8 7 6 5 4 3 2 1
1	MEL	Version (0)	OpCode (LBM = 3)	Flags (0)	TVL offset (4)
5	Transaction ID/Sequence Number			
9 13 17 :	[Optional TLV starts here; otherwise End TLV]			
last				End TLV (0)

Figure 8.21 LBM PDU format.

MSB							LSB
8	7	6	5	4	3	2	1
Reserved (0)							

Figure 8.22 Flags format in LBM PDU.

	1	2	3	4
	8 7 6 5 4 3 2 1	8 7 6 5 4 3 2 1	8 7 6 5 4 3 2 1	8 7 6 5 4 3 2 1
1	Type (3)	Length		
: : : :	Data Pattern			

Figure 8.23 Data TLV format.

	1								2								3								4							
	8	7	6	5	4	3	2	1	8	7	6	5	4	3	2	1	8	7	6	5	4	3	2	1	8	7	6	5	4	3	2	1
1	MEL			version					OpCode (LBR=2)								Flags								TVL offset							
5	Transaction ID/Sequence Number																															
9	[Optional TLV starts here, otherwise end TLV]																															
13																																
17																																
:																																
last																									End TLV (0)							

Figure 8.24   LBR PDU format.

MEP. They allow the transmitting node to discover connectivity data about the path. The PDU used for ETH-LT request information is called LTM, and the PDU used for ETH-LT reply information is called LTR. Frames carrying the LTM PDU are called LTM frames and those carrying the LTR PDU are called LTR frames.

As each network element, containing the MP, needs to be aware of the TargetMAC address in the received LTM frame and associates it to a single egress port, in order for the MP to reply, a unicast ETH-LB to the TargetMAC address could be performed by an MEP before transmitting the LTM frame. This ensures that the network elements along the path to the TargetMAC address have information about the route to the TargetMAC address if the address is reachable in the same MEG.

During a failure condition, the information about the route to the Target-MAC address may age out after a certain time. The ETH-LT function has to be performed before the age out occurs in order to provide information about the route.

If an LTM frame is received by an MP, it forwards the LTM frame to the network element's ETH-LT responder, which performs the following validation:

- Whether the LTM frames have the same MEG level of the receiving MP.
- If the TTL (time to live) field value is 0, the LTM frame (i.e., invalid frame) is discarded.
- Whether the LTM egress identifier TLV is present—the LTM frame without the LTM egress identifier TLV is discarded.

If the LTM frame is valid, the ETH-LT responder performs the following.

- It determines the DA for the LTR frame from the OriginMAC address in the received LTM frame.
- If the LTM frame terminates at the MP, an LTR frame is sent backward to the originating MEP after a random time interval in the range of 0 to 1 s.
- If the above condition applies and LTM frame does not terminate at the MP and the TTL field in the LTM frame is greater than 1, the LTM frame is forwarded toward the single egress port. All the fields of the relayed

LTM frame are the same as the original LTM frame except for the TTL field that is decremented by 1.

Link trace is similar in concept to User Datagram Protocol (UDP) Traceroute. It can be used for retrieving adjacency relationship between an MEP and a remote MP. The result of running ETH-LT function is a sequence of MIPs from the source MEP until the target MIP or MEP where each MP is identified by its MAC address. The link trace also can localize faults when a link and/or a device failure or a forwarding plane loop occurs. The sequence of MPs will likely be different from the expected one. Difference in the sequences provides information about the fault location.

On an on-demand basis, an MEP will multicast LTM on its associated Ethernet connection. The multicast destination MAC address is a class 2 multicast address. The transaction ID, TTL, OriginMAC address, and TargetMAC address are encoded in the LTM PDU. The TargetMAC address is the address of the MEP at the end of the EVC/OVC, which is being traced. It can be learned from CCM. The OriginMAC address is the address of the MEP that initiates the link trace. The transaction ID/sequence number and TargetMAC address are retained for at least 5 s after the LTM is transmitted. This is for comparison with the LTR.

As with CCMs, the sender ID TLV in the LTM/LTR can be used to identify the management address of the MEP to its peers.

After transmitting the LTM frame with a specific transaction number, the MEP expects to receive LTR frames within a specified period of time, usually 5 s. A different transaction number must be used for every LTM frame, and no transaction number from the same MEP may be repeated within 1 min. If an MEP receives an LTR frame with a transaction number that is not in the list of transmitted transaction numbers maintained by the MEP, the LTR frame is invalid. If an MIP receives an LTR frame addressed to it, the MIP should discard it, but an MIP may relay the frame with ETH-LT request information.

The configuration information required by an MEP to support ETH-LT is the following:

- MEG level at which the MEP exists,
- priority of the frames with ETH-LT request information,
- drop eligibility of the frames, and ETH-LT is always marked as drop ineligible,
- TargetMAC address for which ETH-LT is intended, and
- TTL to allow the receiver to determine if frames with ETH-LT request information can be terminated. TTL is decremented every time frames with ETH-LT request information are relayed. Frames with ETH-LT request information with TTL ≤ 1 are not relayed.

The LTR frame contains LTR egress identifier TLV with the Last Egress Identifier field that identifies the network element that originated or forwarded the LTM frame for which this LTR frame is the response. This field takes the same value as the LTM egress identifier TLV of that LTM frame. LTR egress identifier TLV also contains the Next Egress Identifier field that identifies the network element that transmitted this LTR frame and can relay a modified LTM frame to the next hop. This field takes the same value as the LTM egress identifier TLV of the relayed modified LTM frame, if any.

LTM PDU and LTR PDU are depicted in Figures 8.25–8.29.

8.10 ETHERNET ALARM INDICATION SIGNAL (ETH-AIS)

The PDU used for ETH-AIS information is AIS PDU (Fig. 8.30). Frames carrying the AIS PDU are called AIS frames.

An MEP can transmit AIS frames periodically in a direction opposite to its peer MEP(s) immediately following the detection of a defective condition. An MIP is transparent to AIS frames and therefore does not require any information to support ETH-AIS functionality.

The periodicity of AIS frames transmission is based on the AIS transmission period that is encoded as a CCM period. The transmission period can be in the range of 3.3 ms to 10 min; however, 1 s is recommended. An MEP continues to transmit periodic frames with ETH-AIS information until the defective condition is removed.

ETH-AIS is used to suppress alarms following detection of defective conditions at the server (sub)layer. Owing to independent restoration capabilities provided within the STP environments, ETH-AIS is not expected to be applied in the STP environments. Transmission of frames with ETH-AIS information can be enabled or disabled on an MEP (or on a server MEP).

The client layer may consist of multiple MEGs that should be notified to suppress alarms resulting from defective conditions detected by the server layer MEP. The server layer MEP, on detecting the signal fail condition, needs to send AIS frames to each of these client layer MEGs. The first AIS frame for all client layer MEGs must be transmitted within 1 s of detecting the defective condition. If the system is stressed potentially when issuing AIS frames every second across all 4094 VLANs, the AIS transmission period may be configured in minutes. An AIS frame communicates the used AIS transmission period via the Period field.

AIS is generally transmitted with a class 1 multicast DA. A unicast DA is allowed for point-to-point applications.

On receiving an AIS frame, the MEP examines it to ensure that its MEG level corresponds to its own MEG level and detects the AIS defective condition. The Period field indicates the period at which the AIS frames can be expected. Following detection of the AIS defective condition, if no AIS frames are received within an interval of 3.5 times the AIS transmission period indicated in the AIS frames received before, the MEP clears the AIS defective condition.

CoS for AIS frame must be the CoS that yields the lowest frame loss performance for this EVC.

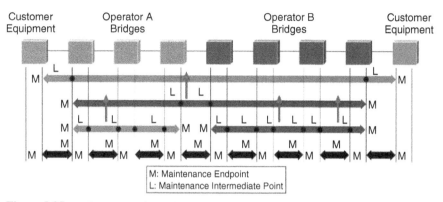

Figure 8.25 Maintenance points and maintenance domains. M, maintenance end point; L, maintenance intermediate point.

	1	2	3	4
	8 7 6 5 4 3 2 1	8 7 6 5 4 3 2 1	8 7 6 5 4 3 2 1	8 7 6 5 4 3 2 1
1	MEL | Version (0)	OpCode (LTM = 5)	Flags	TLV Offset (17)
5	Transaction ID (4 bytes)			
9	TTL (1 bytes)	OriginMAC Address (6 bytes)		
13				
17	TargetMAC Address (6 bytes)			
21	[*Additional TLV starts here*]			
25				
29				
:				
last				End TLV (0)

Figure 8.26 LTM PDU format.

	1	2	3	4
	8 7 6 5 4 3 2 1	8 7 6 5 4 3 2 1	8 7 6 5 4 3 2 1	8 7 6 5 4 3 2 1
1	MEL | Version (0)	OpCode (LTR = 4)	Flags	TVL Offset (6)
5	Transaction ID			
9	TTL	Relay Action	[*TLVs starts here*]	
13				
17				
:				
last				End TLV (0)

Figure 8.27 LTR PDU format.

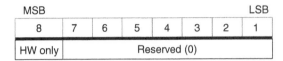

MSB							LSB
8	7	6	5	4	3	2	1
HW only	Reserved (0)						

Figure 8.28 Flags format in LTM PDU.

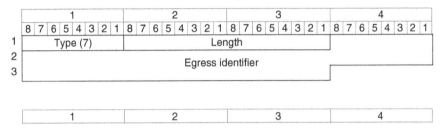

Figure 8.29 LTM egress identifier TLV format.

	1								2								3								4							
	8	7	6	5	4	3	2	1	8	7	6	5	4	3	2	1	8	7	6	5	4	3	2	1	8	7	6	5	4	3	2	1
1	MEL			Version (0)					OpCode (AIS = 33)								Flags								TLV Offset (0)							
5	End TLV (0)																														

Figure 8.30 AIS PDU format.

An MEP detects AIS when it receives an AIS frame that is caused by detection of signal fail condition at a server layer or reception of AIS at a server layer MEP where the MEP does not use ETH-CC function.

AIS state is entered when an MEP receives an AIS and exited when the MEP does not receive AIS frames OR when ETH-CC is used, on clearing of the' LOC defect at the MEP, during an interval equal to 3.5 times the AIS transmission period indicated in the AIS frames received earlier.

Frames with ETH-AIS information can be issued at the client MEG level by an MEP, including a server MEP, on detecting defective conditions. The defective conditions may include

- signal fail conditions when ETH-CC is enabled, and
- AIS condition or LCK condition when ETH-CC is disabled.

Since a server MEP does not run ETH-CC, a server MEP can transmit frames with ETH-AIS information on detection of any signal fail condition. For multipoint ETH connectivity, an MEP cannot determine the specific server layer entity that has encountered defective conditions on receiving a frame with ETH-AIS information. Therefore, on reception of such a frame, the MEP will suppress alarms (i.e., LOC) for all peer MEPs whether there is still connectivity or not.

For a point-to-point ETH connection, an MEP has only a single peer MEP. Therefore, there is no ambiguity regarding the peer MEP for which it should suppress alarms when it receives the ETH-AIS information.

Configuration parameters for an MEP to support ETH-AIS transmission are

- the client MEG level at which the most immediate client layer MIPs and MEPs exist;
- the ETH-AIS transmission period;
- priority of frames with ETH-AIS information.

ETH-AIS is always marked as drop ineligible.

However, for multipoint EVCs/OVCs, a client layer MEP, on receiving an AIS, cannot determine which of its remote peers has lost connectivity. It is recommended that for multipoint, the client layer MEP should suppress alarms for all peer MEPs.

Use of AIS is not recommended for environments that utilize the STP, which provides an independent restoration capability. Owing to the spanning tree and multipoint limitation associated with AIS, the IEEE 802.1 committee has chosen not to support AIS in 802.1ag.

When we have CCM capability, the need for AIS is questioned. The fact is that AIS is generated immediately after the failure is detected. If there is AIS capability, LOC is generated after not receiving three CCMs. Therefore, AIS would provide faster failure reporting than CCM.

8.11 ETHERNET REMOTE DEFECTIVE INDICATION (ETH-RDI)

ETH-RDI is used only when ETH-CC transmission is enabled. When a downstream MEP detects a defective condition, such as receive signal failure or AIS, it will send an RDI in the opposite upstream direction to its peer MEP or MEPs. This informs the upstream MEPs that there has been a downstream failure.

RDI is subject to the same multipoint issue as AIS. An MEP that receives an RDI cannot determine what subset of peer MEPs has experienced a defect. For Y.1711 [14], RDI is encoded as a bit in the Flags field in CCMs. IEEE 802.1ag does not support RDI.

An MEP, on detecting a defective condition with its peer MEP, sets the RDI field in the CCM frames for the duration of the defective condition. CCM frames are transmitted periodically based on the CCM transmission period, when the MEP is enabled for CCM frames transmission. When the defective condition clears, the MEP clears the RDI field in the CCM frames in subsequent transmissions.

The configuration information required by an MEP to support ETH-RDI function is:

- the MEG level at which the MEP exists,
- ETH-RDI transmission period is configured to be the same as ETH-CC transmission period, depending on application,
- the priority of ETH-RDI frames, which is the same as the ETH-CC priority, and

- the drop eligibility of ETH-RDI frames that are always marked as drop ineligible.

An MIP is transparent to frames with ETH-RDI information.

The PDU used to carry ETH-RDI information is CCM. Eight bits in CCM PDU is set to 1 to indicate RDI, otherwise it is set to 0 (Fig. 8.31).

8.12   ETHERNET LOCKED SIGNAL (ETH-LCK)

The LCK condition is caused by an administrative/diagnostic action at a server (sub)layer MEP that results in disruption of the client data traffic.

An MEP continues to transmit periodic frames with ETH-LCK information at the client MEG level until the administrative/diagnostic condition is removed.

An MEP extracts frames with ETH-LCK information (Figure 8.32) at its own MEG level and detects an LCK condition, which contributes to the signal fail condition of the MEP. The signal fail condition may result in the transmission of AIS frames to its client MEPs.

The LCK state is entered when an MEP receives an LCK frame and exits if the MEP does not receive LCK frames during an interval equal to 3.5 times the LCK transmission period indicated in the LCK frames received earlier.

The LCK transmission period is encoded the same as a CCM period. An example of an application that would require administrative locking of an MEP is out-of-service ETH-Test.

The configuration information required by an MEP to support an ETH-LCK transmission is:

- the client MEG level at which the most immediate client layer MIPs and MEPs exist,
- the ETH-LCK transmission period that determines transmission periodicity of frames with ETH-LCK information,
- the priority that identifies the priority of frames with ETH-LCK information, and

MSB							LSB
8	7	6	5	4	3	2	1
RDI	Reserved (0)				Period		

Figure 8.31   Flags format in CCM PDU.

	1		2		3		4	
	8 7 6 5 4 3 2 1	8 7 6 5 4 3 2 1	8 7 6 5 4 3 2 1	8 7 6 5 4 3 2 1
1	MEL	Version (0)	OpCode (LCK = 35)	Flags	TLV Offset (0)
5	End TLV (0)				

Figure 8.32   LCK PDU format.

- the drop eligibility of the ETH-LCK frames that are always marked as drop ineligible.

An MIP is transparent to the frames with ETH-LCK information.

An MEP, when administratively locked, transmits LCK frames in a direction opposite to its peer MEP(s). The periodicity of LCK frame transmission is based on the LCK transmission period, which is the same as the AIS transmission period.

On receiving an LCK frame, an MEP examines its MEG level and detects an LCK condition. If no LCK frames are received within an interval of 3.5 times the LCK transmission period, the MEP clears the LCK condition.

It is used for client layer alarm suppression and enables client MEPs to differentiate between defective conditions and intentional administrative/diagnostic actions at the server layer MEP. This capability is only supported in Y.1731 [5].

8.13 PERFORMANCE MEASUREMENTS

Collecting measurements proactively (i.e., continuously and periodically) or on a on-demand basis for each management entity such as UNI, ENNI, CoS, EVC, and OVC is important in managing networks. These measurements are used in determining if contractual agreements (SLAs) are satisfied, the need for capacity, and the location of network failures. The continuous measurements are coupled with service-level monitoring in accomplishing these tasks.

At the physical port, the following Media Attachment Unit (MAU) Termination Performance Data Set for each MAU Transport Termination is counted [10, 15, 16]:

- number of times the MAU leaves the available state,
- number of times the MAU enters the jabbering state, and
- the number of false carrier events during idle.

Traffic measurements on a per UNI [1] and ENNI bases are counted by NE. They are Octets Transmitted OK, Unicast Frames Transmitted OK, Multicast Frames Transmitted OK, Broadcast Frames Transmitted OK, Octets Received OK, Unicast Frames Received OK, Multicast Frames Received OK, and Broadcast Frames Received OK. In addition, the following anomalies at UNI and ENNI should be counted: Undersized Frames, Oversized Frames, Fragments, Frames with FCS or Alignment Errors, Frames with Invalid CE-VLAN ID, and Frames with S-VLAN ID (at ENNI only).

Measurements on a per UNI, ENNI, CoS per UNI, CoS per ENNI, OVC or CoS per OVC, EVC or CoS per EVC basis for each entity that enforces traffic management at ingress direction (CE to MEN at UNI, and ENNI Gateway to ENNI Gateway at ENNI) are counted:

- green frames sent by the ingress UNI to the MEN,
- yellow frames sent by the ingress UNI to the MEN,

- red (discarded) frames at the ingress UNI,
- green octets sent by the ingress UNI to the MEN,
- yellow octets sent by the ingress UNI to the MEN, and
- red (discarded) octets at the ingress UNI.

At the egress, for the same entities, the following is counted:

- green frames received by the egress UNI from the MEN,
- yellow frames received by the egress UNI from the MEN,
- green octets received by the egress UNI from the MEN, and
- yellow octets received by the egress UNI from the MEN.

At both ingress and egress, for the same entities, the following is counted:

- green frames discarded due to congestion,
- yellow frames discarded due to congestion,
- green octets discarded due to congestion, and
- yellow octets discarded due to congestion.

Counts for the current 15-min interval and the past 32 15-min intervals are stored in NE counters that can store a value of at least $2^{64}-1$. If the counter reaches its maxmium value, it will either keep its maxmium value duration of the interval (interval counter) or wrap around (continuous counter). Timestamp for measurement interval with 1-s accuracy is expected.

Specific threshold values can be assigned to counters by NMS. When a threshold counter exceeds its threshold during a current measurement interval, the NE sends a TCA to the managing system within 60 s of the occurrence of the threshold crossing.

Measurements can be stored in measurement intervals or bins. A measurement interval is an interval of time and stores the results of performance measurements conducted during that interval. A measurement bin is a counter and stores the results of performance measurements falling within a specified range. Figure 8.33 is an example that illustrates the relationship between measurement intervals and measurement bins.

The number of measurement bins per measurement interval depend on application. The minimum number is 2. For example, bins for a measurement of 12-ms end-to-end delay can be as depicted below.

- bin 1 = 0 ms (range is 0 ms ≤ measurement < 4 ms)
- bin 2 = 4 ms (range is 4 ms ≤ measurement < 8 ms)
- bin 3 = 8 ms (range is 8 ms ≤ measurement < 12 ms)
- bin 4 = 12 ms (range is 12 ms ≤ measurement < time-out)

Suspect flag should be set when these measurements are not reliable due to service provisioning or service failure or time-of-day clock adjustment, etc.

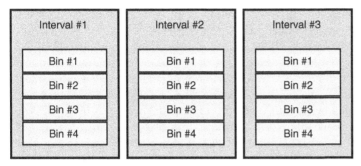

Figure 8.33   Measurement intervals and bins.

8.14   PERFORMANCE MONITORING

There are two types of measurements, periodic and on-demand. Periodic measurements are collected by polling counters of corresponding devices at a polling rate of not slower than 1 per minute. They are available in 15-min bins.

In-service measurements are needed to monitor health of the network as well as user SLAs. Out-of-service measurements are needed before turning up the service, for isolating troubles, and for identifying failed components during failures.

On-demand measurements are mostly needed during failures and service degradation when proactive measurements help to identify possible problems in advance.

As described earlier, each CoS/EVC has its own performance parameters, namely, SLA/SLS. The SLS is usually part of the contract between the service provider and the subscriber. Serviced provider monitors the EVC to ensure contractual obligations are satisfied while subscribers might want to monitor the EVC for the same reasons. SLS parameters to be monitored are delay, jitter, loss, and availability for the given EVC/CoS.

For a point-to-point EVC, the performance counts encompass service frames in both directions of the UNI pair, while for a multipoint EVC, they encompass service frames in one direction over all or a subset of the ordered UNI pairs. Since there are multiple ingress and egress points as well as the potential for frame replication, counters in a multipoint EVC may not be directly correlated. Periodic synthetic frames can be used for delay measurements. By counting and measuring the one-way FLR of uniform synthetic traffic, statistical methods can be used to estimate the one-way FLR of service traffic as well.

In general, the performance measurements can be performed using service frames and synthetic OAM test frames, although the accuracy of the measurements with synthetic frames depends on how closely synthetic frames emulate the service frames.

Delay measurement can be performed using the 1DM or DMM/DMR PDUs. Loss measurement can be performed by counting service frames using the LMM/LMR PDUs as well as by counting synthetic frames via the SLM/SLR (Synthetic Loss Message/Synthetic Loss Reply) PDUs.

For delay measurements, DMM frames are sent from a controller MEP to a responder MEP, which in turn responds with DMR frames. One-way measurements are taken with time-of-day clock synchronization. Two-way measurements (i.e., round-trip measurements) do not require time-of-day clock synchronization.

For loss measurements with synthetic frames, SLMs are sent from an originating MEP to a responder MEP, which in turn responds with SLR frames. One-way and two-way measurements of FLR and availability are always taken with this mechanism.

The test with synthetic frames must generate enough test frames to be statistically valid. The test frames must also be similar to the service frames carried by the EVC, in particular, the OAM test frames must have representative frame length, be generated by a representative arrival process, and be treated by the network elements implementing the network between the MEPs in the same way that service frames are treated.

PM functions are described further in the following sections.

8.15 LOSS MEASUREMENTS

Loss is measured in terms of FLR per UNI, ENNI, EVC, or OVC. For UNI, the FLR can be defined as ((number of frames delivered at the egress of receiving UNI—number of frames transmitted at the ingress of transmitting UNI)/(number of frames transmitted at the ingress of transmitting UNI)) in a given time interval according to the user–service provider contract. This time interval could be a month. An MIP is transparent to frames with ETH-LM (Ethernet Loss Measurement) information and therefore does not require any information to support ETH-LM functionality.

Measuring the one-way FLR of service frames between two measurement points requires transmission and reception counters, where the ratio between them corresponds to the one-way FLR. Collection of the counters needs to be coordinated for accuracy. If counters are allocated for green and yellow frames, it is possible to find the FLR for each color. If there are counters for the total count, it would be possible to identify FLR for the total frames.

Two types of FLR measurement are possible, dual-ended (i.e., one-way) and single-ended (i.e., round-trip or two-way). Dual-ended Loss Measurement can be accomplished by exchanging CCM OAM frames as well as ETH-LM.

When CCM OAM frames are used, they include counts of frames transmitted and frames received, namely, TxFCf, RxFCb, and TxFCb, where

- TxFCf is the value of the local counter TxFCl at the time of transmission of the CCM frame,
- RxFCb is the value of the local counter RxFCl at the time of reception of the last CCM frame from the peer MEP, and

- TxFCb is the value of TxFCf in the last CCM frame received from the peer MEP.

These counts do not include OAM frames at the MEP's MEG level.

When configured for proactive loss measurement, an MEP periodically transmits CCM frames with a period value equal to the CCM transmission period configured for PM application at the transmitting MEP. The receiving MEP detects an unexpected period defect if the CCM transmission period is not the same as the configured value. In this case, frame loss measurement is not carried out. When an MEP detects an LOC defect, it ignores loss measurements during the defective condition and assumes 100% losses.

Near-end and far-end FLRs are calculated across pairs of consecutive frames as

Frame Loss (far end) $= \mid$ TxFCb[tc] $-$ TxFCb[tp] $\mid - \mid$ RxFCb[tc] $-$ RxFCb[tp] $\mid$ Frame Loss (near end) $= \mid$ TxFCf[tc] $-$ TxFCf[tp] $\mid - \mid$ RxFCl[tc] $-$ RxFCl[tp] $\mid$

where

tc $=$ counter values for the current CCM frame.

tp $=$ counter values for the previous CCM frame.

For an MEP, near-end frame loss refers to frame loss associated with ingress data frames, while far-end frame loss refers to frame loss associated with egress data frames.

An MEP maintains the following two local counters for each peer MEP and for each priority class being monitored in a point-to-point ME for which loss measurements are to be performed:

- TxFCl: counter for in-profile data frames transmitted toward the peer MEP.
- RxFCl: counter for in-profile data frames received from the peer MEP.

The TxFCl and RxFCl counters do not count the OAM frames (except proactive OAM frames for protection) at the MEP's MEG level. However, the counters do count OAM frames from the higher MEG levels that pass through the MEPs in a manner similar to the data frames.

Loss measurement can be performed with ETH-LM (Fig. 8.34) as well. The level of accuracy in the loss measurements is dependent on how frames with ETH-LM information are added to the data stream after the counter values are copied in the ETH-LM information. For example, if additional data frames get transmitted and/or are received between the time of reading the counter values and adding the frame with ETH-LM information to the data stream, the counter values copied in ETH-LM information become inaccurate. In order to increase accuracy, a hardware-based implementation that is able to add frames with ETH-LM information to the data stream immediately after reading the counter values is necessary.

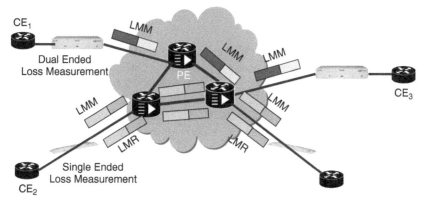

Figure 8.34 Single-ended and dual-ended loss measurements.

MP configuration information to support ETH-LM is the following:

- the MEG level at which the MEP exists,
- the ETH-LM transmission period, which is the frequency of LMM that is configurable to values of 100 ms, 1 s, and 10 s, and 1 s is more common,
- the priority that identifies the priority of the frames with ETH-LM information, and
- the drop eligibility indicating if ETH-LM can be dropped during congestion, and ETH-LM is always marked as drop ineligible.

Whenever a valid LMM frame is received by an MEP, an LMR frame is generated and transmitted to the requesting MEP. An LMM frame with a valid MEG level and a destination MAC address equal to the receiving MEP's MAC address is considered to be a valid LMM frame. An LMR frame contains the following fields and their values:

- TxFCf, which is a 4-octet field. Its value is copied from the LMM frame.
- RxFCf, which is a 4-octet field. Its value is the value of local counter RxFCl at the time of LMM frame reception.
- TxFCb, which is a 4-octet field. Its value is the value of local counter TxFCl at the time of LMR frame transmission.

On receiving an LMR frame, near-end and far-end loss measurements at the MEP can be calculated as:

Frame Loss (far end) $= |TxFCf[tc] - TxFCf[tp]| - |RxFCf[tc] - RxFCf[tp]|$ Frame Loss (near end) $= |TxFCb[tc] - TxFCb[tp]| - |RxFCl[tc] - RxFCl[tp]|$

where

- TxFCf is a 4-octet field of LMR, which carries the value of the TxFCf field in the last LMM PDU received by the MEP from its peer MEP;

- TxFCb is a 4-octet field of LMR, which carries the value of the counter of in-profile data frames transmitted by the MEP toward its peer MEP at the time of LMR frame transmission;
- RxFCf is a 4-octet field of LMR, which carries the value of the counter of data frames received by the MEP from its peer MEP, at the time of receiving the last LMM frame from that peer MEP.

	1	2	3	4
	8 7 6 5 4 3 2 1	8 7 6 5 4 3 2 1	8 7 6 5 4 3 2 1	8 7 6 5 4 3 2 1
1	MEL Version (0)	OpCode (LMM = 43)	Flags (0)	TLV Offset (12)
5	TxFCf			
9	Reserved for RxFCf in LMR			
13	Reserved for TxFCb in LMR			
17	End TLV (0)			

Figure 8.35 LMM PDU format.

	1	2	3	4
	8 7 6 5 4 3 2 1	8 7 6 5 4 3 2 1	8 7 6 5 4 3 2 1	8 7 6 5 4 3 2 1
1	MEL Version	OpCode (LMR = 42)	Flags	TLV Offset
5	TxFCf			
9	RxFCf			
13	TxFCb			
17	End TLV (0)			

Figure 8.36 LMR PDU format.

Single-ended loss measurement is accomplished by the on-demand exchange of LMM and LMR OAM frames. These frames include appropriate counts of frames transmitted and received. Near-end and far-end FLRs are calculated at the end that initiated the loss measurement request.

FLR measurement is performed on a point-to-point basis using a unicast DA. ITU Y.1731 [5] also allows the use of class 1 multicast DA to support multipoint testing.

LMM and LMR PDU formats used by an MEP to transmit LMM information are shown in Figures 8.35 and 8.36.

8.16 AVAILABILITY

Service availability is one of the customer SLA parameters. It is usually calculated per month or per year. An availability of 99.999% (i.e., five 9s) per year, which is equivalent to approximately 5 min downtime in a year, for carrier-class systems and services is common in the industry.

Service availability is the availability of the EVC for a given CoS between UNI-C at location A and UNI-C at location Z. If both the locations A and Z are on one network, the EVC consists of

- EVC segment over the Access Network of location A;
- EVC segment over Backbone Network;
- EVC segment over the Access Network of location Z.

If location A is an operator's footprint while the location Z is in another operator's footprint, the EVC consists of

- OVC1, OVC2, ... OVCn of EVC segments in MEN1, MEN2, ..., MENn, respectively;
- EVC segment over the link between ENNI1 and ENNI2, EVC segment over the link between ENNI2 and ENNI3 ..., and EVC segment over link between $ENNI_{n-2}$ and $ENNI_{n-1}$.

The service availability is the product of the availabilities of all the components making the EVC. In order to determine the availability of the EVC, availability of each of these components needs to be monitored and correlated.

A number of consecutive FLR measurements are needed to evaluate the availability/unavailability status of an entity. The default number is 10 [17].

A configurable availability threshold Ca and a configurable unavailable threshold Cu are used in evaluating the availability/unavailability status of an entity. The thresholds can range from 0 to 1 where the availability threshold is less than or equal to the unavailability threshold. However, MEF 17 [18] recommends Ca = Cu = C.

Whenever a transition between availability and unavailability occurs in the status, an adjacent pair of availability indicators should be reported to management systems (i.e., EMS and NMS).

Availability is calculated via sliding windows based on a number of consecutive FLR measurements [17–19]. The availability for a CoS/EVC/OVC of a point-to-point connection during Δt_k is based on the FLR during the short interval and each of the following $n - 1$ short intervals and the availability of the previous short time interval (i.e., sliding window of width $n \Delta t$).

Figure 8.37 presents an example of the determination of the availability for the small time intervals Δt_k within T with a sliding window of 10 small time intervals, as we determine the availability of an EVC from UNI i to UNI j for a time interval T, excluding the small time intervals that occur during a maintenance interval.

Each Δt_k in T is defined to be either "Available" or "Unavailable," and this is represented by $A_{\langle i,j \rangle}(\Delta t_k)$, where $A_{\langle i,j \rangle}(\Delta t_k) = 1$ means that Δt_k is available and $A_{\langle i,j \rangle}(\Delta t_k) = 0$ means that Δt_k is unavailable. The definition of $A_{\langle i,j \rangle}(\Delta t_k)$ is based on the FLR function discussed in the following.

Let $I_{\Delta t}^{\langle i,j \rangle}$ be the number of ingress service frames at UNI i to be delivered to UNI j for a given CoS where the first bit of each service frame arrives at UNI i

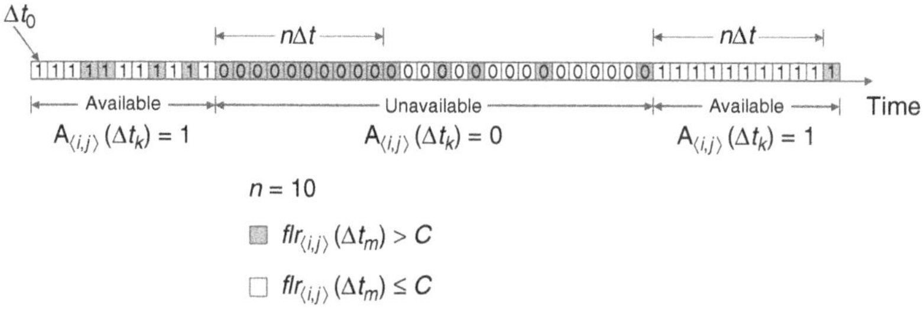

Figure 8.37 Determining availability via sliding window. Δt, a time interval much smaller than T; C, a loss ratio threshold, which if exceeded suggests unavailability; n, the number of consecutive small time intervals, Δt, over which to assess availability.

within the time interval Δt and is subject to an ingress bandwidth profile with either no color identifier or a color identifier that corresponds to green.

Let $E_{\Delta t}^{\langle i,j \rangle}$ be the number of unique (not duplicate) unerrored egress service frames where each service frame is the first egress service frame at UNI j that results from a service frame counted in $I_{\Delta t}^{\langle i,j \rangle}$.

Then,

$$\text{flr}_{\langle i,j \rangle}(\Delta t) = \begin{cases} \left(\dfrac{I_{\Delta t}^{\langle i,j \rangle} - E_{\Delta t}^{\langle i,j \rangle}}{I_{\Delta t}^{\langle i,j \rangle}} \right) & \text{if } I_{\Delta t}^{\langle i,j \rangle} \geq 1 \\ 0 & \text{otherwise} \end{cases}$$

Δt_0 is the first short time interval agreed by the service provider and subscriber at or after turnup of the EVC. $A_{\langle i,j \rangle}(\Delta t_k)$ for $k = 0$ is

$$A_{\langle i,j \rangle}(\Delta t_0) = \begin{cases} 0 \text{ if } \text{flr}_{\langle i,j \rangle}(\Delta t_m) > C, \forall m = 0, 1, \ldots n - 1 \\ 1 \text{ otherwise} \end{cases}$$

and for $k = 1, 2, \ldots$ is

$$A_{\langle i,j \rangle}(\Delta t_k) = \begin{cases} 0 \text{ if } A_{\langle i,j \rangle}(\Delta t_{k-1}) = 1 \text{ and } \text{flr}_{\langle i,j \rangle}(\Delta t_m) > C, \\ \quad \forall m = k, k+1, \ldots k + n - 1 \\ 1 \text{ if } A_{\langle i,j \rangle}(\Delta t_{k-1}) = 0 \text{ and } \text{flr}_{\langle i,j \rangle}(\Delta t_m) \leq C, \\ \quad \forall m = k, k+1, \ldots k + n - 1 \\ A_{\langle i,j \rangle}(\Delta t_{k-1}) \quad \text{otherwise} \end{cases}$$

Multipoint-to-multipoint EVC availability may be calculated by a scheme agreed by the service provider and customer. Theoretically, one may have to monitor simultaneously all possible point-to-point connections for the given multipoint connection and multiply availability of all the point-to-point connections. However, this is impractical to monitor and calculate in the real world. Instead, it may be practical to monitor and calculate FLR using synthetic frames for a subset of point-to-point connections simultaneously.

FLR-based availability calculation is still not widely practiced in the industry due to difficulty with loss calculations. Instead, failure intervals are measured and the availability is calculated by dividing unavailability time by the total service time.

8.17 FRAME DELAY MEASUREMENTS

There are two types of FD (frame delay) measurements, one-way and two-way (or round-trip delay-RTD). RTD is defined as the time elapsed since the start of transmission of the first bit of the frame by a source node until the reception of the last bit of the loopbacked frame by the same source node, when the loopback is performed at the frame's destination node.

One-way FD is measured by MEPs periodically sending 1DM frames (Fig. 8.38), which include appropriate Transmit Time Stamps. The FD is calculated at the receiving MEP by taking the difference between the Transmit Time Stamp and a Receive Time Stamp, which is created when the 1DM frame is received.

1DM frame can be sent periodically with a period of 100 ms, 1 s, and 10 s, and 1 s is more common. The 1DM frames are discard ineligible if tagging is used. Before initiating 1DMs between NEs, a Boolean clock synchronization flag must indicate that the clock synchronization is in effect.

Two-way DM measures round-trip delay and does not require synchronized clocks. It is accomplished by MEPs exchanging DMM and DMR frames. Each of these DM OAM frames includes Transmit Time Stamps. Y.1731 [5] allows an option for inclusion of additional time stamps such as a Receive Time Stamp

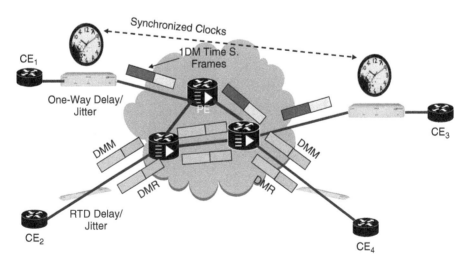

Figure 8.38 One-way delay and RTD measurements.

and a return Transmit Time Stamp. These additional time stamps compensate for DMR processing time.

The one-way delay is calculated by Frame Delay = RxTimef − TxTimeStampf, where RxTimef is the time that the 1DM PDU was received and TxTimeStampf is the time stamp at the time the 1DM PDU was sent.

Two-way delay measurement avoids the clock synchronization issue but could incur inaccuracy due to the DMM to DMR processing in the target MEP. Consequently, ITU-T Y.1711 [14] allows for two options in the measurement of two-way delay. If the target MEP turnaround delay is not considered significant then the round-trip delay can be calculated by Frame Delay = RxTimeb − TxTimeStampf, where RxTimeb is the time that the DMR PDU is received by the initiating MEP.

A more accurate two-way delay measurement can be achieved if the target MEP turnaround delay is subtracted out. In this cast, the round-trip delay can be calculated by Frame Delay = (RxTimeb − TxTimeStampf) − (TxTimeStampb − RxTimeStampf), where TxTimeStampb is the time that the DMR PDU is sent by the target MEP and RxTimeStampf is the time that the DMM PDU is received by the target MEP.

Unicast destination MAC address for point-to-point services is configurable.

The priority of the DMM frame transmission is configurable so that its priority corresponds to the EVC CoS to be monitored. The period values for DMM are 100 ms, 1 s, and 10 s, and 1 s is more common. Also, DMR frame pairs for FDV (frame delay variation) calculation can be configured. An offset range of 1 through 10 can be set.

Frame size for DMM and 1DM frame transmissions is configurable. The range of frame sizes from 64 through MTU size in 4-octet increments is recommended.

8.18 INTERFRAME DELAY VARIATION (IFDV) MEASUREMENTS

Similar to delay measurements, there are two IFDV measurements, round-trip and one-way IFDVs. Round-trip IFDV is calculated exactly as the difference between two consecutive two-way FD measurements, while one-way IFDV is the difference between two consecutive 1DMs. It is measured in microseconds.

For one-way Ethernet FD measurement, only the receiver MEP collects statistics. For two-way Ethernet FD measurement, only the initiator MEP collects statistics.

DMM and DMR PDUs are given in Figures 8.39 and 8.40, respectively.

8.19 TESTING

Testing is one of the areas in Carrier Ethernet that is still being worked [23, 24]. RFC 2544 [10] defines a mechanism to perform out-of-service throughput

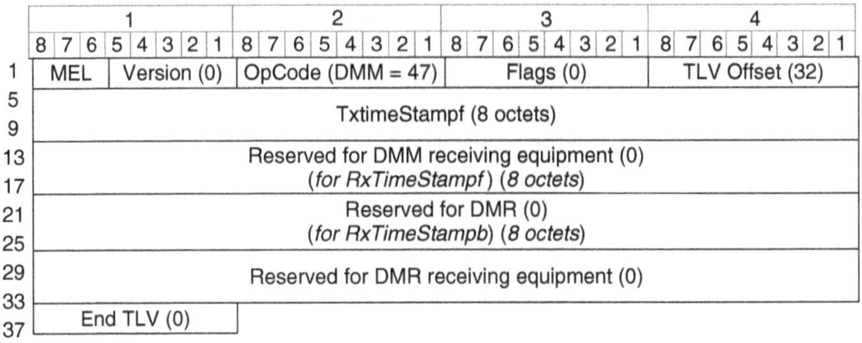

	1								2								3								4							
	8	7	6	5	4	3	2	1	8	7	6	5	4	3	2	1	8	7	6	5	4	3	2	1	8	7	6	5	4	3	2	1
1	MEL			Version (0)					OpCode (DMM = 47)								Flags (0)								TLV Offset (32)							
5 9	TxtimeStampf (8 octets)																															
13 17	Reserved for DMM receiving equipment (0) *(for RxTimeStampf)* (8 octets)																															
21 25	Reserved for DMR (0) *(for RxTimeStampb)* (8 octets)																															
29 33	Reserved for DMR receiving equipment (0)																															
37	End TLV (0)																															

Figure 8.39   DMM PDU format.

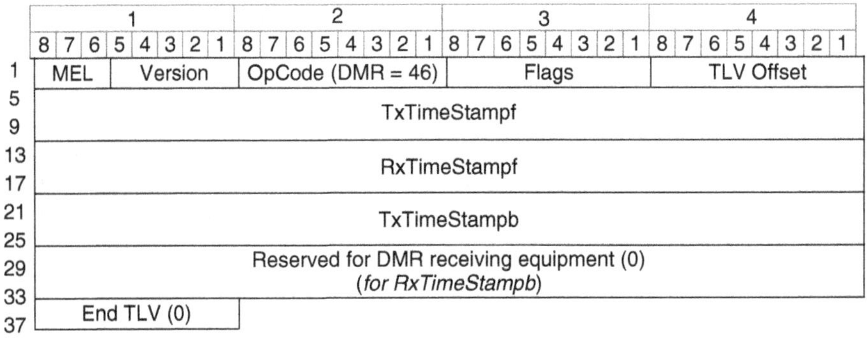

	1								2								3								4							
	8	7	6	5	4	3	2	1	8	7	6	5	4	3	2	1	8	7	6	5	4	3	2	1	8	7	6	5	4	3	2	1
1	MEL			Version					OpCode (DMR = 46)								Flags								TLV Offset							
5 9	TxTimeStampf																															
13 17	RxTimeStampf																															
21 25	TxTimeStampb																															
29 33	Reserved for DMR receiving equipment (0) *(for RxTimeStampb)*																															
37	End TLV (0)																															

Figure 8.40   DMR PDU format.

and delay measurements, while Y.1731 [5] defines an incomplete mechanism to perform in-service testing. In this section, we cover testing defined in References 5, 10, 11.

Testing is concerned with the testing of equipment and their resources, and transport facilities. Testing may be carried out for the purpose of testing connecting facilities in preparation for installation of new equipment, accepting newly installed interfaces or service assignments, validating trouble reports, supporting fault localization and verification of repair.

The equipment provides diagnostics that can examine the state of each significant element of hardware and that can identify faults and isolate failures within the smallest replaceable unit of hardware. A management system initiates these diagnostics.

The equipment runs diagnostics on hardware, checks on software, and reports the result to the managing system.

8.19.1   Y.1731 Testing

Test signal (TST) is an OAM message exchanged between MEPs in one way only, which includes test pattern to test throughput, measure bit errors, or detect frames delivered out of sequence. In general, tests can be performed in service or out of service.

TST OAM messages are generally sent with a unicast DA. The use of a class 1 multicast DA is also allowed for multipoint testing. TST is a one-way diagnostic function. If the diagnosis is performed in service, the repetition rate must not be disruptive of client layer traffic. If the diagnosis is performed out of service, the affected MEPs will initiate LCK messages. The TST OAM PDU includes a transaction ID/sequence number and also typically includes pseudorandom test data, which is checked for bit errors by the receiving MEP.

When an in-service ETH-Test function is performed, data traffic is not disrupted and the frames with ETH-Test information are transmitted in such a manner that limited part of the service bandwidth is utilized. MEP attributes to be configured to support ETH-Test are the following.

- MEG level.
- Unicast MAC address of the peer MEP for which ETH-Test is intended.
- Data pattern with an optional checksum. The length and contents of the data pattern are configurable at the MEP. The contents can be a test pattern and an optional checksum. Test patterns can be PRBS $(231-1)$ and all "0" pattern, etc. At the transmitting MEP, a test signal generator is configured. At a receiving MEP, a test signal detector is configured.
- Priority, which is configurable per operation, identifies the priority of frames with ETH-Test information.
- Drop eligibility that identifies the eligibility of frames with ETH-Test information to be dropped when congestion conditions are encountered.

Configuration attributes such as the transmission rate of ETH-Test information and the total interval of ETH-Test are undefined.

An MIP is transparent to the frames with ETH-Test information and therefore does not require any configuration information to support ETH-Test functionality.

An MEP inserts frames with ETH-Test information toward a targeted peer MEP at its configured MD level. The receiving MEP detects these frames with ETH-Test information and makes the intended measurements if the MD level matches. The PDU used for ETH-Test information is called TST, which supports bidirectional testing.

Each TST frame is transmitted with a specific sequence number. A different sequence number must be used for every TST frame, and no sequence number from the same MEP may be repeated within 1 min. When an MEP is configured for an out-of-service test, the MEP also generates LCK frames at the immediate client MEG level in the same direction where TST frames are transmitted.

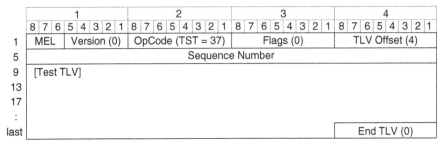

Figure 8.41   TST PDU format.

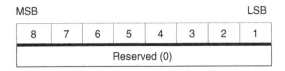

Figure 8.42   Flags format in TST PDU.

Information elements present in TST PDU (Figs 8.41 and 8.42) are:

- sequence number, which is a 4-octet field and
- test, which is an optional field whose length and contents are determined at the transmitting MEP. The contents of Test field indicate a test pattern and also carry optional checksum, as mentioned above.

8.19.2   RFC 2544

RFC 2544 [10] and RFC 1242 [20] define a procedure to test out-of-service throughput, latency (delay), and frame loss rate for a given port or an EVC.

Test setup consists of a tester with both transmitting and receiving ports. Connections are made from the sending ports of the tester to the receiving ports of the DUT (device under test) and from the sending ports of the DUT back to the tester (Fig. 8.43).

Frame sizes to be used in testing should include the maximum and minimum sizes and enough sizes in between to be able to get a full characterization of the DUT performance. At least five frame sizes are recommended to be used in each test condition.

For an MTU of 1522 bytes, frame sizes of 64, 128, 256, 512, 1024, 1280, and 1522 bytes should be considered. For larger frame sizes such as 4000 bytes and 9200 bytes, additional frame sizes should be used in the testing.

The stream of test frames should be augmented with 1% frames addressed to the hardware broadcast address to determine if there is any effect on the forwarding rate of the other data in the stream. The broadcast frames should be evenly distributed throughout the data stream, for example, every hundredth frame.

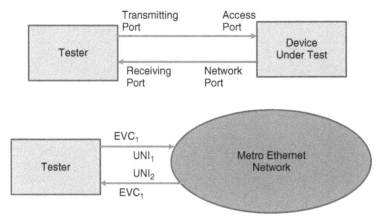

Figure 8.43 Out-of-service testing.

In order to test the bidirectional performance of a DUT, the test series should be run with the same data rate being offered from each direction.

The test equipment should include sequence numbers in the transmitted frames and check for these numbers on the received frames to identify the frames that were received out of order, duplicate frames received, the number of gaps in the received frame numbering sequence, and the number of dropped frames.

In order to simulate real traffic scenarios, some of the tests described below should be performed with both steady-state traffic and traffic consisting of repeated bursts of frames. The duration of the test should be at least 60 s.

For testing of the throughput, which is the fastest rate at which the count of test frames transmitted by the DUT, a specific number of frames at a specific rate through the DUT are sent and the frames transmitted by the DUT are counted. If the count of offered frames is equal to the count of received frames, fewer frames are received than were transmitted, the rate of the offered stream is reduced by a percentage of maximum rate and the test is rerun. The throughput test results can be plotted as frame size versus frame rate, as depicted in Figure 8.44.

The procedure for latency is first the throughput for DUT at each of the frame sizes mentioned above should be determined by sending a stream of frames at a particular frame size through the DUT at the determined throughput rate to a specific destination. The stream should be at least 120 s in duration.

An identifying tag should be included in one frame after 60 s. The time for frame transmission and receival is recorded. The latency is the difference between the frame arrival timer and transmission time.

The test must be repeated at least 20 times, with the reported value being the average of the recorded values. This test is performed for each of the test frames addressed to a new destination network.

For frame loss rate, the percentage of offered frames that are dropped are reported. The first trial is run for the frame rate that corresponds to 100% of the

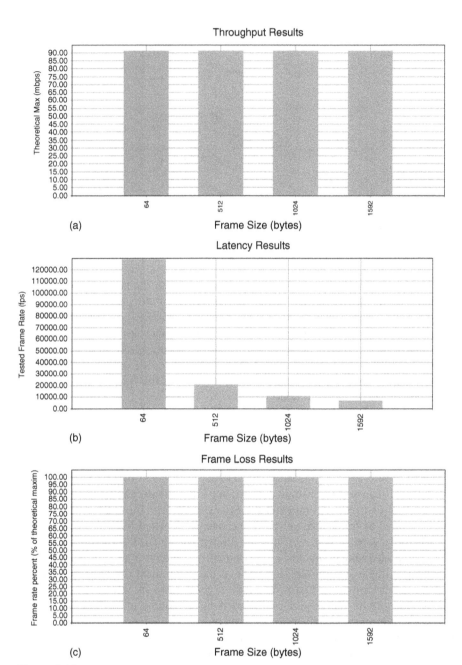

Figure 8.44   Examples of RFC 2544. (a) RFC 2544 throughput report for 100 Mbps EVC.
(b) RFC 2544 latency report for 100 Mbps point-to-point EVC. (c) RFC 2544 frame loss report for
100 Mbps point-to-point EVC.

maximum rate for the input frame size. The procedure is repeated for the rate that corresponds to 90% of the maximum rate used and then for 80% of this rate. This sequences is continued at reducing 10% intervals until there are two successive trials in which no frames are lost. Depending on the amount of time to be spent on testing, the maximum granularity of the trials can be chosen to be smaller than the 10%.

The frame loss rate at each point is calculated by ((input_count − output_count) × 100)/input_count.

The results of the frame loss rate test can be plotted as a graph where the x-axis is the input frame rate as a percentage of the theoretical rate for the given frame size and the y-axis is the percentage loss at the particular input rate.

In order to test DUT for back-to-back frames [20], a burst of frames with minimum interframe gaps to the DUT is sent and the number of frames forwarded by the DUT is counted. If the count of transmitted frames is equal to the number of frames forwarded, the length of the burst is increased and the test is rerun. If the number of forwarded frames is less than the number transmitted, the length of the burst is reduced and the test is rerun.

The back-to-back value is the number of frames in the longest burst that the DUT will handle without the loss of any frames. The trial length must be at least 2 s and should be repeated at least 50 times, with the average of the recorded values being reported.

At present, there are various applications that can generate back-to-back frames. Remote disk servers and data transfers between databases during disaster recovery are among these applications. Therefore, it is necessary to test the service for back-to-back frames.

8.19.3 Complete Service Testing

As given above, if a service provider wants to test an EVC before turning it up for the service, the likely test scenarios involved are the following:

- Verifying physical connectivity between CPE and the network.
- Looping back frames remotely at each user port of CPE, network port, and VLAN to ensure swapping of SA and DA in order to make loopback work.
- Running RFC 2544 testing per EVC basis to determine throughput, delay, and jitter, and tests must be run for various packet sizes and MTUs.
- Collecting PM parameters for EVC/OVC and ENNI/UNI.
- Setting threshold for frame loss and generate alarms.
- Verifying ENNI/UNI and EVC/OVC alarms:
  - User ports of CPE shall be disconnected and reconnected. Link Down and Link Up events shall be reported with appropriate severity such as critical for Link Down event.

- Network ports of CPE shall be disconnected and reconnected. Link Down and Link Up events should be reported with appropriate severity.
- EVC shall be taken down administratively and corresponding event shall be reported.
- Power cord for CPE shall be disconnected and reconnected. Dying Gasp alarm with appropriate severity should be reported.
- Tracing EVC path using ETH-LTM.
- In order to verify Y.1731/802.1ag capabilities for the given EVC, MD, MA, and Up MEP and Down MEP at customer interfaces and network interfaces of the CPE, verifying CCM capability by configuring CCM for 1-s interval for the EVC.
- Testing DMM/DMR and LMM/LMR capabilities.
- For TDM circuit, verifying data integrity using a standard 220—1 PRBS pattern.

8.20 SECURITY

Security of physical access for Metro Ethernet services is the same as that for any other wireline services. Users must identify themselves with their unique user IDs and password before performing any actions on the NE and NMS. The NE maintains the identity of all users logged on and logged off (or locks) a user ID after a specified time interval of inactivity, for example, 15 min.

If the NE can accept remote-operation-related commands, the user can be authenticated via public/private key technology.

At EVC level, SOAM traffic is not permitted to transit the MEPs of an MA at the same MD level or higher, while SOAM traffic at a higher MD level is transparent to the MA. This domain hierarchy provides a mechanism for protecting MPs from other MPs with which the MP has not been designed to communicate. However, it is possible for one MP to flood another with OAM PDUs, potentially resulting in a denial of service attack on an MP.

In order to make sure that network elements are not susceptible to a denial of service attack via OAM PDUs, it is necessary to rate-limit OAM PDU traffic at the local node.

8.21 OAM BANDWIDTH

An adequate OAM bandwidth for each EVC and each NE should be allocated to ensure timely response to problems in the network. MEF SOAM PM draft requires an NE that can process at least 10 times its EVC capacity in OAM PDUs per second.

In other words, for example, if an NE claims to support 1000 EVCs, it must be able to process at least 10,000 OAM PDUs per second, with any legal or illegal OAM PDU content.

Our rule of thumb is to allocate 1% bandwidth for OAM.

8.22   CONCLUSION

SOAM tools and procedures defined for Metro Ethernet services are the key to the broader deployment of the services, by saving substantial amount of operational expenses for service providers and allowing users to monitor its service in real time.

In this chapter, we covered most of these tools and procedures. New tools and procedures are being standardized in ITU and MEF.

REFERENCES

1. MEF 15, Requirements for Management of Metro Ethernet Phase 1 Network Elements, November 2005.
2. MEF 7.1, Phase 2 EMS-NMS Information Model, October, 2009.
3. IEEE Std 802.3ah-2004 IEEE Standard for Local and Metropolitan Area Networks—Specific requirements—Part 3: Carrier Sense Multiple Access With Collision Detection (CSMA/CD) Access Method and Physical Layer Specifications Amendment: Media Access Control Parameters, Physical Layers, and Management Parameters for Subscriber Access Networks.
4. IEEE 802.1ag-2007, Virtual Bridged Local Area Networks—Amendment 5: Connectivity Fault Management, December 2007.
5. ITU-T Recommendation G.8013/Y.1731, OAM Functions and Mechanisms for Ethernet based Networks, 2011.
6. MEF 31, Service OAM Fault Management Definition of Managed Objects, January 2011.
7. MEF 35 Service OAM Performance Monitoring Implementation Agreement, April 2012.
8. MEF 16, Ethernet Local Management Interface (E-LMI), January 2006.
9. RFC 2281, Cisco Hot Standby Router Protocol (HSRP), March 1998.
10. RFC 2544, Benchmarking Methodology for Network Interconnect Devices, 1999.
11. ITU-T Y.1564 Ethernet service activation test methodology, 3/25/2011
12. MEF 30, Service OAM Fault Management Implementation Agreement, January 2011.
13. ITU-T G.8010, Architecture of Ethernet Layer Networks, 2004.
14. ITU-T Y.1711, Operation & Maintenance Mechanism for MPLS networks, 2004.
15. RFC 3636, Definitions of Managed Objects for IEEE 802.3 Medium, September 2003.
16. ITU-T Q.840.1, Requirements and analysis for NMS-EMS management interface of Ethernet over Transport and Metro Ethernet Network (EoT/MEN), 2007.
17. MEF 10.2, Ethernet Services Attributes- Phase 2, October 2009
18. MEF 10.2.1, Performance Attributes Amendment to MEF 10.2, January 2011.
19. ITU-T Y.1563, Ethernet frame transfer and availability performance, 2009.
20. RFC 1242, Benchmarking Terminology for Network Interconnection Devices, 1991.
21. MEF 17, Service OAM Requirements and Framework—Phase 1, April 2007.
22. MEF 20, UNI Type II Implementation Agreement, June 2008.
23. MEF, Carrier Ethernet Service Activation Testing-Draft, July 2012.
24. MEF, Service Activation Testing Protocol Data Unit-Draft, July 2012.

Chapter 9

Circuit Emulation
Services (CES)

9.1 INTRODUCTION

A pseudowire is an emulation of a native service over a packet switched network (PSN). The native service may be low rate TDM, SDH/SONET, while the PSN may be Ethernet, MPLS, or IP. Especially, in mobile backhaul locations, equipment with TDM and SONET/SDH interfaces are common. The CESoETH (CES over Ethernet) is the solution to provide low cost Ethernet transport to these locations.

The CESoETH is offered at UNI. As operators begin offering CESoETH at locations outside their footprints, the service will be offered at ENNI as well.

In CES, data streams are converted into frames for transmission over Ethernet. At the destination site, the original bit stream is reconstructed when the headers are removed, payload is concatenated, and clock regenerated, while ensuring very low latency. The MEN behaves as a virtual wire that is an application layer function as described in Chapter 5.

When the TDM data is delivered at a constant rate over a dedicated channel, the native service may have bit errors, but data is never lost in transit. All PSNs suffer to some degree from packet loss, and this must be compensated when delivering TDM over a PSN.

The standard TDM pseudowire operating modes are CESoPSN, Structure-Agnostic *TDM* over Packet (SAToP), TDMoIP, and HDLCoPSN (HDLC emulation over PSN).

In SAToP [3], TDM is typically unframed. When it is framed or even channelized, the framing and channelization structure are completely disregarded by the transport mechanisms. In such cases, all structural overhead must be transparently transported with the payload data, and the employed encapsulation method provides no mechanisms for its location or utilization. On the other hand, structure-aware TDM transport may explicitly safeguard TDM structure.

Networks and Services: Carrier Ethernet, PBT, MPLS-TP, and VPLS, First Edition. Mehmet Toy.
© 2012 John Wiley & Sons, Inc. Published 2012 by John Wiley & Sons, Inc.

   TDM pseudowire over Ethernet (Ethernet CES) provides emulation of TDM services, such as $N \times 64$ kbps, T1, T3, and OC-n, across a MEN, but transfers the data across MEN, instead of a traditional circuit-switched TDM network. From the customer perspective, this TDM service is the same as any other TDM service.

   From the provider's perspective, two CES interworking functions (IWFs) are provided to interface the TDM service to the Ethernet network. The CES IWFs (Fig. 9.1) are connected via the MEN using point-to-point Ethernet Virtual Connections (EVCs) [1].

   The basic CES is a point-to-point constant bit rate (CBR) service, similar to the traditional leased line type of TDM service. However, service multiplexing can be performed (Fig. 9.2) ahead of the CES IWFs, such as aggregation of multiple emulated T1 lines into a single T3 or OC-3 link, creating a multipoint-to-point or even a multipoint to multipoint configuration.

   In the unstructured emulation mode (SAToP), a service is provided between two service-end-points that use the same interface type. Traffic entering the MEN on one end point leaves the network at the other and vice versa, transparently. The MEN must maintain the bit integrity, timing, and other client-payload-specific characteristics of the transported traffic, without causing any degradation. All the management, monitoring, and other functions related to that specific flow must be performed without changing or altering the service payload information or capacity.

   The unstructured emulation mode is suitable for leased line services or any other transfer-delay-sensitive (real-time) applications. For example, the DS1

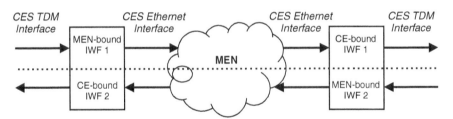

Figure 9.1   CES architecture.

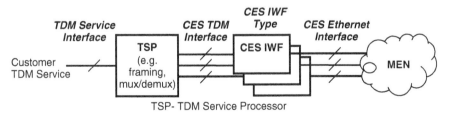

Figure 9.2   CES architecture with multiplexing.

unstructured CBR service is intended to emulate a point-to-point DS1 circuit, transparently carrying any arbitrary 1.544 Mbps data stream. The end-user timing source for these interface signals is not necessarily traceable to a *primary reference source* (PRS).

In structured emulation mode, a service provided between two service-endpoints use the same interface type. Traffic entering the MEN on one end-point is handled as overhead and payload. The MEN maintains the bit integrity, timing information, and other client-payload-specific characteristics of the transported traffic without causing any degradation that would exceed the requirements for the given service. All the management, monitoring, and other functions related to that specific flow must be performed without changing or altering the service payload information or capacity.

A fractional DS1 service, where the framing bit and unused channels are stripped, and the used channels transported across the MEN as an $N \times 64$ kbps service, is an example of structured CES.

MEF 8 [1] defines the CES IWF, which allows communication between non-Ethernet circuits (such as T1, E1, E3, and T3) and Ethernet UNI interfaces. CES is typically implemented on NID, but it can be implemented on edge routers as well.

In the simplest case, a pair of IWFs connected by an EPL (Ethernet Private Line) service. However, in many cases, several IWFs may share an Ethernet UNI and are connected via EVPL (Ethernet Virtual Private Line) service (Fig. 9.3).

CES is widely used in Metro Ethernet Backhaul where TDM/SONET is still a popular interface for base stations (BSs) and radio network controllers (RNCs), as depicted in Figure 9.4.

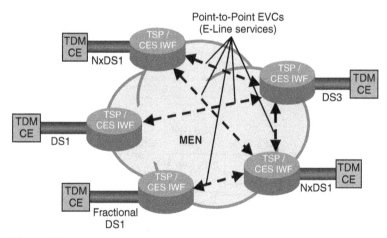

Figure 9.3   TDM virtual private line configurations.

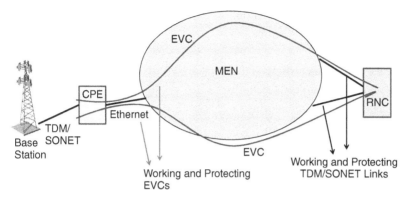

Figure 9.4 TDM over Ethernet for mobile backhaul.

9.2 CIRCUIT EMULATION FUNCTIONS

CES functional blocks are depicted in Figure 9.5. The TDM Service Processor (TSP) may terminate framing overhead or multiplex several customer TDM services into a single service to be emulated.

One or multiple CES IWFs may use one Ethernet interface. When there are multiple CES IWFs, the flows of the IWFs are multiplexed and demultiplexed using the ECDX (Emulated Circuit De/Multiplexing Function). The output of ECDX, which is in the packet domain, interfaces with the EFTF (Ethernet Flow Termination Function) that handles the MAC layer functions.

The CES IWF is the adaptation function that interfaces the CES application to the Ethernet layer, by handling emulation of the service presented at the CES TDM interface. The CES IWF is responsible for all the following functions required for the emulated service:

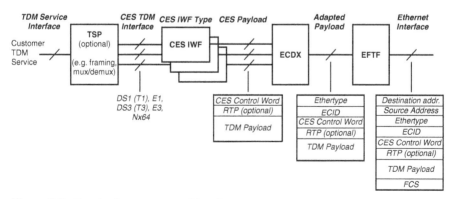

Figure 9.5 Functional components and interface types.

- encapsulation and decapsulation
- payload formation and extraction
- synchronization
- carriage of TDM signaling and alarms
- error response and defect behavior
- TDM performance monitoring.

The interfaces to the CES IWF can be DS1, E1, DS3, E3, or $N \times 64$ kbps TDM interfaces or packetised TDM payload, CES control word, and RTP header.

The ECDX prepends an Emulated Circuit Identifier (ECID) and assigns the Ethertype field (0x88d8) to every Ethernet frame sent to the MEN. In the TDM-bound direction, the ECDX determines the destination CES IWF of each Ethernet frame from the ECID value and strips the Ethertype and ECID fields, before handing off the CES payload to the CES IWF.

The interfaces to the ECDX consist of the following:

- CES payload (i.e., packetized TDM payload, CES control word, optional RTP header) [4] and
- Adapted payload (i.e., the CES Payload, plus the ECID and Ethertype fields).

EFTF takes an adapted payload from the ECDX and then creates an Ethernet Service Frame by adding MAC Destination and Source addresses, optional VLAN tag (if required) and associated Tag ID and User Priority information, any padding required to meet the minimum Ethernet frame size, and the frame check sequence (FCS).

In the TDM-bound direction, the EFTF takes in an Ethernet frame from the MEN and checks the FCS, discarding the frame if it is incorrect. It determines whether it contains CES payload from the Ethertype field, and forwards it to its associated ECDX function, for passing to the appropriate CES IWF.

9.3 ADAPTATION FUNCTION HEADERS

CES over Ethernet adaptation function operates directly on top of the Ethernet layer. There are three components in the adaptation function header [1]:

- **ECID** that identifies the emulated circuit being carried is added by ECDX. This separates the identification of the emulated circuit from the Ethernet layer, allowing the MEN operator to multiplex several emulated circuits across a single EVC where required.

  The ECID consists of a single, 20-bit unsigned binary field containing an identifier for the circuit being emulated. This is added by the ECDX and shown in Figure 9.6.

  ECIDs have local significance only and are associated with a specific MAC address. Therefore, an emulated circuit may be given different

0 19	20 31
Emulated Circuit Identifier (ECID) (20 bits)	Reserved (set to 0x102)

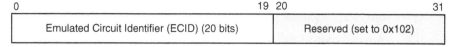

Figure 9.6 Emulated circuit identifier (ECID).

ECIDs for each direction of the circuit. ECIDs are selected during the creation of an emulated circuit.

Length of ECID is the same as an MPLS label size to ease interworking with an MPLS-based circuit emulation service.

- **CESoETH control word** providing sequencing and signaling of defects such as alarm indication signal (AIS) of the TDM circuit, or packet loss detected in the MEN, is added by the CES IWF. The CESoETH control word added by the TDM-bound IWF (Fig. 9.7) consists of the following.
  - 'L' bit indicating Local TDM failure impacting TDM data when set. For structure-agnostic emulation, an MEN-bound IWF sets 'L' bit to one when loss of signal (LOS) is detected on the TDM service interface. For structure-aware emulation, an MEN-bound IWF sets the 'L' bit to one where the TDM circuit indicates LOS, AIS, or loss of frame alignment (LOF). An MEN-bound IWF clears the 'L' bit as soon as the defect condition is rectified.

    For structure-agnostic emulation, on reception of CESoETH frames marked with the 'L' bit set to one, the TDM-bound IWF discards the payload and plays out AIS code for the scheduled duration of the CESoETH frame.

    For structure-aware emulation, on reception of CESoETH frames marked with the 'L' bit set to one, the TDM-bound IWF discards the payload and plays out AIS for the scheduled duration of the CESoETH frame.
  - 'R' bit representing **Remote Loss of Frames** (LOFs) *indication* when set by a MEN-bound IWF. In other words, it indicates that its local TDM-bound IWF is not receiving frames from the MEN and consequently, has entered a Loss of Frames State (LOFS), perhaps due to congestion or other network-related faults.

    A TDM-bound IWF enters a LOFS following the detection of a locally preconfigured number of consecutive lost (including late frames that are discarded) CESoETH frames and exits LOFS following reception of a locally preconfigured number of consecutive CESoETH frames. An MEN-bound IWF sets the 'R'

0 1 2 3	4	5	6 7	8 9	10 15	16 31
Reserved set to zero	L	R	M (2 bits)	FRG (2 bits)	LEN (Length) (6 bits)	SN (Sequence Number) (16 bits)

Figure 9.7 Structure of the CESoETH control word.

bit to 1 on all frames transmitted into the MEN, while its local TDM-bound IWF is in the LOFS.

On detection of a change in state of the 'R' bit in incoming CESoETH frames, a TDM-bound IWF reports it to the local management entity.

– 'M' bits that are called **Modifier bits and** set by the MEN-bound IWF to supplement the meaning of the L bit, as shown in Table 9.1: When an MEN-bound IWF is not capable of detecting the conditions described in Table 9.1, it clears the 'M' field to zero on frames to be transmitted into the MEN. On the other hand, a TDM-bound IWF discards a CESoETH frame, where the 'M' field is set to a value that it does not support.

– FRG **(Fragmentation) bits** that are used for fragmenting multiframe structures into multiple CESoETH frames (Table 9.2).

– LEN **(Length)** is an unsigned binary number indicating the length of the payload that is the sum of the size of the CESoETH control word (4 octets), size of the optional RTP header (12 octets, if present),

Table 9.1 Meaning of the Local TDM Failure Modification Bits

L Bit 4	M		Indicates
	Bit 6	Bit 7	
0	0	0	No local TDM defect detected.
0	1	0	Receipt of RDI at the TDM input to the MEN-bound IWF. When this indication is received, a TDM-bound IWF may generate RDI in the local TDM circuit. This is applicable only to structure-aware emulation.
0	1	1	CESoETH frames containing non-TDM data such as signaling frames.
1	0	0	A TDM defect that triggers AIS generation at the TDM-bound IWF, such as LOS or LOF in structure-aware operation.

Table 9.2 Meaning of the Fragmentation Bits

FRG Bit 8	Bit 9	Indicates
0	0	Entire multiframe structures are carried in a single CESoETH frame or no multiframe structure is being indicated
0	1	Packet carrying the first fragment of the multiframe structure
1	0	Packet carrying the last fragment of the multiframe structure
1	1	Packet carrying an intermediate fragment of the multiframe structure

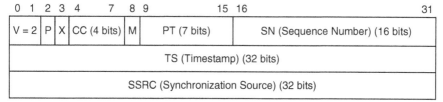

0 1 2 3 4   7 8 9   15 16   31
V = 2 | P | X | CC (4 bits) | M | PT (7 bits) | SN (Sequence Number) (16 bits)
TS (Timestamp) (32 bits)
SSRC (Synchronization Source) (32 bits)

Version number (V) is always be set to 2
Padding field (P) is always be set to 0
Header extension (X) is always be set to 0
CSRC (contributing source) count (CC) field is always be set to 0
Marker field (M) is always be set to 0.
PT = Payload Type

Figure 9.8   RTP header structure.

and size of the TDM payload. The LEN does not include the size of ECID.

A nonzero LEN indicates the presence of padding. LEN is set to zero where the length equals or exceeds 42 octets.

−SN (**sequence number**) is a 16 bit unsigned binary number that increments by one for each frame transmitted into the MEN with the same ECID, including those frames that are fragments of multiframe structures. The receiving IWF primarily uses it to detect frame loss and to restore frame sequence. The initial value of the SN is random.

• *Optional RTP* (real-time transport protocol) **header** providing timing and sequencing where appropriate. CESoETH uses the fields of the RTP header as described in Figure 9.8.

The RTP header in CESoETH can be used in conjunction with at least the following modes of timestamp generation:

−**Absolute mode**: The MEN-bound IWF sets timestamps using the clock recovered from the incoming TDM circuit. As a consequence, the timestamps are closely correlated with the sequence numbers. All CESoETH implementations supporting RTP support this mode.

−**Differential mode**: The two IWFs at either end of the emulated circuit have access to the same high quality synchronization source, and this synchronization source is used for timestamp generation.

9.4 SYNCHRONIZATION

In any circuit emulation scheme, the clock that is used to play out the data at the TDM-bound IWF must be the same frequency as that used to input the data at the MEN-bound IWF. Unsynchronized clocking will result in frame slips.

For clocking, the TDM-bound IWF can use

• TDM line timing, which is the clock from the incoming TDM line;

- external timing, which is the clocking from an external reference clock source;
- free run timing, which is the clocking from a free-running oscillator;
- Ethernet line timing, which is recovering the clock from the Ethernet interface.

Ethernet line timing covers all methods where information is extracted from the Ethernet, including

- adaptive timing, where the clock is recovered from data in the CESoETH frames, and the arrival time of the frames;
- differential timing, where the clock is recovered from a combination of data contained in the CESoETH frames, and knowledge of a reference clock common to both the MEN-bound and TDM-bound IWFs. Such a reference may be distributed by a variety of means (Fig. 9.9).

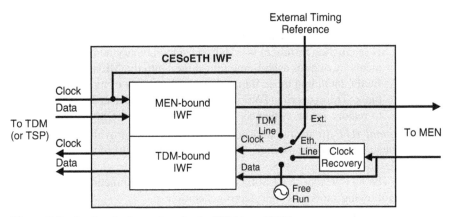

Figure 9.9 Synchronization options for the TDM-bound IWF.

9.5 TDM APPLICATION SIGNALING

Circuit Emulation applications interconnected using a CESoETH service may exchange signaling in addition to TDM data. The typical example is telephony applications that exchange their off-hook/on-hook state in addition to TDM data carrying PCM-encoded voice.

With structure-agnostic emulation, it is not required to intercept or process CE signaling. Signaling is embedded in the TDM data stream, and hence it is carried end-to-end across the emulated circuit.

With structure-aware emulation, transport of common channel signaling (CCS) may be achieved by carrying the signaling channel with the emulated service such as channel 23 for DS1. However, Channel associated signaling (CAS) such as DS1 robbed bit signaling requires knowledge of the relationship of the timeslot to the interface multiframe structure, which is indicated by the framing bits.

A generic method for extending the $N \times 64$ kbps basic service by carrying CE signaling (CAS or CCS) in separate signaling packets is independent of the TDM circuit type.

CESoETH data frames and their associated signaling frames have the same destination MAC address, Ethertype, and Usage of the RTP header. CESoETH data frames and their associated signaling frames use different ECID values. On the other hand, CESoETH data frames and their associated signaling frames use a separate sequence number space.

When the RTP header is used,

- data frames and associated signaling frames use a different payload type value (both allocated from the range of dynamically allocated RTP payload types);
- data frames and associated signaling frames use a different SSRC value;
- the timestamp values for the data and associated signaling frames **are** identical at any given time.

In the case where the CE application interconnected by a basic $N \times 64$ kbps CESoETH service is a telephony application using CAS, the payload of each signaling frame consists of N 32-bit words, where N is the number of timeslots in the service. The i-th word of the payload contains the current "ABCD" value of the CAS signal corresponding to the i-th timeslot (Fig. 9.10).

Signaling frames **are** sent three times at an interval of 5 ms on any of the following events:

- Setup of the emulated circuit.
- A change in the signaling state of the emulated circuit.
- LOFs defect has been cleared.
- Remote Loss of Frames indication has been cleared.

A signaling frame **is** sent every 5 s in the absence of any of the events, except when there is a failure in the local TDM circuit leading to the L flag being set in the associated data frames.

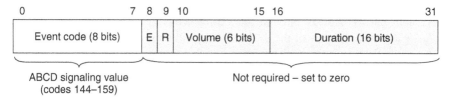

Figure 9.10   Encoding format for CAS "ABCD" values.

9.6   CESoETH DEFECTS AND ALARMS

CESoETH defects can be caused by the TDM interface interworking or adaptation section and EVC section provided by MEN. Defects caused by the TDM interface are LOS, AIS, and LOF. Defects caused by IWF are malformed frames, jitter buffer overrun and underrun. Defects caused by MEN are misconnection of CESoETH frames, reordering and LOFs, and late arriving CESoETH frames.

Misconnection is caused by wrongly directed frames. They are discarded. If the percentage of stray frames persists above a defined level for a configurable period of time, 2.5 s by default, the misconnection alarm is reported.

Detection of out-of-sequence or lost CESoETH frames is accomplished by use of the sequence number, from either the RTP header or the CESoETH control word. Out-of-sequence frames are reordered where they arrive in time to be played out. Out-of-sequence CESoETH frames that cannot be reordered **are** discarded.

If loss of one or more CESoETH frames is detected by the TDM-bound IWF, it generates exactly one "replacement octet" for every lost octet of TDM data. If the FLR (frame loss ratio) is above a defined threshold for a configurable period of time, 2.5 s by default, a LOF alarm is generated. The LOF alarm **is** cleared if FLR remains below a defined threshold for a configurable period of time, 10 s by default.

When the frames arrive after their scheduled playout time (i.e., Late Arriving Frames), even the jitter buffer is not full, they are considered as lost frames.

If the percentage of frames arriving too late to be played out exceeds a defined level for a configurable period of time, 2.5 s by default, a Late Frame alarm is generated.

CESoETH frames can be malformed if the PT value in its RTP header does not correspond to one of the PT values allocated for this direction of the emulated circuit. If a malformed frame is received by the TDM-bound IWF in time to be played out to the TDM interface, it discards the malformed frame and generates exactly one "replacement octet" for every lost octet of TDM data. If the percentage of malformed CESoETH frames is above a defined level for a configurable period of time, 2.5 s by default, a Malformed Frames alarm is generated. The Malformed Frames alarm **is** cleared if no malformed packets have been detected for a configurable period of time, 10 s by default.

The TDM-bound IWF contains a "jitter buffer" that accumulates data from incoming CESoETH frames to smooth out variation in arrival time of the CESoETH frames. Data is played out of the jitter buffer to the TDM service at a constant rate. The delay through this buffer needs to be as small as possible, in order to reduce latency of the TDM service, but large enough to absorb IFDV (interframe delay variation). The *jitter buffer overrun* condition occurs when the jitter buffer at the TDM-bound IWF cannot accommodate the newly arrived valid CESoETH frame because of insufficient buffer space.

The *jitter buffer underrun* condition occurs when there is no correctly received CESoETH payload ready to be played out on the TDM interface as a result of frames lost in MEN or discarded because of error conditions, and replacement data must be played out instead.

If a CESoETH frame arrives that cannot be stored in the jitter buffer because of a jitter buffer overrun condition, the TDM-bound IWF discards the frame. If the percentage of frames causing jitter buffer overruns is above a defined level for a configurable period of time, 2.5 s by default, a jitter buffer overrun alarm is generated. The jitter buffer overrun alarm **is** cleared if no cases of jitter buffer overrun have been detected for a configurable period of time, 10 s by default.

9.7 PERFORMANCE MONITORING OF CESoETH

CESoETH service supporting DS1 circuit using ESF framing may monitor messages carried in the FDL (Facility Data Link).

All events in the TDM-bound IWF, which lead to replacement data being played out (except when a result of receiving a CESoETH packet with an AIS or Idle Code indication), result in errored seconds or potentially severely errored seconds.

The collective sum of all these errors can be aggregated into Frame Error Ratio (FER), as the number of errored (i.e., lost or discarded) CESoETH frames, over the total number of CESoETH frames sent. This includes situations where the CESoETH frame fails to arrive at the TDM-bound IWF (i.e., lost in MEN), arrives too late to be played out, arrives too early to be accommodated in the jitter buffer, and arrives with bit errors causing the frame to be discarded.

An MEN-bound CES IWF maintains the number of CESoETH frames transmitted and number of payload octets transmitted. A TDM-bound CES IWF maintains the number of CESoETH frames received, number of payload octets received, number of lost frames detected, number of frames received that are out-of-sequence but successfully reordered, number of transitions from the normal to the LOFS, number of malformed frames received, number of jitter buffer overruns, and number of jitter buffer underruns.

9.8 CESoETH SERVICE CONFIGURATION

CESoETH is offered at UNI and ENNI interfaces via E-Line services. For the CESoETH, UNI, EVC, ENNI, and OVC attributes are configured. In addition, the following parameters need to be assigned when an emulated circuit is set up.

- Service type such as structure-agnostic DS1 and unstructured DS1.
- TDM Bit Rate that is applicable to structure-aware $N \times 64$ kbps services. The bit rate is defined by N. It is the same for each direction of the emulated circuit.
- *Payload size* that is defined as the number of octets of TDM payload carried in each CESoETH frame. It is the same for each direction of the emulated circuit.

- ECID.
- TDM clocking option for the TDM-bound IWF.
- Whether the RTP header has to be used.
- Whether the generic TDM application signaling method has to be used.
- Whether the control word has to be extended from 32 bits to 64. When an extended control word is selected, the extension applies to both directions of the emulated circuit.

If RTP is used, the following additional parameters need to be assigned:

- Payload type.
- Timestamp mode, absolute or differential.
- Timestamp resolution parameter that encodes the bit rate of the clock used for setting the timestamps in RTP headers as a multiple of the basic 8 kHz rate.
- SSRC value.

If generic TDM signaling is used, ECID for signaling frames and payload type are to be set.

In addition to parameters mentioned above, control word and defect and alarm parameters need to be configured.

9.9 CONCLUSION

CESoETH is one of the key Metro Ethernet Services that is widely deployed at mobile backhaul networks. Synchronization of both ends is necessary for the service to work. Delay, jitter, and loss for MBH [2] are also tight.

In this chapter, we described the service, how it works, and the service parameters.

REFERENCES

1. MEF 8. Implementation Agreement for the Emulation of PDH Circuits over Metro Ethernet Networks; 2004 Oct.
2. MEF 22. Mobile Backhaul Implementation Agreement Phase 1; 2009 Jan.
3. RFC 5086. Structure-Aware Time Division Multiplexed (TDM) Circuit Emulation Service over Packet Switched Network (CESoPSN); 2007 Dec.
4. RFC 3550. A Transport Protocol for Real-Time Applications; 2003 July.

Chapter 10

Carrier Ethernet Local Management Interface

10.1 INTRODUCTION

Ethernet local management interface (ELMI) protocol defined in Reference 1 operates between the customer edge (CE) device and network edge (NE) of service provider. It defines the protocol and procedures allowing autoconfiguration of the CE (customer equipment) device by the service provider's NE device (Fig. 10.1). The ELMI protocol also provides the means of notification of the status of an EVC and UNI, and enables the CE to request and receive status and service attributes information [2] from the MEN, so that it can configure itself to access Metro Ethernet services.

ELMI is a protocol that is terminated by the UNI-C on the CE side of the UNI and by the UNI-N on the MEN side of the UNI (Fig. 10.2).

The ELMI protocol includes the following procedures:

- Notification to the CE device of the addition of an EVC: the respective CPEs are informed of the availability of a new EVC once the service provider turns on the service. Furthermore, the service end points are notified of the corresponding CE-VLAN ID to be used by a given service.
- Notification to the CE device of the deletion of an EVC.
- Notification to the CE device of the availability (active/partially active) or unavailability (inactive) state of a configured EVC, so that the CE device can take some corrective action, such as rerouting traffic to a different EVC or other WAN service, when informed that an EVC has become inactive.
- Notification to the CE device of the availability of the Remote UNI so that the CE device can take some corrective action, such as rerouting traffic to a different EVC or other WAN service, when informed that the remote UNI is down.

Networks and Services: Carrier Ethernet, PBT, MPLS-TP, and VPLS, First Edition. Mehmet Toy.
© 2012 John Wiley & Sons, Inc. Published 2012 by John Wiley & Sons, Inc.

- Communication of UNI and EVC attributes to the CE device.
  - **(a)** The network informs the CE device as to which VLAN ID is used to identify each EVC to avoid the possibility of a VLAN mismatch between the service provider and CE.
  - **(b)** Remote UNI identification to ensure that the right end points have been connected by an EVC
  - **(c)** Autoconfiguration of bandwidth profiles by CE device.

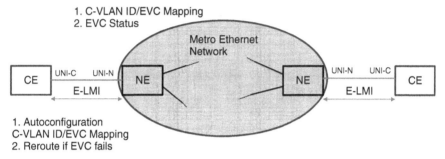

Figure 10.1 Ethernet local management interface.

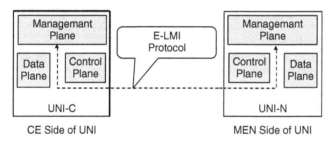

Figure 10.2 ELMI termination in functional components.

Destination Address	Source Address	E-LMI Ethertype	E-LMI PDU (message)	CRC
6 Octets	6 Octets	2 Octets	46–1500 Octets (Data + Pad)	4 Octets

Figure 10.3 ELMI framing structure.

In order to transfer ELMI messages between the UNI-C and the UNI-N, a framing or encapsulation mechanism is needed. The ELMI frame structure is based on the IEEE 802.3 untagged MAC-frame format where the ELMI messages are encapsulated inside Ethernet frames. The ELMI framing structure is presented in Figure 10.3.

The destination address and ELMI Ethertype must be 01-80-C2-00-00-07 and 88-EE, respectively. (01-80-C2-00-00-07) DA can be used if there is no 802.1Q complaint component between UNI-C and UNI-N.

EVC Status can be New, Active, Not Active, and Partially Active:

- **New**—EVC has just been added to the CE-VLAN ID/EVC Map.
- **Active**—EVC is in the CE-VLAN ID/EVC Map and fully operational between the UNIs in the EVC.
- **Not Active**—EVC is in the CE-VLAN ID/EVC Map but not capable of transferring traffic among any of the UNIs in the EVC.
- **Partially Active**—It is applicable for Multipoint-to-Multipoint EVCs. When a Multipoint-to-Multipoint EVC is "Partially Active," it is capable of transferring traffic among some but not all of the UNIs in the EVC.

Tables 10.1 and 10.2 detail the possible combinations of point-to-point EVC status and Multipoint-to-Multipoint EVC status, respectively.

The ELMI protocol employs two messages, STATUS and STATUS ENQUIRY. Every message of the ELMI protocol consists of (i) Protocol Version, (ii) Message Type, (iii) Report Type, and (iv) Other information elements (IEs) and subinformation elements. The ELMI message parts 0, 0, and 0 are common to all the ELMI messages where each message may have additional information and subinformation elements. The example of E LMI message organization and ELMI protocol IEs are shown in Figure 10.4 and Table 10.3, respectively.

The values of the subinformation elements used for the ELMI protocol are shown in Table 10.4.

The coding of the IEs other than Protocol Version and Message Type is detailed in Reference 1.

Table 10.1   Possible Status Combinations for a Point-to-Point EVC

New	Active	Not Active
✓	✓	
✓		✓
	✓	
		✓

Table 10.2   Possible Status Combinations for a Multipoint-to-Multipoint EVC

New	Active	Not Active	Partially Active
✓	✓		
✓		✓	
✓			✓
	✓		
		✓	
			✓

8	7	6	5	4	3	2	1	Octet
Protocol Version information element								1
Message Type information element								2
Report Type								3, 4, 5

Other information and sub-information elements as required · 6, ...

Figure 10.4   Example of General ELMI message organization.

Table 10.3   Information Element Identifiers

Information Element	Identifier
	Bits
	8 7 6 5 4 3 2 1
Protocol Version	Not applicable
Message Type	Not applicable
Report Type	0 0 0 0 0 0 0 1
Sequence Numbers	0 0 0 0 0 0 1 0
Data Instance (DI)	0 0 0 0 0 0 1 1
UNI status	0 0 0 1 0 0 0 1
EVC status	0 0 1 0 0 0 0 1
CE-VLAN ID/EVC map	0 0 1 0 0 0 1 0

Table 10.4   Subinformation Element Identifiers

Subinformation element	Identifier
	Bits
	8 7 6 5 4 3 2 1
UNI identifier	0 1 0 1 0 0 0 1
EVC parameters	0 1 1 0 0 0 0 1
EVC identifier	0 1 1 0 0 0 1 0
EVC map entry	0 1 1 0 0 0 1 1
Bandwidth profile	0 1 1 1 0 0 0 1

10.2 ELMI MESSAGES

STATUS message is sent by the UNI-N to the UNI-C in response to a STATUS ENQUIRY message to indicate the status of EVCs or for the exchange of sequence numbers. It may be sent without a STATUS ENQUIRY to indicate the status of a single EVC.

The STATUS message described in Table 10.2 can include EVC Service Attributes and Parameters that enable automatic configuration of CE devices based on the network configuration.

EVC Status IEs are arranged in the message in ascending order of EVC Reference IDs; that is, the EVC with the lowest EVC Reference ID is first, the second lowest EVC Reference ID second, and so on. If all IEs cannot be sent in a single Ethernet frame, more STATUS messages **are** sent with Type of Report *Full Status Continued*. The asynchronous STATUS message **must** contain a Single EVC Status IE that precedes the CE-VLAN ID/EVC IE.

The IEs carried in the STATUS message are Sequence Number, Data Instance (DI), UNI Status, EVC Status, and CE-VLAN ID/EVC Map. The STATUS message can be FULL or asynchronous. In a full STATUS message, the UNI-C uses the EVC Status IE to detect a change in status of configured EVCs. When the UNI-C sends a STATUS ENQUIRY message with a Report Type of *Full Status*, the UNI-N responds with a STATUS message containing an EVC Status IE for each EVC configured at that UNI. Each EVC Status IE contains an active bit and a partially active bit indicating the availability or unavailability of that EVC. For a Point-to-Point EVC, the EVC status is "Active" if and only if the active bit is set to 1. For a Multipoint-to-Multipoint EVC, the status of the EVC is defined as in Table 10.5.

If the UNI-C receives an EVC Status IE indicating that the EVC is Not Active, the CE stops transmitting frames on the EVC until the UNI-C receives an EVC Status IE for that EVC indicating a status of Active or Partially Active.

Since there is a delay between the time the MEN makes an EVC available and the time the UNI-N transmits an EVC Status IE notifying the UNI-C, there is a possibility of the CE receiving frames on an EVC marked as Not Active.

Similarly, since there is a delay between the time the MEN detects that an EVC has become Not Active or Partially Active and the time UNI-N transmits an EVC Status IE notifying the UNI-C, there is a possibility of the MEN receiving frames on a Not Active or Partially Active EVC and drop of frames.

Asynchronous Status is used to notify the UNI-C that the EVC has changed status without waiting for a request from UNI-C. The UNI-N uses the EVC Status IE to inform the UNI-C about a change in status of a configured EVC. This STATUS message may be sent when the UNI-N detects the EVC status change, and the Report Type is set to *Single EVC Asynchronous Status*.

STATUS ENQUIRY is sent by the UNI-C to request status or to verify sequence numbers. The UNI-C must send a STATUS message in response to a STATUS ENQUIRY message. The STATUS ENQUIRY message consists of Protocol Version, Message Type, Report Type, Sequence Numbers, and DI. Brief descriptions of each message element are given in Section 10.3.

Table 10.5 Status of a Multipoint-to-Multipoint EVC

Active Bit	Partially Active Bit	Status
1	0	Active
0	0	Not Active
0	1	Partially Active
1	1	Not Defined

10.3 ELMI MESSAGE ELEMENTS

Protocol Version, which is 1-octet field, indicates the version supported by the sending entity (UNI-C or UNI-N).

Message Type, which is the second part of every message, identifies the function of the ELMI message being sent. Its coding is coded as in Table 10.6. Bit 8 is reserved for possible future use as an extension bit.

Report Type IE indicates the type of enquiry requested when included in a STATUS ENQUIRY message or the contents of the STATUS message. The length of this IE is 3 octets (Table 10.7).

Sequence Numbers IE exchanges sequence numbers between the UNI-N and the UNI-C on a periodic basis. This allows each protocol entity to detect if it has not received messages and to acknowledge receipt of messages to the other entity. The length of this IE is 4 octets (Table 10.8).

CE-VLAN ID/EVC Map IE conveys how CE-VLAN IDs are mapped to specific EVCs (Table 10.9). The maximum number of bytes needed to carry this IE depends on the number of VLAN IDs mapped to an EVC. When the number of octets needed exceeds the maximum length that can be specified in the TLV length octet (255), this IE can be repeated for the same EVC.

The EVC Reference ID allows the UNI-C to correlate information received in the CE-VLAN ID/EVC Map IE and the EVC Status IE to the same EVC. It is a binary encoded number in the range 0–65,535. The EVC Reference ID is

Table 10.6 Message Type Coding

Bits	Message Type
8 7 6 5 4 3 2 1	
0 1 1 1 1 1 0 1	STATUS
0 1 1 1 0 1 0 1	STATUS ENQUIRY

Table 10.7 Report Type Coding

Bits	Report Type
8 7 6 5 4 3 2 1	
0 0 0 0 0 0 0 0	Full Status
0 0 0 0 0 0 0 1	ELMI Check
0 0 0 0 0 0 1 0	Single EVC Asynchronous Status
0 0 0 0 0 0 1 1	Full Status Continued

Table 10.8 Sequence Numbers Information Element

8	7	6	5	4	3	2	1	Octet
Sequence Numbers information element identifier per Table 10.3								1
Length of **Sequence Numbers** contents (= 00000010)								2
Send sequence number								3
Receive sequence number								4

Table 10.9 CE-VLAN ID/EVC Map Information Element

8	7	6	5	4	3	2	1	Octet
CE-VLAN ID/EVC Map information element identifier per Table 10.3								1
Length of **CE-VLAN ID/EVC Map** information element								2
EVC reference ID								3
EVC reference ID—continue								4
Reserve 0	Last IE	CE-VLAN ID/EVC Map Sequence #						5
Reserve 0						Untagged /priority tagged	Default EVC	6
EVC map entry Subinformation element								7–10

Table 10.10 UNI Status Information Element

8	7	6	5	4	3	2	1
UNI Status information Element Identifier as per Table 10.3							
Length of **UNI Status** Information Element Contents							
CE-VLAN ID/EVC Map Type							
Bandwidth Profile Sub-Information Element							
UNI Identifier Sub-Information Element							

Table 10.11 CE-VLAN ID/EVC Map Type Coding

Bits	CE-VLAN ID/EVC Map Type
8 7 6 5 4 3 2 1	
0 0 0 0 0 0 0 1	All to one bundling
0 0 0 0 0 0 1 0	Service multiplexing with no bundling
0 0 0 0 0 0 1 1	Bundling

locally significant, which means that a given EVC can have a different value of EVC Reference ID at each UNIs in the EVC. If the sequence number exceeds 6 bits counter, it rolls over to zero.

UNI Status IE conveys the status and other relevant UNI service attributes of the UNI as defined in MEF UNI 10.2 [2]. This IE cannot be repeated in a STATUS message. The length of this IE depends on the number and size of UNI identifier subinformation element (Tables 10.10 and 10.11).

EVC Status IE (Tables 10.12 and 10.13) conveys the status and attributes of a specific EVC on the UNI. This IE can be repeated as necessary to indicate the status of all configured EVCs on the UNI.

DI IE (Table 10.14): DI reflects the current state of UNI and EVC information that is active on the UNI-N and UNI-C. Whenever there is a mismatch in DI, it is time to exchange UNI and EVC information between UNI-N and UNI-C. The format is shown in Table 10.14.

Bandwidth profile subinformation element (Table 10.15) conveys the characterization of the length and arrival of a sequence of the Service Frames at a

Table 10.12 EVC Status Information Element

8	7	6	5	4	3	2	1	Octet
EVC Status information element identifier per								
Information element				**Identifier**				
				Bits 8 7 6 5 4 3 2 1				
Protocol Version				Not applicable				
Message Type				Not applicable				1
Report Type				0 0 0 0 0 0 0 1				
Sequence Numbers				0 0 0 0 0 0 1 0				
Data Instance (DI)				0 0 0 0 0 0 1 1				
UNI Status				0 0 0 1 0 0 0 1				
EVC Status				0 0 1 0 0 0 0 1				
CE-VLAN ID/EVC Map				0 0 1 0 0 0 1 0				
Table								
Length of **EVC Status** information element								2
EVC Reference ID								3
EVC Reference ID (continue)								4
EVC Status Type								
Reserve 0				Reserve 0	Partially Active	Active	New	5
EVC parameters subinformation element								6
EVC ID subinformation element								7
Bandwidth Profile subinformation element								8

Table 10.13 EVC Status Coding

Bits 3 2 1	EVC Status
0 0 0	Not Active
0 0 1	New and Not Active
0 1 1	New and Active
0 1 0	Active
1 0 0	Partially Active
1 0 1	New and Partially Active

reference point UNI. This subinformation element is included in the UNI Status and EVC Status IEs. It can be repeated, up to eight times in an EVC Status IE, when there are eight per CoS identifier bandwidth profiles.

The coding of the various fields in the bandwidth profile subinformation element are shown in Table 10.16.

Table 10.14   Data Instance Information Element Format

8	7	6	5	4	3	2	1	Octet
Data Instance information element identifier								1
Length of **Data Instance** information element contents (00000101)								2
Reserved 0								3
Data Instance								4
Data Instance—continue								5
Data Instance—continue								6
Data Instance—continue								7

Data Instance: Any integer value packed in 4 bytes. Value "0x00000000" is unique and will be used by UNI-C to send its first message to UNI-N. Value "0x00000000" is never sent by the UNI-N.

Table 10.15   Bandwidth Profile Subinformation Element

8	7	6	5	4	3	2	1	Octet
Bandwidth profile subinformation element								1
Length of **Bandwidth profile** subinformation element contents (=00001100)								2
Reserve 0						CM	CF	Per CoS bit	3
CIR magnitude								4
CIR multiplier in binary. CIR = (CIR multiplier) * $10^{(\text{CIR magnitude})}$ (kbps).								4.1 / 4.2
CBS magnitude								5
CBS multiplier CBS = (CBS multiplier) * $10^{(\text{CBS magnitude})}$ (KB).								5.1
EIR magnitude								6
EIR multiplier in binary. EIR = (EIR multiplier) * $10^{(\text{EIR magnitude})}$ (kbps)								6.1 / 6.2
EBS magnitude								7
EBS multiplier EBS = (EBS multiplier) * $10^{(\text{EBS magnitude})}$ (KB)								7.1
user_priority bits 111	user_priority bits 110	user_priority bits 101	user_priority bits 100	user_priority bits 011	user_priority bits 010	user_priority bits 001	user_priority bits 000	8

Table 10.16 Coding in Bandwidth Profile Subinformation Element

Field Name	Value	Meaning
Per CoS bit (octet 3, bit 1)	0	user_priority bit values are ignored and not processed
	1	user_priority bit values are significant
Coupling Flag (CF) (octet 3, bit 2)	0	Coupling Flag not set
	1	Coupling Flag set
Color Mode Flag (CM) (octet 3, bit 3)	0	Color Mode Flag is not set
	1	Color Mode Flag is set
user_priority bits 000 (octet 8, bit 1)	0	Bandwidth profile does not apply to frames with user_priority = 000
	1	Bandwidth profile applies to frames with user_priority = 000
user_priority bits 001 (octet 8, bit 2)	0	Bandwidth profile does not apply to frames with user_priority = 001
	1	Bandwidth profile applies to frames with user_priority = 001
user_priority bits 010 (octet 8, bit 3)	0	Bandwidth profile does not apply to frames with user_priority = 010
	1	Bandwidth profile applies to frames with user_priority = 010
user_priority bits 011 (octet 8, bit 4)	0	Bandwidth profile does not apply to frames with user_priority = 011
	1	Bandwidth profile applies to frames with user_priority = 011
user_priority bits 100 (octet 8, bit 5)	0	Bandwidth profile does not apply to frames with user_priority = 100
	1	Bandwidth profile applies to frames with user_priority = 100
user_priority bits 101 (octet 8, bit 6)	0	Bandwidth profile does not apply to frames with user_priority = 101
	1	Bandwidth profile applies to frames with user_priority = 101
user_priority bits 110 (octet 8, bit 7)	0	Bandwidth profile does not apply to frames with user_priority = 110
	1	Bandwidth profile applies to frames with user_priority = 110
user_priority bits 111 (octet 8, bit 8)	0	Bandwidth profile does not apply to frames with user_priority = 111
	1	Bandwidth profile applies to frames with user_priority = 111

Table 10.17   EVC Map Entry Subinformation Element

8	7	6	5	4	3	2	1	Octet
EVC map entry subinformation element identifier per								
Subinformation element				**Identifier**				
				Bits 8 7 6 5 4 3 2 1				
UNI identifier				0 1 0 1 0 0 0 1				
EVC parameters				0 1 1 0 0 0 0 1				1
EVC identifier				0 1 1 0 0 0 1 0				
EVC map entry				0 1 1 0 0 0 1 1				
Bandwidth profile				0 1 1 1 0 0 0 1				
Table								2
Length of **EVC map entry** contents								3
CE-VLAN ID								4
CE-VLAN ID								5
								6

Table 10.18   UNI Identifier Subinformation Element

8	7	6	5	4	3	2	1	Octet
UNI identifier subinformation element identifier per Table 10.3								1
Length of **UNI identifier** contents								2
ASCII Octet								3
ASCII Octet								4
. . .								. . .
. . .								. . .

EVC map entry Subinformation Element (Table 10.17) is to specify one or more CE-VLAN ID values of **1, 2, . . . , 4095**.

UNI identifier Subinformation Element (Table 10.18) is to convey the value of UNI identifier.

EVC identifier Subinformation Element (Table 10.19) is to convey the value of EVC identifier.

EVC parameters subinformation element: The EVC parameters subinformation element conveys the service attributes of an existing EVC on the UNI. This subinformation element can be repeated, if necessary, in a STATUS message to indicate the service attributes of all configured EVCs on the UNI. The format and coding of this IE are shown in Tables 10.20 and 10.21.

10.4   ELMI SYSTEM PARAMETERS AND PROCEDURES

The behavior of the ELMI protocol is defined by a set of procedures that need to be carried out based on the events at the CE and the MEN, and the

Table 10.19 EVC Identifier Subinformation Element

8	7	6	5	4	3	2	1	Octet
EVC identifier subinformation element identifier per								

Subinformation element	Identifier	
	Bits 8 7 6 5 4 3 2 1	
UNI identifier	0 1 0 1 0 0 0 1	1
EVC parameters	0 1 1 0 0 0 0 1	
EVC identifier	0 1 1 0 0 0 1 0	
EVC map entry	0 1 1 0 0 0 1 1	
Bandwidth profile	0 1 1 1 0 0 0 1	

Table	
Length of **EVC identifier** contents	
ASCII Octet	2 / 3
ASCII Octet	4

Table 10.20 EVC Parameters Subinformation Element

8	7	6	5	4	3	2	1	Octet
EVC parameters subinformation element identifier								1
Length of **EVC parameters** subinformation element contents								2
Reserve 0				EVC Type				3

Table 10.21 EVC Type Coding

Bits 3 2 1	EVC Type
0 0 0	Point-to-Point EVC
0 0 1	Multipoint-to-Multipoint EVC

received ELMI messages or PDUs (Protocol Data Units) by the UNI-C and the UNI-N.

The ELMI procedures are characterized by a set of ELMI messages that will be exchanged at the UNI. These message exchanges can be asynchronous or periodic.

Periodic message exchanges are governed by timers, status counters, and sequence numbers.

Polling Timer (T391) is the timer for UNI-C to poll status of UNI and EVCs (full status) via STATUS ENQUIRY messages. The timer starts with transmission of STATUS ENQUIRY message. If STATUS message is not received, an error is recorded.

T391 can take values from 5 to 30 s. The recommended default value for T391 is 10 s.

Polling Verification Timer (T392) is the timer for UNI-N to respond STATUS ENQUIRY messages. T392 starts with transmission of STATUS message and stops with receiving STATUS ENQUIRY. It is greater than T391 and takes values of 5–30 s with a recommended default of 15 s. If STATUS ENQUIRY message is not received, an error is recorded.

Polling Counter (N391) is a counter of UNI-C for full STATUS polling counts. It takes values in the range of 1–65,000 with a recommended default of 360.

Status Counter (N393) is a counter for both UNI-C and UNI-N for consecutive error counts.

Sequence Numbers allow the UNI-N and the UNI-C to determine the status of the ELMI process including correlating STATUS ENQUIRY messages with STATUS messages.

The UNI-C and the UNI-N maintain the send and receive internal counters. The send sequence counter maintains the value of the send sequence number field of the last Sequence Numbers IE sent. The receive sequence counter maintains the value of the last received send sequence number field in the Sequence Numbers IE and maintains the value to be placed in the next transmitted received sequence number field.

The value zero in the receive sequence number indicates that the receive sequence number field contents are undefined; this value is normally used after initialization. Figure 10.5 shows an example of the use of the send and receive sequence numbers.

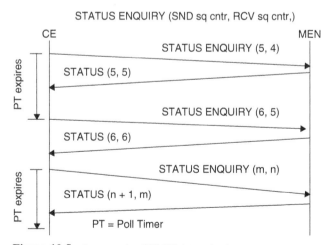

Figure 10.5 An example of ELMI status check.

10.4.1 Periodic Polling Process

The periodic polling process can be summarized as follows.

1. At least, every **T391** seconds, the UNI-C sends a STATUS ENQUIRY message to the UNI-N.
2. At least every **N391** polling cycles, the UNI-C sends a *Full Status* STATUS ENQUIRY.
3. The UNI-N responds to each STATUS ENQUIRY message with a STATUS message and resets the Polling Verification Timer, which is used by the UNI-N to detect errors. If the UNI-C sends a STATUS ENQUIRY requesting full status, the UNI-N responds with a STATUS message with the Report Type specifying *Full Status*. The STATUS message sent in response to a STATUS ENQUIRY contains the Sequence Numbers and Report Type IEs.

   If the UNI-N cannot fit EVC status IEs and service attributes and parameters for all EVCs into a single *Full Status* STATUS message, the UNI-N responds with a *Full Status Continued* STATUS message, containing as many EVC Status IEs as allowed by the Ethernet frame size.
4. If the STATUS message is a *Full Status*, or *Full Status Continued* STATUS message, the UNI-C updates its configuration according to the status of the UNI and the status and service attributes of each configured EVC.

   On receipt of a *Full Status Continued* STATUS message, the UNI-C continues to request EVC status by sending a *Full Status Continued* STATUS ENQUIRY message (without waiting for the Polling Timer to expire). The UNI-C S restarts Polling Timer with value **T391** each time it transmits a *Full Status Continued* STATUS ENQUIRY message. When the UNI-N responds with a *Full Status* STATUS message, it is an indication that all information has been sent.
5. The UNI-C compares the EVC reference ID sent in the full status with the previously reported EVC reference ID.

In addition to triggering *Full Status* and *Full Status Continued* reports every N391 polling cycles, the DI is used to trigger such reports each time there is a change in EVC or UNI information.

10.5 UNI-C AND N PROCEDURES

When the UNI-C comes up for first time or is restarted, the UNI-C sets its DI to 0 and sends the ELMI STATUS ENQUIRY with Report Type *Full Status*.

The UNI-C will then receive *Full Status* or *Full Status Continued* reports, including the latest UNI and EVC information and update its local database. After the reports, the UNI-C's local DI **is** set to the UNI-N DI that is received in the *Full Status* report. The DI value will stay the same until the Status procedure is complete.

When the UNI-N comes up first, it sets its DI value to a nonzero value that is different from the DI value received in the first message received from the UNI-C. Any change in information related to UNI or EVC including status change results in incrementing DI value to reflect the change in data.

For ELMI Check, on receipt of ELMI STATUS ENQUIRY, with ELMI Check, UNI-N responds with ELMI STATUS and include the current value of DI.

ELMI notifies the UNI-C of a newly added EVC using a *Full Status* STATUS message.

In addition, the UNI-N and the UNI-C use the information provided by periodic polling for error monitoring and detect reliability errors (i.e., nonreceipt of STATUS/STATUS ENQUIRY messages or invalid sequence numbers in a Sequence Numbers IE) and Protocol errors. The UNI-N and the UNI-C ignore messages (including their sequence numbers) containing these errors.

Unrecognized information and subinformation elements are ignored by both the UNI-C and UNI-N. When an error is detected, the appropriate management entity is notified.

10.6 CONCLUSION

LMI is a protocol terminated by the UNI-C of the CE and by the UNI-N of the MEN. It provides the status of an EVC and UNI, and enables the CE to configure itself to access Carrier Ethernet services. E-LMI messages, STATUS and STATUS ENQUIRY, are encapsulated inside Ethernet frames. This chapter described ELMI message elements and procedures between UNI-C and UNI-N.

ELMI is not widely used in the industry today. Demarcation devices supporting ELMI are few. Service providers are not eager to provide ELMI services either. This might change as Operation Support Systems that can manage ELMI become available and user demand for ELMI picks up.

REFERENCES

1. MEF 16. Ethernet Local Management Interface; January 2006.
2. MEF 10.2. Ethernet Services Attributes Phase 2; October 2009.

Chapter 11

PB (Provider Bridges), PBB (Provider Backbone Bridges), and PBT (Provider Backbone Transport)

11.1 INTRODUCTION

Carrier Ethernet is very close to prove itself as an alternative technology for MEN in addition to enterprise networks. In order for Carrier Ethernet to be an alternative for backbone, reliability and scalability need to be supported in addition to its QoS and ENNI (external network-to-network interface) capabilities. End-to-end Ethernet networks is expected to be 30–40% cheaper than its alternatives [1, 2], as a result of its simplicity compared to alternatives, although as more features added to Carrier Ethernet, keeping the cost low is going to be a challenge.

Core networks connect networks of different cities, regions, countries, or continents. The complexity of these technologies imposes substantial financial burdens on network operators, both in the area of capital expenditures (CAPEX) and operational expenditures (OPEX). Low cost Ethernet is very attractive. New extensions to the Ethernet protocols [3–6] are developed to transform Ethernet to a technology ready for use in MANs/WANs.

Services supported in LAN/MEN such as E-LAN and E-LINE will be supported end to end. This results in no changes to the customer's LAN equipment, providing end-to-end usage of the technology, contributing to wider interoperability and low cost. Service level agreements (SLAs) provide end-to-end performance, based on rate, frame loss, delay, and jitter and enable traffic engineering (TE) to fine tune the network flows.

A high degree of scalability is needed for handling different traffic types and for user separation inside the network, especially, the scalability in terms

Networks and Services: Carrier Ethernet, PBT, MPLS-TP, and VPLS, First Edition. Mehmet Toy.
© 2012 John Wiley & Sons, Inc. Published 2012 by John Wiley & Sons, Inc.

of address space, maximum transmission speed, and maximum transmission distance. Recent research and standardization efforts aim at speeding up Ethernet to 100 Gbps. High bandwidth and various tagging and tunneling capabilities help Ethernet to support scalability.

Network errors need to be detected and repaired, without the users noticing it. It has to be possible to monitor, diagnose, and centrally manage the network with standard, vendor-independent implementations and to provide services rapidly. Carrier-class OAM needs to be implemented. A 50 ms recovery time during failures, which is inherited from SONET, is also considered one of the parameters of reliability for Ethernet networks.

If Ethernet is ever going to be a credible wide area layer-2 network transport protocol, it needs to be able to transparently transport any of the major protocols used in a converged network. This capability is provided in IP through the use of pseudowires and MPLS. The PBB-TE (provider backbone bridging traffic engineering) could provide an alternative to the **MPLS** core. Layer-2 Ethernet-based virtual private networks (VPNs) are being offered in parallel to layer-3 MPLS-based IP-VPNs.

The PBT (provider backbone transport) is a group of enhancements to Ethernet that are defined in the IEEE's (PBB-TE). The PBB-TE separates the Ethernet **service layer** from the **network layer** thus enabling the development carrier-grade public Ethernet services. Its capabilities may be summarized as follows:

- Enabling TE in Ethernet networks. PBT tunnels reserve appropriate bandwidth and support the provisioned QoS metrics that guarantee SLAs will be met without having to overprovision network capacity.
- Multiplexing of any service inside PBT tunnels, including Ethernet and MPLS services. This flexibility allows service providers to deliver native Ethernet initially and MPLS-based services (VPWS, VPLS) if and when they require.
- Protecting a point-to-point Ethernet tunnel by allowing to provision an additional backup tunnel to provide resiliency.
- Scalability by turning off MAC learning features, the undesirable broadcast functionality that creates MAC flooding and limits the size of the network is removed. In addition, PBT offers a full 60-bit addressing scheme that enables virtually limitless numbers of tunnels to be set up in the service provider network.
- Better service management by having more effective alarm correlation, service-fault correlation, and service-performance correlation, as a result of the fact that the network knows both the source and destination address in addition to the route since each packet is self identifying.

Underlying protocols for PBT are as follows:

- 802.1AB [7] link layer discovery protocol (LLDP) is used to discover the network layout and forwarding this information to the control plane or management layer;

- 802.1ag [3] protocol to monitor the links and trunks in the PBT layout;
- PBT is an expansion of the PBB (provider backbone bridge) protocol, IEEE 802.1ah [4];
- 802.1Qay [5], PBBs with TE or PBT protocol.

In the following sections, we describe 802.1AB and 802.1ah protocols first and then PBT-TE.

11.2   IEEE 802.1AB

802.1AB standards (Station and Media Access Control Connectivity Discovery) defines "LLDP" that is designed for advertising information to stations attached to the same 802 LAN, to populate physical topology and device discovery management information databases. The protocol facilitates the identification of stations connected by IEEE 802 LANs/MANs, their points of interconnection, and access points for management protocols.

The protocol can operate with all IEEE 802 access protocols and network media. Network management information schema and object definitions of LLDP are suitable for storing connection information about adjacent stations and compatible with the IETF PTOPO management information base (MIB) [8].

An IEEE 802.1AB-enabled device advertises its system information to its neighbors. The neighbors save the system information in an MIB, which can then be used by a management protocol such as SNMP. With the MIB, the network management system knows the network topology, where systems are connected to it, port status, etc.

Normally, a network management system would discover the network status. The purpose of allowing the network to discover itself and then tell it to a management station is as follows:

- to facilitate multivendor interoperability and use of standard management tools to discover and make available physical topology information for network management;
- to make it possible for network management to discover certain configuration inconsistencies or malfunctions that can result in impaired communication at higher layers;
- to provide information to assist network management in making resource changes and/or reconfigurations that correct configuration inconsistencies or malfunctions.

In the following sections, we describe 802.1AB architecture, LLDP protocol, and MIB.

11.2.1   Architecture

The LLDP is a link layer protocol and optional part of 802 LAN protocol stack. It is a media-independent protocol intended to be run on all IEEE 802 devices,

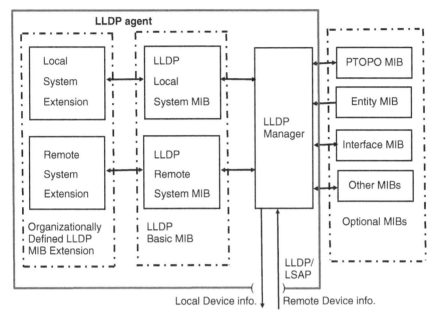

Figure 11.1 LLDP components.

allowing a LLDP agent to learn higher layer management reachability and con-nection endpoint information from adjacent devices. It runs on all IEEE 802 media and over the data-link layer only, allowing two systems running different network layer protocols to learn about each other.

Architecturally, the LLDP runs on top of the uncontrolled port of an IEEE 802 MAC client. It may be run over an aggregated MAC client as described in Reference 9. It may be desirable for stations to prohibit the transmission of LLDP PDUs over the uncontrolled port until the controlled port has been authorized. The spanning tree state of a port does not effect the transmission of LLDP PDUs.

Current applications use ports at transport (TCP/UDP) protocol layer to identify points of access in a machine and to define connections using end-point service identifiers (pair protocol, port). At LLC (logical link control) over Ethernet, this function is performed using link service access points (LSAP) that provide interface ports for users above LLC sublayer (Fig. 11.1).

Each LLDP agent (Fig. 11.2) is responsible for causing the following tasks to be performed:

- maintaining current information in the LLDP local system MIB;
- extracting and formatting LLDP local system MIB information for trans-mission to the remote port;
- provide LLDP agent(s) status at regular intervals or whenever there is a change in the system condition or status;
- recognizing and processing received LLDP frames;

- maintaining current values in the LLDP remote system MIB;
- using the procedure somethingChangedLocal() and the procedure some-thingChangedRemote() to notify the PTOPO MIB manager and MIB managers of other optional MIBs whenever a value status change has occurred in one or more objects in the MIBs depicted in Figure 11.2.

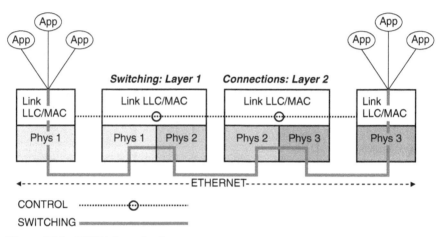

Figure 11.2 LLDP information flow.

MIB objects are generally accessed through SNMP. Objects in the MIB are defined using the mechanisms of the Structure of Management Information (SMI). The LLDP MIB manager is responsible for updating and maintaining each LLDP remote systems MIB associated with the LLDP agent.

11.2.2 Principles of Operation

The LLDP runs over the data-link layer; therefore, devices that use different network layer protocols can still identify each other. An 802-based LAN with LLDP contains agents, which use the services of the LLC and the MAC address to transmit and receive information from and to other LLDP agents. It uses a special multicast MAC address (Section 11.2.3) to send the information to other agents.

On Ethernet networks, LLDP header and messages are encapsulated in an 802 slow protocol frame, which includes a multicast destination address and the source MAC address of the transmitting interface. Although LLDP is defined as an Ethernet protocol, any layer 2 protocol can use it.

The transmission of the LLDPDU (link layer discovery protocol data unit) can be triggered by the following two factors:

- The expiration of a timer, which is the transmit countdown timing counter, and

- Status or value changes in one of the information elements in the local system.

The LLDP agent can transmit information and receive information about the remote systems. It is possible to turn either the receiving or the sending functions of the agent off. This makes it possible to configure an implementation, which restricts a local LLDP agent either to transmit or receive only, or to do both.

When the transmission of the LLDPDU has been started, the local system will put its new information from its MIB in a TLV (type, length, and value). The TLVs are inserted into an LLDPDU, which will be transmitted using the special multicast address.

When an LLDP receive module receives the incoming LLDPDU, it will recognize LLC entity and MAC address. It will then use the information from different TLVs to fill the LLDP remote systems MIB.

The LLDP protocol is essentially a one-way protocol. Each device configured with an active LLDP agent sends periodic messages to the slow protocols multicast MAC address [9]. The device sends the periodic messages on all physical interfaces enabled for LLDP transmission and listens for LLDP messages on the same set on interfaces. Each LLDP message contains information identifying the source port as a connection endpoint identifier. It also contains at least one network address that can be used by an NMS (network management system) to reach a management agent on the device via the indicated source port. Each LLDP message contains a configurable time-to-live value, which tells the recipient LLDP agent when to discard each element of learned topology information. Additional optional information may be contained in LLDP PDUs to assist in the detection of configuration inconsistencies.

The LLDP protocol is designed to advertise information useful for discovering pertinent information about a remote peer and to populate topology management information as defined in Reference 8. It is intended to neither act as a configuration protocol for remote devices nor as a mechanism to signal control information between peers.

During the operation of LLDP, it may be possible to discover configuration inconsistencies between devices on the same physical LAN. This protocol does not provide a mechanism to resolve those inconsistencies, rather a means to report discovered information (RDI) to higher layer management entities.

11.2.3   802.1AB Frame Format

The frame format of the 802.1AB standard is derived from a normal Ethernet frame and contains the following:

- Destination address is the special LLDP multicast address 01-80-C2-00-00-0E,
- Source address contains the MAC address of the sending port, and
- Ethertype is the LLDP ethertype, 88-CC.

The information fields in each LLDP frame are contained in a LLDPDU as a sequence of short, variable length, information elements known as TLVs that each include TLV fields where the following occur:

- Type identifies what kind of information is being sent
- Length indicates the length of the information string in octets
- Value is the actual information that needs to be sent (e.g., a binary bit map or an alphanumeric string that can contain one or more fields).

Each LLDPDU includes four mandatory TLVs and optional TLVs as selected by network management:

- **Chassis ID TLV** identifies the chassis that contains the LAN station associated with the transmitting LLDP agent. It is also used for the MAC service access point (MSAP), which is used to identify a system.
- **Port ID TLV** contains the port component of the MSAP identifier associated with the transmitting LLDP agent.
- **Time-to-Live TLV** contains an integer value in the range of 0–65535 s, which is used to indicate the Time to Live of the LLDPDU. When this time reached 0, the LLDPDU should be thrown away. Normally, the LLDPDU is renewed before this exceeding of time.
- **From 0 to *n* optional TLVs**, as allowed by the space limitation of the LLDPDU. The optional TLVs can be inserted in any order.
- **End-of-LLDPDU TLV** is the last TLV in the LLDPDU and used to mark the end of the LLDPDU.

The chassis ID and the port ID values are concatenated to form a logical MSAP identifier that is used by the recipient to identify the sending LLDP agent/port. Both the chassis ID and port ID values can be defined in a number of convenient forms such as port 1 (portID) and IP-Phone (ChassisID) as defined in LLDP MIB [10].

A 0 value in the TTL field of Time to Live TLV tells the receiving LLDP agent how long all information pertaining to this LLDPDUs MSAP identifier will be valid so that all the associated information can later be automatically discarded by the receiving LLDP agent if the sender fails update it in a timely manner.

Currently defined optional TLVs may be used to describe the system and/or to assist in the detection of configuration inconsistencies associated the MSAP identifier: IEEE 802.1 Organizationally Specific TLV set contains Port VLAN-ID TLV, Port and Protocol VLAN-ID TLV, VLAN Name TLV, and Protocol Identity TLV. IEEE 802.3 Organizationally Specific TLV set contains the following:

- MAC/PHY Configuration/Status TLV (indicates the autonegotiation capability and the duplex;
- speed of IEEE 802.3 MAC (PHYs);

- power via MDI TLV indicating the capabilities and current status of IEEE 802.3 PMDs that either require or are able to provide power over twisted-pair copper links;
- link aggregation TLV indicating the current link aggregation status of IEEE 802.3 MACs;
- maximum frame size TLV indicating the maximum supported IEEE 802.3 frame size.

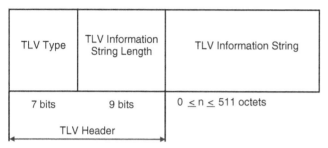

Figure 11.3 Organizationally specific TLV format.

Organizationally Specific TLVs (Fig. 11.3) can be defined by either the professional organizations or the individual vendors that are involved with the particular functionality being implemented within a system. The basic format and procedures for defining Organizationally Specific TLVs are provided below.

Information for constructing the various TLVs to be sent is stored in the LLDP local system MIB. Selection of which particular TLVs to send is under control of the network manager. Information received from remote LLDP agents is stored in the LLDP remote systems MIB. The LLDP remote systems MIB is updated after all TLVs have been validated.

The LLDPDU is checked to ensure that it contains the correct sequence of mandatory TLVs and then each optional TLV is validated in succession. LLDP-DUs and TLVs that contain detectable errors are discarded.

The LLDP is designed to operate in conjunction with MIBs defined by IETF, IEEE 802, and others. The LLDP agents automatically notify the managers of these MIBs whenever there is a value or status change in an LLDP MIB object.

The LLDP managed objects for the local system are stored in the LLDP local system MIB. Information received from a remote LLDP agent is stored in the local LLDP agent's LLDP remote system MIB. LLDP MIBs are as follows:

- IETF Physical Topology in Reference 8 allows an LLDP agent to expose learned physical topology information, using a standard MIB. The LLDP is intended to support the PTOPO MIB.
- IETF Entity MIB in Reference 11 allows the physical component inventory and hierarchy to be identified. Chassis IDs passed in the LLDPDU may identify entPhysicalTable entries. The SNMP agents that implement the

LLDP MIB should implement the entPhysicalAlias object from the Entity MIB version 2 or higher.

- IETF Interfaces MIB in Reference 10 provides a standard mechanism for managing network ports. Port IDs passed in the LLDPDU may identify ifTable (or entPhysicalTable) entries. The SNMP agents that implement the LLDP MIB, should also implement the ifTable and ifXTable for the ports that are represented in the Interfaces MIB.

11.3 PROVIDER BACKBONE BRIDGES (PBB)

The PBB, namely, MAC-in-MAC encapsulation, is based on IEEE 802.1ah [4]. The PBB supports complete isolation of individual client-addressing fields and isolation from address fields used in the operator backbone. Client provider bridge (PB) frames are encapsulated and forwarded in the backbone network based on new B-DA (backbone destination address), B-SA (backbone source address), and B-VID (backbone VLAN-ID).

Although Q-in-Q supports a tiered hierarchy (i.e., no tag, C-Tag/C-VLAN ID, and S-Tag/S-VLAN ID), the service provider can create 4094 customer VLANs, which is insufficient for large metropolitan and regional networks. 802.1ah introduces a new 24-bit tag field (I-SID), service instance identifier, to overcome 12-bit S-VID (S-VLAN ID) defined in PB. This 24-bit tag field is proposed as a solution to the scalability limitations encountered with the 12-bit S-VID defined in PBs.

The PBB service is connectionless. Multiple Spanning Tree Protocol (MSTP) is used to prevent loops. The MSTP differs from STP (spanning tree protocol) in that it is capable to allow frames assigned to different VLANs to follow separate paths through the network. This is based on different MSTP instances, within various regions of LANs. These regions are all connected into a single spanning tree, the common spanning tree (CST). The VLAN tags are reserved on a network, rather than a per-port basis. Flooding is used when destination MAC addresses are not recognized.

The PBBs [4] operate the same way as traditional Ethernet bridges. IEEE 802.1ag [3] Connectivity Fault Management (CFM) addresses the end-to-end OAM such as loopback at specific MAC, link trace, and continuity check.

The PBB implementation uses the MAC addresses of the Ethernet UNIs (ingress ports), rather than customer MACs in the switch forwarding tables. This eliminates the MAC address explosion by greatly reducing the number of MAC addresses that must be learned and maintained by switches in the service provider's core infrastructure.

Keeping the number of MAC addresses to a minimum reduces the aging out and relearning of MAC addresses, thus enhancing end-to-end performance and making network forwarding far more stable.

The MAC learning process observes the MAC addresses of frames received at each port and updates filtering data base conditionally on the state of receiving port.

The PBB networks are managed via SNMP. Separation of addresses of both customer and service providers creates a secure environment. When one side needs to make changes to its network, this side should not have to worry about the overlapping MAC addresses or VLANs. This leads to a simpler operation. There are no forwarding loops and broadcast storms between customer and service providers, which lead to robustness. By not trading addresses between different tiers would reduce memory and processing power, eventually reduces CAPEX.

11.3.1   802.1ah Frame Format

Evolution of headers from 802.1 to 802.1ah is depicted in Figure 11.4. The upper side of the frame (payload to customer-DA) is the part that is used in a PBs [12] enabled network. The lower part is the part that is added by a PBEB (Provider Backbone Edge Bridge) when the frame travels through a backbone network.

Extra MAC header added by the 802.1ah protocol can operate as a normal Ethernet frame. Therefore, it can be used by Ethernet devices that do not have PBB enabled.

Backbone source and destination address have the same function as they would have in normal Ethernet, just as the Ethertype and Backbone-VLAN ID.

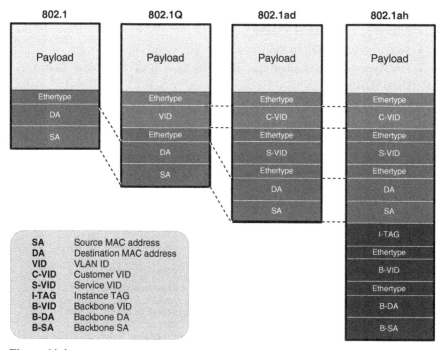

Figure 11.4 Evolution of frame headers.

The next field contains the instance-tag (I-TAG) that comes in two flavors, which are used by the PBEBs:

- Short-service instance TAG
- Long-service instance TAG

Short-service instance TAG (32 bits) consists of the 24-bit I-SID, which identifies the service in the provider's network, and 4 control fields as follows:

- Priority indicates the customer frame's priority,
- Drop eligible is a 1-bit field that indicates the customer drop eligibility,
- Res1 is a 2-bit field that is used for any future format variations, and
- Res2 is a 2-bit field that is reserved for future format variations.

Long-service instance TAG (128 bits) consists of the short tag, including the customer source and destination address. The purpose of the long-service instance TAG is to indicate that an Ethernet frame is encapsulated inside the PBB header, while the short-service instance TAG is intended for multiprotocol encapsulation. The fields are distinguished by a different Ethertype. The Ethertype can be used by PBEBs to know what kind of frame is encapsulated (DIX, 802.3, etc.).

The frame format is depicted in Figure 11.5. Each field is described below:

- Payload
- Ethertype (802.1Q = 81-00) (16 bits)
- Customer-VLAN ID (16 bits)
- Ethertype (802.1ad = 88-a8)(16 bits)
- Service-VLAN ID (16 bits)
- Ethertype (standard ethertype, e.g., ethertype IPv4 (08-00) or IPv6 (86-DD)) (ts)16 bits
- Customer-SA (MAC) (48 bits)
- Customer-DA (MAC) (48 bits)
- I-TAG (Short I-TAG: 32 bits, Long I-TAG: 128 bits)
- Ethertype (can be used to specify what is encapsulated) (16 bits)

B-DA	B-SA	B-TAG	I-TAG	L/T	User data	FCS

B-TAG - Backbone Tag which is identical to S-TAG and optional in the frame

I-TAG - Service Instance Tag which is optional in frame, that encapsulates customer addresses, introduces service instance identifier (I-SID) that allows each BEB (Backbone Edge Bridges) to support a number of backbone service instances and permits the unambiguous identification up to 224 backbone service instances within a single **Provider Backbone Bridge Network (PBBN)**.

Figure 11.5 PBB frame format.

- Backbone-VLAN ID (16 bits)
- Ethertype (specifies whether long or short I-TAG) (16 bits)
- Backbone-SA (MAC) (48 bits)
- Backbone-DA (MAC) (48 bits)

The **B-TAG**-backbone tag, which is identical to S-TAG and optional in the frame, and **I-TAG**-service instance tag, which is optional in frame, that encapsulates customer addresses, introduces service instance identifier (I-SID) that allows each BEB (Backbone Edge Bridges) to support a number of backbone service instances, and permits the unambiguous identification up to 224 backbone service instances within a single **Provider Backbone Bridge Network** (**PBBN**).

Long-service instance tag indicates that an Ethernet frame is encapsulated inside the PBB header, while the short-service instance tag is intended for multi-protocol encapsulation. The fields are distinguished by different EtherType. The Ethertype (before the I-Tag field) can be used by PBEBs to know the type of frame that is encapsulated, such as DIX and 802.3. I-Tag frame consists of the following fields:

- *Priority Code Point (I-PCP)*—This 3-bit field encodes the priority and drop eligible parameters of the service request primitive associated with this frame using the same encoding as specified for VLAN tags in Chapter 6. The Provider Backbone Bridged Network operates on the priority associated with the B-TAG.
- *Drop Eligible Indicator (I-DEI)*—This 1-bit field carries the drop eligible parameter of the service request primitive associated with this frame. The Provider Backbone Bridged Network operates on the drop eligibility associated with the B-TAG.
- *No Customer Addresses (NCA)*—This 1-bit field indicates whether the C-DA (customer destination address) and C-SA (customer source address) fields of the tag containing valid addresses. A value of 0 indicates the C-DA and C-SA fields containing valid addresses. A value of 1 indicates that the C-DA and C-SA fields do not contain valid addresses.
- *Reserved 1 (Res1)*—This 1-bit field is used for any future format variations. The Res1 field contains a value of 0 when the tag is encoded and is ignored when the tag is decoded.
- *Reserved 2 (Res2)*—This 2-bit field is used for any future format variations. The Res2 field contains a value of 0 when the tag is encoded. The frame will be discarded if this field contains a nonzero value when the tag is decoded.
- *Service Instance Identifier (I-SID)*—This 24-bit field carries the service instance identifier of the backbone service instance.
- *Encapsulated Customer Destination Address (C-DA)*—If the NCA (no customer address) bit is 0, this 48-bit field carries the address in the destination address parameter of the service request primitive associated with

this frame. If the NCA bit is 1, this field has a value of 0 when the tag is encoded and is not used when the tag is decoded.

- *Encapsulated Customer Source Address (C-SA)*—If the NCA bit is 0, this 48-bit field carries the address in the source address parameter of the service request primitive associated with this frame. If the NCA bit is 1, this field has a value of 0 when the tag is encoded and is not used when the tag is decoded. The service instance identifier is encoded in a 24-bit field. A BEB may not support the full range of I-SID values but supports the use of any I-SID values in the range 0 through a maximum N, where N is a power of 2 and specified for that implementation.

11.3.2   PBB Principles of Operation

The PBB located at the backbone of the PBT network is called *Backbone Core Bridge* (BCB). The bridge located at the edge of PBT network is called *BEB*. The BCB is an S-VLAN Bridge used within the core of a PBBN. The BEB is a system that encapsulates customer frames for transmission across a PBBN.

A PBEB connects the customers' 802.1ad provider-bridges-enabled networks to the provider network. A customer Ethernet frame coming from the customer network arrives at a PBEB to be transmitted to the backbone. The PBEB adds a service provider MAC header to the customer Ethernet frame, thus allowing customer (C-MAC) addresses and VLANs to be independent of the backbone address (B-MAC) and backbone VLANs administered by the PBBN operator to relay those frames across the backbone. The S-VLANs used to encapsulate customer frames are known as backbone VLANs (B-VLANs). The resources supporting those VLANs are considered to be part of the PBBN.

The PBEB checks the service provider MAC address against its forwarding tables and forwards Ethernet frame as if it is an ordinary frame. As a result, in the core of the network, the switches do not have to be PBB enabled. The frame travels through the network and arrives on the other edge at another Edge Bridge. This PBEB will deencapsulate the extra MAC header, so the customer's frame is again present on the other side of the network. The PBEB forwards the Ethernet frame by using the internal forwarding table and the frame arrives at its destination.

Spanning trees can be employed in the PB and BB networks. Customer spanning trees may extend over provider network. However, the PB network and BB network spanning trees must be decoupled to scale the provider network (Fig. 11.6).

11.3.3   Provider Backbone Bridge Network (PBBN)

The BEB is of three types, I type Backbone Edge Bridge (I-BEB), B type Backbone Edge Bridge (B-BEB), and IB type Backbone Edge Bridge (IB-BEB).

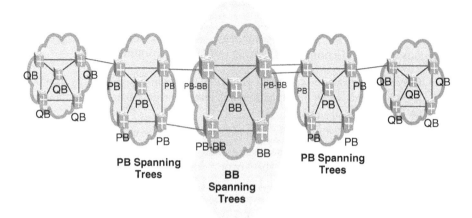

Figure 11.6 Spanning trees in PB and BB networks.

I-component is responsible for encapsulating frames received from customers, assigning each to a backbone service instance and destination identified by a backbone destination address, backbone source address, and a service instance identifier (I-SID) (Fig. 11.7).

I-Component allows each BEB to support more than one such virtual medium for each combination of backbone address and VLAN identifier. If the **I-component** does not know which of the other BEBs provides connectivity to a given customer address, it uses a provisioned default encapsulating B-MAC address that reaches all the other BEBs logically attached to that virtual medium.

Each **I-component** learns the association between customer source addresses received (encapsulated) from the backbone and the parameters that identify the

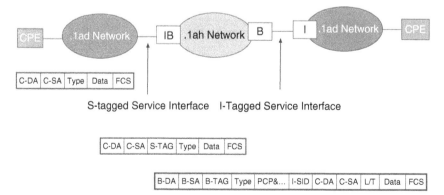

Figure 11.7 Multiproviders Ethernet service delivery.

backbone connection, so subsequent frames to that address can be transmitted to the correct BEB.

B-Component is an S-VLAN component with one or more customer backbone ports (CBPs). **B-component** is responsible for the following:

- relaying encapsulated customer frames to and from I-components or other B-components when multiple domains interconnect, either within the same BEB or externally connected,
- checking that ingress/egress is permitted for frames with that I-SID,
- translating the I-SID (if necessary) and using it to assign the supporting connection parameters (backbone addresses if necessary and VLAN identifiers) for the PBBN, and
- relaying the frame to and from the provider network port(s) that provide connectivity to the other bridges within and attached to the backbone.

The number of I-components (0 or more) and B-components (0 or 1) within a given BEB is a direct consequence of the type and number of external interfaces supported by that BEB. The I-component and B-component may be in the same bridge or may be in different bridges. The types of BEBs may be classified as I-BEB, B-BEB, and IB-BEB.

The PBBN provides multipoint tunnels between Provider Bridged Networks (PBNs) (Fig. 11.8), where each B-VLAN carries many S-VLANs. S-VLANs

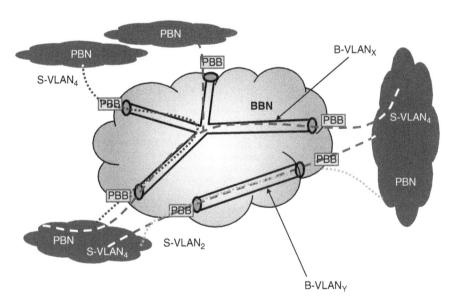

Figure 11.8 PBBN consisting of multiple tunnels between PBNs.

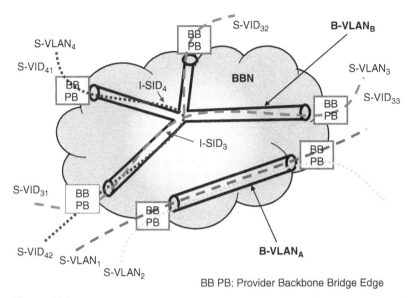

Figure 11.9   Single I-SID per S-VLAN. An I-SID uniquely identifies an S-VLAN within the Backbone; the MAP Shim translates between S-VID and I-SID; the I-SID to (from) S_VID mapping is provisioned when a new service instance is created.

may be carried on a subset of a B-VLAN. For example, all point-to-point S-VLANs could be carried on a single multipoint B-VLAN providing connection to all end points (Fig. 11.9). An I-SID uniquely identifies an S-VLAN within the backbone. When a new service instance is created, I-SID to/from S-VID mapping is provisioned.

Regardless of the I-SID address size, the map tables only have 4096 entries since only one I-SID exists per S-VLAN and only 4096 S-VLANs exist per PB. A different S-VID in each PBN maps to the I-SID.

Backbone POP MAC address, B-MAC addresses, identify the Edge Provider Backbone Bridges (BB PB) (Fig. 11.10). B-MAC addresses are learned by other Edge-Backbone-Edge Bridges. The backbone edge MAC address determines which edge on the B-VLAN will receive the frame. Frames may be flooded by sending with broadcast or multicasts DA B-MACs to the B-VLAN.

The PBB allows customer's MAC addresses to overlap with the provider's MAC addresses, because the customers' service frames are tunneled by PBBs and are not used when switching frames inside the provider's network. As a result, customers are free to assign identifier and class of service values to their VLANs, without concern that the service provider will alter them. Meanwhile, the service provider does not need to worry about coordinating VLAN administration with its customers.

Service provider's switches only use the provider MAC header. There is no need for them to maintain visibility of customers' MAC addresses, reducing the burden on the forwarding tables in the provider's network. This also ensures

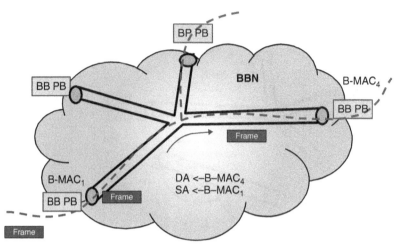

Figure 11.10   BB PB.

that changes to the customers' networks do not impact the provider network, improving the stability of the service provider's network.

Customer security is improved, because the service provider switches are no longer inspecting the customer MAC header.

The PBBs tunnels the customers' Service Frames and control frames, such as STP frames through the provider's network. This allows control protocols, such as STP, to be used separately by the customers' networks and the service providers' network. However, the STP that is used in the customers' networks, should not interact with the STP used in the service provider's part of the network, which is PBB enabled.

11.4   PBT (PROVIDER BACKBONE TRANSPORT)

Traffic-engineered provider backbone bridging (PBB-TE), PBT, is an Ethernet derivative intended to bring connection-oriented characteristics and deterministic behavior to Ethernet. It turns off Ethernet's spanning tree and media access control address flooding and learning characteristics. That lets Ethernet behave more similar to a traditional carrier transport technology.

The frame format of PBT is exactly as the format used for implementing PBB. The difference is in the meaning of the frame's fields. The VID and the B-DA fields together form a 60-bit globally unique identifier.

The control plane is used to manage the forwarding tables of the switches; to create PBT tunnels, all switches need to be controlled from one (PBT) domain. This is necessary for the control plane to fill the forwarding tables of the switches and thus set up a path. Since the control plane fills the forwarding tables, learning and flooding/broadcasting, and spanning tree have become obsolete for PBT. This technique enables circuit switching on an Ethernet.

There is a total of 4094 VLANs, which can be set up according to the 802.1Q standard. When using, for example, 64 VIDs for PBT, the other 4030 can be used for "ordinary" connectionless Ethernet.

The PBT is a combination of Ethernet extensions, created to establish carrier Ethernet. IEEE 1Qay [5] is based on extensions of References 3, 4, 7 and 12. It is developed with scalability and QoS in mind. To use Ethernet as a transport technique in MAN/WAN, OAM functionality is required. The PBT is able to use standards in References 3, 6, 7 and 13 and developing MEF specification in Reference 14. The PBT can take full advantage of the MEF work on ENNI specification [14] to allow peering of Ethernet services.

When using PBT, it becomes possible to set up a path with QoS and a backup path over an Ethernet-based network (Fig. 11.11). The PBT tunnels are created, appropriate bandwidth is reserved to guarantee SLA without having an overprovisioned network capacity. This provisioning needs to be handled by network management layer. The precise implementation of the provisioning is not described yet in Reference 5. A single physical part of the network can have PBT, PBB, and normal Ethernet frames on different VLANs.

The PBT packets will be switched on VID and destination MAC address. There are two VLANs when path protection is enabled: one VLAN provides the current work path and the other provides the backup path. Fast Reroute uses reserved set of B-VID tags.

The PBB-TE reuses existing implementations of VLANs and double tagging and combines them with the network separation and layering principles of PBB.

In the PBB model, the B-VID identifies a packet flooding domain that inter-connects different PB networks. In the PBB-TE model, the B-VID identifies a specific path through the network, in combination with the B-DA address.

Forwarding Model PBB-TE is intended to be deployed as a connection-oriented packet-switched network layer. The PBB-TE runs on independent VLAN learning (IVL) capable switches, which allows packets to be forwarded based on

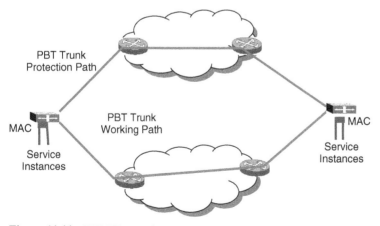

Figure 11.11   PBT-TE protection.

the 60-bit concatenation of the B-VID and destination MAC address. Broadcast frames are discarded. Packets with unknown VID+MAC entries are discarded before being flooded.

End-to-end OAM and protection mechanism are based on References 3 and 6. A 12-bit VID field is used to identify alternate paths to the associated destination MAC address, for protection switching. One VID is assigned for the working path and another VID is assigned for the protection path. Multiple VIDs could support k-shortest path routing or be assigned to protection paths assigned to different failure scenarios.

Any service with a defined EtherType can be transported over PBB-TE. For example, E-Line and E-LAN, pseudowires, and OAM packets can run over PBB-TE.

The PBT can use 802.1AB for network management autodiscovery of network nodes. Bridge forwarding tables are populated using a network management system or GMPLS control plane. Control-plane management system directly provision packet forwarding into the forwarding tables. The GMPLS can be used to populate forwarding tables by GMPLS-**controlled** Ethernet label switching (GELS) [15]. The PLSB (provider link state bridging) using IS-IS protocol helps PBB-TE support multipoint, multicast, and connectionless services. Protocols used by PBT is depicted in Figure 11.12.

11.4.1 Principles of Operation

The PBT ignores spanning tree states; however, STP can be used on the non-PBT-VLANs. 802.1AB tells the management how the network layout looks like.

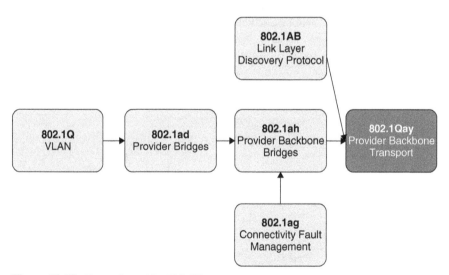

Figure 11.12 Protocols used by PBB-TE.

In turn, the network management fills the forwarding tables of the layer 2 devices. This makes provisioning and adding nodes much simpler.

Forwarding of the packets is based on explicit routes through the network. When the continuity check messages are lost in the network, resulting a fault condition, the VLAN-ID (VID) at the sending host will be changed to the backup VID. This process should take place within 50 ms.

Let us consider the example in Figure 11.13, where customer A and B networks are LANs. LAN A is connected to the provider via a PBEB (A). The provider network contains several other Provider Backbone Core Bridges—PBCB (C) and (D). Customer B is connected to the provider network through PBEB (B).

A management layer or a suitable control plane manages the core network with the PBEBs and PBCBs. This control plane fills the forwarding database in the PBCBs with the correct entries, so they are able to forward traffic through the path/tunnel. For example, the PBCB C has an entry for PBEB B in combination with a certain Backbone VLAN-ID, the 60-bit identifier, to forward traffic on a certain port in the direction of PBEB D. These tunnels are created by the operator through the management layer.

When a customer from site A sends traffic to a customer at site B, its destination will be reached as following:

- The traffic originating from customer A is sent through the customer's network to the PBEB A.
- PBEB A is configured to add the 24-bit I-SID of a certain value to the I-TAG field, based on the S-VID of the customer's frame.
- With the use of this value, a tunnel and its backup tunnel can be selected. These tunnels, identified by the 60-bit identifier, are reserved before the tunnel becomes operational.

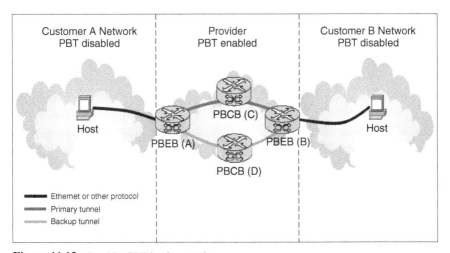

Figure 11.13   Provider PBT implementation.

- The PBEB A also adds the backbone destination address of PBEB B and its own backbone source address and the backbone VLAN-ID.
- The encapsulated frame will be sent through the tunnel and will arrive at PBCB C.

The PBCB C has a forwarding entry, which it selects based on the B-VID. It will forward the frame to PBEB B. The PBEB B will deencapsulate the frame and forward the frame based on the I-SID, which is associated with a certain VLAN-ID. The frame will be delivered to customer B.

Paths are being monitored by the 802.1ag protocol, so when an error occurs, the backup tunnel is automatically switched.

11.4.2   End-to-End Carrier Ethernet with Multiple PBT Domains and Single Domain

Implementing end-to-end carrier Ethernet connection with QoS, protection, and scalability requires implementing PBT inside the customer's network and in provider's network. Each network can have its own domain.

This network has multiple PBT domains (Fig. 11.14), three control planes exist, in order to ensure the PBT tunnels to be created and ended correctly. Every time a frame travels from one domain to another, the frame has to be deencapsulated and encapsulated.

PBEBs create single points of failure in the connections. This could be solved by installing additional PBEBs between the domains. Each domain is monitored by CFM per domain. When a link error occurs on one domain, the other domains will not know that the problem has occurred. To solve this issue, the management layer should be handled on another (OSI) layer.

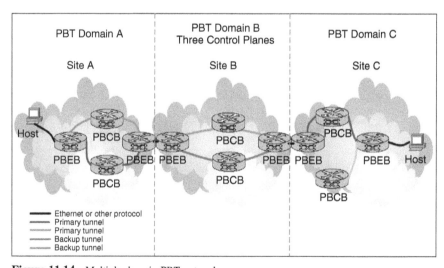

Figure 11.14 Multiple domain PBT network.

Customer and provider networks can be under one PBT domain (Fig. 11.15). This example is more applicable to corporate networks connecting offices of multiple locations.

A single-domain solution also has its specific advantages and disadvantages. For instance, one of the advantages is that only one control plane is required, but a disadvantage is that this solution has less scalability for the different partners.

Another example to multiple PBT domain networks is where access networks implementing PBT while backbone connecting them with non-PBT network (Fig. 11.16). This example is also very much applicable to enterprise networks.

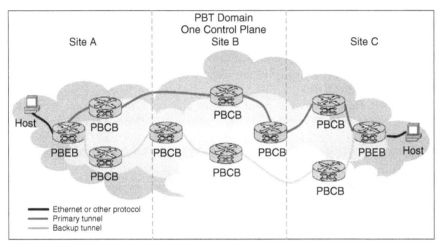

Figure 11.15 Single PBT domain network.

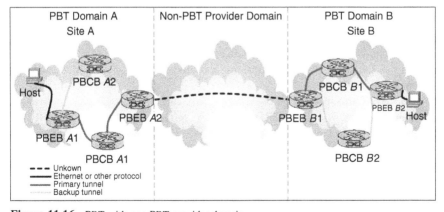

Figure 11.16 PBT with non-PBT provider domain.

11.4.3   PBB-TE Network

The PBB-TE provides a method for full TE of Ethernet switched path (ESP) in a PBBN network. To do this, the PBB-TE replaces the MSTP control plane with either a management plane or an external control plane and then populates the bridge filtering tables of the corresponding B-components and BCBs by creating static filtering table entries for the ESP-MACs and ESP-VIDs.

The external PBB-TE management/control plane is responsible for maintaining and controlling all the topology information to support point-to-point or point-to-multipoint unidirectional **ESPs** over the PBBN.

The PBB-TE topology can coexist with the existing active MSTP or with the shortest path bridging (SPB) topology by allocating B-VID spaces to PBB-TE, MSTP, or SPB, or PBB-TE can be stand alone.

The PBB-TE takes control of the ESP-VID range from the BCBs and backbone edge bridges (BEB) of the PBBN. It will maintain routing tree(s) that provides corouted reverse path(s) from the B-DA to the B-SA. The ESP-VID(s) used in this reverse ESP does not have to be the same one used for the forward ESP. Each of the provisioned ESPs are identified by a 3-tuple: <ESP-MAC DA, ESP-MAC SA, ESP-VID>.

The PBB-TE can provide point-to-point and point-to-multipoint services. A point-to-point PBB-TE service instance, which is called a *PBB-TE trunk*, is provided by a pair of corouted unidirectional point-to-point ESPs and by a pair of 3-tuples: <DA1, SA1,VID1>, <SA1, DA1, VID2>. The VIDs can be the same or different.

A point-to-multipoint PBB-TE service is provided by one multipoint ESP and n unidirectional point-to-point ESPs, routed along the leaves of the multicast ESP and correspondingly by $n+1$3-tuples: <DA, SA, VID>, <SA, SA1, VID1>, ..., <SA, SAn, VIDn>.

In a PBB-TE region, backbone edge bridges (BEBs) mark the demarcation between the provider backbone bridged network (PBBN) and the networks attached to it. The protection domain is defined to be the area between the CBPs of BEBs.

ESPs are provisioned from one CBP to the other, each one identified by the tuple <B-DA,B-SA, B-VID>. Each ESP represents a unidirectional path and the ESP pairs that form the bidirectional path are defining a PBB-TE trunk. The ESPs belonging to the same PBB-TE trunk are corouted but can also be identified by different B-VIDs.

Two PBB-TE trunks can be provisioned, working trunk and protecting trunk, each identified one 3-tuple: <CBP-B, CBP-A, VID-1> and <CBP-A, CBP-B, VID-2>. The working trunk consists of two ESPs identified by the 3-tuples <CBP-B, CBP-A, VID-3> and <CBP-A, CBP-B, VID-4>.

On the terminating BEBs, the VLAN membership of each participating port has to be configured for the B-components.

Each of the PBB-TE trunks is monitored by an independent Maintenance Association. One MA is set to monitor PBB-TE trunk-1 and a second to monitor

PBB-TE trunk-2. Each of these two MAs is associated with a pair of ESPs and identified by their corresponding 3-tuples. The MA that monitors the PBB-TE trunk-1 contains VID-1 and VID-2 in its VID list. Two Up MEPs, associated with this MA, are configured on the CBPs that terminate each PBB-TE trunk.

Each of these MEPs has its own primary VID, VID-1 for the MEP on the West B and VID-2 for the MEP on the East B. Each MEP receives frames that are tagged with any of the VIDs in the MA list but sends frames that are tagged only with that MEPs primary VID.

The MEP for the working trunk on the West B and the corresponding MEP on the East B exchange CCM frames over primary VIDs. If a fault occurs at any of the ESPs, the MEP on the receiving end will be notified. In particular, if a fault on the working ESP occurs, the MEP on the East B-component will declare a remote MEP defect. A bridge does not set remote MEP CCM defect within 3.25 * CCMinterval seconds of the receipt of a CCM and sets remote MEP CCM defect within 3.5*CCMinterval seconds after the receipt of the last CCM.

All subsequent CCMs sent by MEP will have the RDI field set for as long as proper CCMs are not received by the MEP. A reception of a CCM frame with the RDI field set will cause a change of the B-VID parameter of the CBP on West B-component, to the preconfigured value of the protection ESP. The resulted behavior will be to move the specific service instance to the protection PBB-TE trunk.

11.4.4   PBT – MPLS Interworking

The MPLS is the primary technology in the core of IP-based converged networks. The MPLS uses **pseudowire tunnels** for the transport of services across a core network. At the edge, the PBB-TE can support access to the core and provide Metro Ethernet Services, **E-Line**, E-LAN, and E-Tree (Figs. 11.17–11.19).

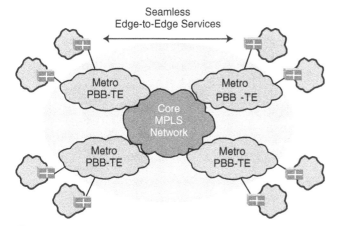

Figure 11.17   A network of PBB-TE and MPLS.

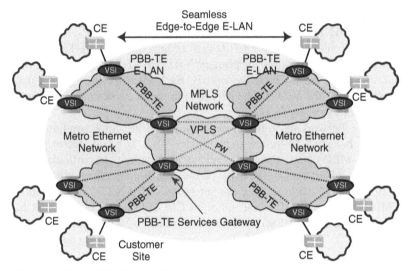

Figure 11.18 A PBB-TE E-LAN service.

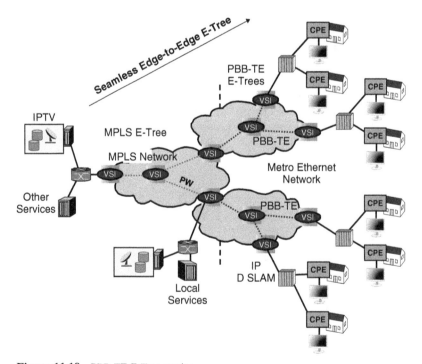

Figure 11.19 PBB-TE E-Tree service.

11.5 CONCLUSION

The PBT is the technology to make end-to-end Ethernet feasible. With scalable addressing, QoS, service OAM, G.8031 [13] protection methods, and MEF defined services, the PBT is a low cost alternative for carrier-class networking.

In fact, some service providers are in the process of implementing PBT. We expect to see more deployment in the coming years.

REFERENCES

1. Cummings J. The T-MPLS vs. PBT debate, Network World 2007 June 18.
2. Lunk P. Traffic engineering for Ethernet: PBT vs. T-MPLS, Lightwave, 2007 May 3.
3. IEEE 802.1ag -2007, Virtual Bridged Local Area Networks—Amendment 5: Connectivity Fault Management; 2007 Dec.
4. IEEE 802.1ah 2008, IEEE Standard for Local and metropolitan area networks—Virtual Bridged Local Area Networks Amendment 7: Provider Backbone Bridges; 2008.
5. IEEE 802.1Qay-2009, Provider Backbone Bridge Traffic Engineering; 2009.
6. ITU-T Y.1731, OAM Functions and Mechanisms for Ethernet based Networks, 5/ 2006.
7. IEEE 802.1AB, Station and Media Access Control Connectivity Discovery; 2005.
8. RFC 2922, Physical Topology MIB; 2000 Sept.
9. IEEE Std. 802.3–2005, Carrier sense multiple access with collision detection (CSMA/CD) access method and physical layer specifications.
10. IETF RFC 2863, The Interfaces Group MIB; 2000 Jun.
11. IETF RFC 2737, Entity MIB (Version 2, 1999 Dec).
12. IEE 802.1ad, OAM Functions and Mechanisms for Ethernet based Networks, 5/ 2006.
13. ITU-T G.8031, Ethernet linear protection switching; 2011.
14. MEF 35, Service OAM Performance Monitoring Implementation Agreement, April 2012.
15. RFC 5828, Generalized Multiprotocol Label Switching (GMPLS) Ethernet Label Switching Architecture and Framework; 2010 Mar.

Chapter 12

T-MPLS (Transport MPLS)

12.1 INTRODUCTION

The MPLS (multiprotocol label switched system) was originally developed by the IETF in order to address core IP router performance issues. It became a dominant technology for carriers' converged IP/MPLS core networks and as a platform for data services such as IP-VPN. With increasing packet networking, the ITU-T became interested in adapting MPLS to make it a "carrier-class" network.

The MPLS first appeared on the transport networking scene in the form of GMPLS (generalized multiprotocol label switching), a set of protocols or protocol extensions offered up by the IETF, which could be implemented as a distributed control plane for the setup and teardown of connections based on circuits, wavelengths, or whole fibers in a transport network.

GMPLS's focus is solely on improving the management of core optical resources and not on packet transport efficiency. Some carriers have taken advantage of GMPLS to support topology and neighbor discovery in order to improve provisioning of wavelengths or circuits in their networks. However, this is limited to typically core optical switches and large multiservice provisioning platform (MSPPs).

The T-MPLS (transport multiprotocol label switched system) offers packet-based alternatives to SONET circuits and promises operators much greater flexibility in how packet traffic is transported through their metro and core optical networks. In T-MPLS, a new profile for MPLS is created so that MPLS label switched paths (LSPs) and pseudowires (PWs) can be engineered to behave as TDM circuits or layer 2 virtual connections, replacing SONET section, line, and path. SONET will remain as a physical interface (PHY) along with gigabit and 10-Gb Ethernet.

The T-MPLS is a connection-oriented packet-switched transport layer network technology based on the MPLS bearer plane modeled in Reference 1. Its architectural principles are as follows:

- A T-MPLS layer network can operate independently of its clients and its associated control networks.

Networks and Services: Carrier Ethernet, PBT, MPLS-TP, and VPLS, First Edition. Mehmet Toy.
© 2012 John Wiley & Sons, Inc. Published 2012 by John Wiley & Sons, Inc.

- Transport connections can have very long holding times. Therefore, T-MPLS includes features traditionally associated with transport networks, such as protection switching and operation and maintenance (OAM) functions.

- A T-MPLS network will not peer directly with an IP/MPLS network. An LSP initiated from an IP/MPLS network element will be encapsulated before it transits a T-MPLS network. Similarly if IP/MPLS is used as a server layer for T-MPLS, then an LSP initiated from a T-MPLS network element will be encapsulated before it transits an IP/MPLS network. This results in control-plane independence between T-MPLS and IP/MPLS.

The T-MPLS is intended to be a separate layer network with respect to MPLS. However, T-MPLS will use the same data link protocol ID (e.g., Ether-Type), frame format, and forwarding semantics as defined for MPLS frames. The semantics used within a label space is defined in Reference 2.

The T-MPLS may be described as a transport network profile of IETF RFCs and Reference 1. Reference 3 defines architecture of T-MPLS layer network, while Reference 4 defines data plane and UNI/NNI. Equipment functionalities and management are based on References 5 and 6, respectively.

The OAM, which is based on References 7 and 8, is specific to the transport network, and functionality is referenced from Reference 9. This provides the same OAM concepts and methods (e.g., connectivity verification, alarm suppression, remote defect indication) already available in other transport networks, without requiring complex IP data plane capabilities.

The result is the T-MPLS, a connection-oriented packet transport network based on MPLS that provides managed point-to-point connections to different client layer networks such as Ethernet. However, unlike the MPLS, it does not support a connectionless mode and is intended to be simpler in scope, less complex in operation, and more easily managed. Layer 3 features have been eliminated and the control plane uses a minimum of IP to lead to low cost equipment implementations.

As an MPLS subset, T-MPLS abandons the control protocol family that IETF defines for MPLS. It simplifies the data plane, removes unnecessary forwarding processes, and adds ITU-T transport style protection switching and OAM functions.

Forwarding behavior of T-MPLS is a subset of IETF defined MPLS. This common data/forwarding plane retains the essential nature of MPLS and ensures that interoperability and interworking will be readily achievable. The features are end-to-end LSP protection, survivability methods mimicking the current linear [10], ring [11] and shared mesh options, per layer monitoring, fault management, and operations that leverage existing SONET OAM mechanisms.

The T-MPLS has strict data and control-plane separation. It does not use layer 3 and IP-related features, such as LSP merge, penultimate hop popping (PHP), and equal cost multiple path (ECMP), eliminating the need for packet transport network elements to process IP packets, while also improving OAM.

The changes allow establishment of bidirectional end-to-end LSPs, configured and monitored via a central network management system. Control-plane traffic via three types of signaling communication channels, namely, in-fiber/in-band via native IP packets, in-fiber/in-band via dedicated LSP, in- or out-fiber/out-of-band, is supported.

The management plane will be used for manual/automated provisioning in the same way as SDH and OTN/WDM networks. However, the control plane for T-MPLS is envisaged to be ASON (automatically switched optical network)/GMPLS and will thus enable more dynamic and intelligent operation.

MPLS Fast ReRoute (FRR) capability requires the use of LSP merge that is excluded from T-MPLS. Since no control plane is involved, protection switching performance can in principle be fast.

Bidirectional T-MPLS LSPs are supported by pairing the forward and backward directions to follow the same path (i.e., the same nodes and links). The pairing relationship between the forward and the backward directions is known in each node traversed by the bidirectional LSP. Both LSP and E-LSP are supported.

Both TTL and EXP as defined in Reference 12 use either the pipe[1] or the short-pipe model. QoS within each T-MPLS sublayer is decoupled from any other one and from client QoS. Diff-Serv as defined in Reference 12 is supported for only pipe and short-pipe models.

Both per-platform and per-interface label space are supported. The T-MPLS will not reserve labels for its own use independently of the MPLS. This helps to ensure that interoperability and interworking will be readily achievable.

Transport MPLS networks can be used to implement Metro Ethernet E-Line services and E-LAN services.

12.2   DIFFERENCES FROM MPLS

In order to define a subset of the MPLS that is connection oriented and to use the established transport OAM model, several MPLS protocol features are excluded from T-MPLS. Key differences of T-MPLS compared with MPLS include the following:

- *Use of Bidirectional LSPs*. T-MPLS pairs the forward and backward LSPs that follow the same nodes and links to provide bidirectional connections, while MPLS LSPs are unidirectional.

---

[1] Three tunnel modes are defined in Reference 12, Uniform, Short-Pipe, and Pipe. In **Uniform mode**, any changes made to the EXP value of the top label on a label stack are propagated both *upward*, as new labels are added, and *downward*, as labels are removed. **In Short-Pipe mode**, the IP Precedence bits on an IP packet are propagated upward into the label stack as labels are added. When labels are swapped, the existing EXP value is kept. If the topmost EXP value is changed, this change is propagated downward only within the label stack and not to the IP packet. **Pipe mode** is just similar to Short-Pipe mode, except the PHB on the MPLS to IP link is selected based on the removed EXP value rather than the recently exposed DSCP value.

- *No PHP Option*. This mechanism pops the MPLS label at the penultimate node, sending IP packets to the last node. This eases the processing at the final node and also means that MPLS OAM packets cannot reach this node.
- *No LSP Merging Option*. LSP merge means that all traffic forwarded along the same path to the same destination may use the same MPLS label. Although this improves scalability, it makes OAM and performance monitoring (PM) difficult because the traffic source becomes ambiguous and unknown. In FRR MPLS link protection, it must be possible to merge two LSPs into one at a node. However, this can create problems in maintaining OAM integrity.
- *No ECMP Option*. ECMP allows traffic within one LSP to be routed along multiple network paths. This not only requires additional IP header processing, as well as MPLS label processing, but also makes OAM more complex, as continuity check (CC) and PM flows may follow different paths. The ECMP allows MPLS packets to be sent over multiple LSPs to the same endpoint.

The T-MPLS and MPLS systems connectivity requires interworking between IP/MPLS PWs and T-MPLS PW at a very preliminary stage. The interworking requirements have been defined in References 7 and 8. There is a working assumption that T-MPLS OAM should reuse where possible the recently developed OAM standards for Ethernet [13, 14].

12.3 ARCHITECTURE

Transport MPLS is a connection-oriented packet transport technology, based on MPLS frame formats (Fig. 12.1). It profiles MPLS, avoiding the complexity and need for IP routing capability and deeper packet inspection. The T-MPLS allows for guaranteed SLAs and defines protection switching and restoration. Fault localization and multioperator service offering are possible.

A Transport MPLS layer network is operated with network management and/or by a control plane. The control plane inside the T-MPLS network is ASON [15, 16]/GMPLS [17].

The T-MPLS leverages MPLS as service layer, as depicted in Figures 12.2 and 12.3. Its characteristic information, client/server associations, the topology,

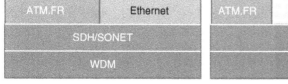

Figure 12.1 Transport network alternatives.

Client Services

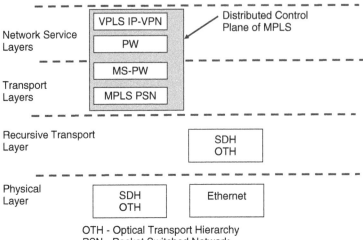

OTH - Optical Transport Hierarchy
PSN - Packet Switched Network

Figure 12.2   Layers of MPLS networks.

Client Services

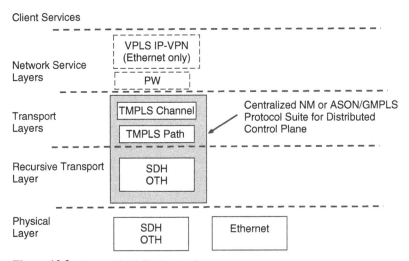

Figure 12.3   Layers of T-MPLS network.

and partitioning of T-MPLS transport networks [3] use a subset of the MPLS architecture [1].

The T-MPLS architecture defines one layer network that is a path layer network. The MPLS layer network characteristic information can be transported through MPLS links supported by trails in other path layer networks such as Ethernet MAC layer network.

The T-MPLS is intended to be a separate layer network with respect to MPLS. However, T-MPLS will use the same data link protocol ID such as EtherType, frame format, and forwarding semantics as defined for MPLS frames. The semantics used within a label space are defined in Reference 2.

The T-MPLS is conceived within the ITU-T's layered network architecture, such that layer networks (i.e., servers) operate independently of their clients (networks or services) and their associated management and signaling networks. As a result, T-MPLS networks will be able to carry customer traffic completely transparently and securely, without unwanted interaction (Fig. 12.4). In other words, the T-MPLS can transport any type of packet service such as Ethernet, IP, and MPLS. The precise adaptation or client handling needs to be defined on a service-by-service basis.

The OAM methodology for T-MPLS is based on Reference 9. However, Reference 9 does not include the complete set of tools that transport networks conventionally provide, such as tools for PM or on-demand OAM.

The control plane for T-MPLS has not been defined. It is expected that the ITU-T's ASON control model will be applied, using GMPLS protocols.

12.3.1   T-MPLS Interfaces

The T-MPLS, similar to MPLS, defines UNI interface, which is the interface between a client and service node, and NNI, which is between two service nodes (Fig. 12.5).

As depicted in Figure 12.6, the T-MPLS NNI may carry informational elements of data plane, which may include a data communication network (DCN), supporting management-plane and control-plane communications, control plane

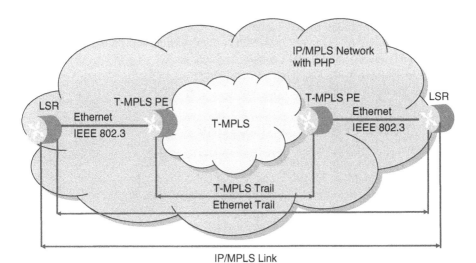

Figure 12.4   IP/MPLS via Ethernet over T-MPLS network.

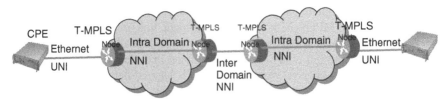

Figure 12.5   Multioperator T-MPLS network.

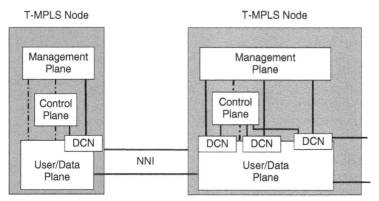

Figure 12.6   T-MPLS NNI planes.

providing signaling and routing, and management plane. The ENNI itself may consists of informational elements of three planes as data plane, including the OAM, control plane, and management plane.

The T-MPLS network node interfaces for T-MPLS-over-ETH (MoE), T-MPLS-over-synchronous digital hierarchy (SDH) (MoS), T-MPLS-over-optical transport hierarchy (OTH) (MoO), T-MPLS-over-plesiosynchronous digital hierarchy (PDH) (MoP), and T-MPLS-over-resilient packet ring (RPR) (MoR) are defined in Reference 4. The payload bandwidths of PDH, SDH, and OTH for MoR are given in Tables 12.1, 12.2, and 12.3, respectively.

12.4   T-MPLS FRAME STRUCTURE

The T-MPLS pushes an additional header structure to packet. Figure 12.7 depicts the frame format of a T-MPLS packet for Ethernet physical medium, while Figure 12.8 illustrates the relationship between network layers and T-MPLS header.

Table 12.1 The Bandwidth of the Payload of PDH Path Signals

PDH Type	PDH Payload, kbps
P11s	$1,536-(64/24)$ H $1,533$
P12s	1,980
P31s	33,856
P32e	$4,696/4,760 \times 44,736 = 44,134$
P11s-Xv, $X = 1-16$	$= 1,533$ to $= 24,528$
P12s-Xv, $X = 1-16$	1,980–31,680
P31s-Xv, $X = 1-8$	33,856–270,848
P32e-Xv, $X = 1-8$	$= 44,134$ to $= 353,072$

Table 12.2 The Bandwidth of the Payload of SDH VCs

VC (Virtual Container) Type	VC Payload, kbps
VC-11	1,600
VC-12	2,176
VC-2	6,784
VC-3	48,384
VC-4	149,760
VC-4-4c	599,040
VC-4-16c	2,396,160
VC-4-64c	9,584,640
VC-4-256c	38,338,560
VC-11-Xv, $X = 1-64$	1,600–102,400
VC-12-Xv, $X = 1-64$	2,176–139,264
VC-2-Xv, $X = 1-64$	6,784–434,176
VC-3-Xv, $X = 1-256$	48,384–12,386,304
VC-4-Xv, $X = 1-256$	149,760–38,338,560

Table 12.3 The Bandwidth of OTH Optical Channel Data Units (ODUs)

ODU (Optical Data Unit) Type	OPU (Optical Payload Unit) Payload, kbps
ODU1	2,488,320
ODU2	$238/237 \times 9,953,280 = 9,995,277$
ODU3	$238/236 \times 39,813,120 = 40,150,519$
ODU1-Xv, $X = 1-256$	2,488,320–637,009,920
ODU2-Xv, $X = 1-256$	$= 9,995,277$ to $= 2,558,709,902$
ODU3-Xv, $X = 1-256$	$= 40,150,519$ to H $10,278,532,946$

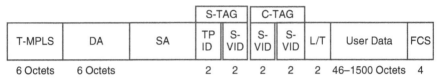

T-MPLS	DA	SA	S-TAG		C-TAG		L/T	User Data	FCS
			TP ID	S-VID	S-VID	S-VID			
6 Octets	6 Octets		2	2	2	2	2	46–1500 Octets	4

Figure 12.7   Frame format of T-MPLS.

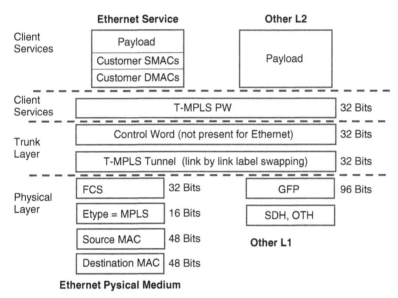

Figure 12.8   Mapping of T-MPLS frame structure into network layers.

12.5   T-MPLS NETWORKS

A typical T-MPLS network configuration is given in Figure 12.9, where the T-MPLS network provides a primary LSP and backup LSP. The switching between the primary and secondary LSP tunnels can take place within 50 ms. These T-MPLS tunnels can support both layer 3 IP/MPLS traffic flows and layer 2 traffic flows via PWs.

In Figure 12.9, the T-MPLS tunnel is transported over MPLS LSPs. Various services can be transported over these T-MPLS tunnels as depicted in Figure 12.10.

12.5.1   T-MPLS Protection

The T-MPLS protection can be linear or ring. The linear protection switching for the T-MPLS as defined in Reference 10 can be 1 + 1 or 1:1. The ring protection switching is defined in Reference 11.

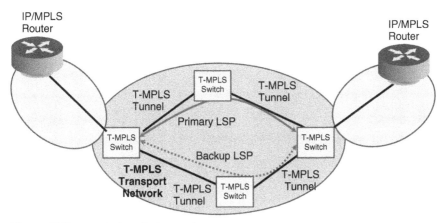

Figure 12.9 A typical T-MPLS network.

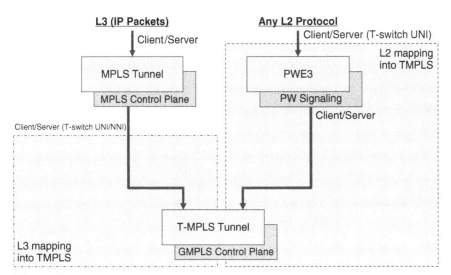

Figure 12.10 Services over T-MPLS tunnels.

The 1 + 1 architecture operates with unidirectional switching, while the 1:1 architecture operates with bidirectional switching. In the 1 + 1 architecture, a protection transport entity is dedicated to each working transport entity. The normal traffic is copied and fed to both working and protection transport entities with a permanent bridge at the source of the protected domain. The traffic on working and protection transport entities is transmitted simultaneously to the sink of the protected domain, where a selection between the working and protection transport entities is made based on some predetermined criteria, such as server defect indication.

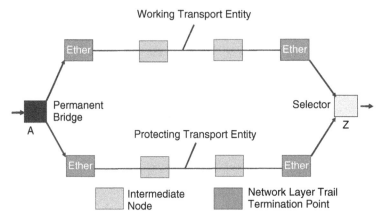

Figure 12.11 Unidirectional 1 + 1 trail protection switching architecture.

Figure 12.11 depicts a unidirectional 1 + 1 trail protection scheme, where protection switching is performed by the selector at the sink side of the protection domain based on purely local protection sink information. The working (protected) traffic is permanently bridged to the working and protection connection at the source side of the protection domain. If connectivity check packets are used to detect defects of the working and protection connection, they are inserted at the source of the protection domain of both working and protection side and detected and extracted at the sink side of the protection domain.

For example, if a unidirectional defect in the direction of transmission from node A to node Z occurs for the working connection as in Figure 12.11, this defect will be detected at the sink of the protection domain at node Z and the selector at node Z will switch to the protection connection.

In the 1:1 architecture, the protection transport entity is dedicated to the working transport entity. However, the normal traffic is transported either on the working transport entity or on the protection transport entity using a selector bridge at the source of the protected domain. The selector at the sink of the protected domain selects the entity that carries the normal traffic. Since source and sink need to be coordinated to ensure that the selector bridge at the source and the selector at the sink select the same entity, an APS protocol is necessary.

In 1:1 bidirectional protection switching operation, the protection switching is performed by both the selector bridge at the source side and the selector at the sink side of the protection domain based on local or near-end information and the APS protocol information from the other side or far end.

If connectivity check packets are used to detect defects of the working and protection connection, they are inserted at the both working and protection side.

The protection operation mode can be nonrevertive or revertive. In nonrevertive types, the service will not be switched back to the working connection if the switch requests are terminated. In nonrevertive mode of operation, when the failed connection is no longer in an SF (signal fail) or an SD (signal degrade) condition, and no other externally initiated commands are present, a No Request state is entered. During this state, switching does not occur.

In revertive types, the service will always return to the working connection if the switch requests are terminated. In revertive mode of operation, under conditions where working traffic is being transmitted via the protection connection and when the working connection is restored, if local protection switching requests have been previously active and now become inactive, a local Wait to Restore state is entered. This state normally times out and becomes a No Request state after the Wait to Restore timer has expired. Then, reversion back to select the working connection occurs. The Wait to Restore timer deactivates earlier if any local request of higher priority preempts this state.

Except for the case of $1 + 1$ unidirectional switching, an APS signal is used to synchronize the action at the A and Z ends of the protected domain. Request/State type, Requested signal, Bridged signal, Protection configuration are communicated between A and Z.

The APS payload structure in a T-MPLS OAM frame is undefined. The field values for the APS octets are given in Table 12.4.

APS protocol must prevent misconnections and minimize the number of communication cycles between A and Z ends of the protected domain, in order to minimize the protection switching time. The communication may be once

Table 12.4   APS Channels Field Value Description

1111 Lockout of Protection (LP)	1110 Signal Fail for Protection (SF-P)	1101 Forced Switch (FS)
1100 Signal Fail (SF)	1010 Signal Degrade (SD)	1000 Manual Switch (MS)
0110 Wait to Restore (WTR)	Request/State	0100 Exercise (EXER)
0010 Reverse Request (RR)	0001 Do Not Revert (DNR)	0000 No Request (NR)
Others Reserved for future international standardization	0 No APS Channel	A
1 APS Channel	0 $1 + 1$ (Permanent Bridge)	B
1 (1:1)n (Selector Bridge) ($n \in 1$)	0 Unidirectional switching	D
1 Bidirectional switching	0 Nonrevertive operation	Protection Type
R	1 Revertive operation	0 Null Signal
Requested Signal 1–254 Normal Traffic Signal 1–254	255 Unprotected Traffic Signal	0 Null Signal
Bridged Signal 1–254 Normal Traffic Signal 1–254	255 Unprotected Traffic Signal	

Z → A, twice Z → A and A → Z, or three times (Z → A, A → Z, and Z → A). This is referred to as *1-phase, 2-phase*, and *3-phase protocols*.

For example, in a bidirectional 1:1 trail protection configuration, if a defect in the direction of transmission from node Z to node A occurs for the working connection Z to A, this defect will be detected at the node A. The APS protocol initiates the protection switching. The protocol is as follows:

- node A detects the defect;
- the selector bridge at node A is switched to protection connection A to Z, and the merging selector at node A switches to protection connection Z to A;
- the APS command sent from node A to node Z requests a protection switch;
- after node Z validates the priority of the protection switch request, the merging selector at node Z is switched to protection connection A to Z, and the selector bridge at the node Z is switched to protection connection Z to A (i.e., in the Z to A direction, the working traffic is sent on both working connection Z to A and protection connection Z to A);
- then the APS command sent from node Z to A is used to inform node A about the switching;
- now the traffic flows on the protection connection.

The T-MPLS subnetwork connection protection is used to protect a section of a connection, where two separate routes are available. The section can be within an operator's network or multiple operators' networks. Two independent subnetwork connections exist, which act as working and protection transport entities for the protected traffic.

The T-MPLS sublayer trail termination functions provide OAM functions to determine the status of the working and protection T-MPLS sublayer trails.

The protection switching types can be a unidirectional switching type or a bidirectional switching type. In unidirectional switching, only the affected direction of the connection is switched to protection, the selectors at each end are independent. Unidirectional 1 + 1 SNC protection can be either revertive or nonrevertive. This type is applicable for 1 + 1 T-MPLS trail and SNC/S protection.

In bidirectional switching, both directions of the connection, including the affected direction and the unaffected direction, are switched to protection. In this case, the APS needs to coordinate the two end points. Bidirectional 1:1 SNC/S protection should be revertive. This type is applicable for 1:1 T-MPLS trail and SNC/S protection.

Protection switching can be triggered by the following:

- Manually via Clear, Lockout of Protection, Forced Switch, Manual Switch commands. **Clear** command clears all of the externally initiated switch commands. **Lockout of Protection (LP)** fixes the selector position to the working connection, prevents the selector from switching to the protection connection when it is selecting the working connection, and switches the selector from the protection to the working connection when it is selecting the protection connection. **Forced Switch** switches the selector from the working connection to the protection connection, unless a higher priority switch request, LP, is in effect. **Manual Switch (MS)** switches the selector from the working connection to the protection connection, unless an equal or higher priority switch request (i.e., LP, FS, SF, or MS) is in effect.

- Automatically via Signal Fail or Signal Degrade associated with a protection domain.
- Protection switching function of Wait to Restore, Reverse Request, Do Not Revert, and No Request. **Wait to Restore** state is only applicable for the revertive mode and applies to a working connection. This state is entered by the local protection switching function in conditions where working traffic is being received via the protection connection when the working connection is restored, if local protection switching requests have been previously active and now become inactive. It prevents reversion back to select the working connection until the Wait to Restore timer has expired. The Wait to Restore time may be configured by the operator in 1 min step between 5 and 12 min. An SF or SD condition will override the WTR. **No Request** state is entered by the local protection switching function under all conditions where no local protection switching requests (including Wait to Restore) are active.

Target switching time for trail and SNC protection is 50 ms. The switching time excludes the detection time necessary to initiate the protection switch and the hold off time. A hold off timer is started when a defect condition is declared and runs for a nonresettable period, which is provisionable from 0 to 10 s in steps of 100 ms. When the timer expires, protection switching is initiated if a defect condition is still present at this point.

12.6 CONCLUSION

The T-MPLS is a connection-oriented carrier-grade packet transport technology. With OAM&P based on traditional transport concepts, the T-MPLS aimed at achieving full OAM capabilities. The T-MPLS is able to converge any L2 and L3 protocol on a common packet-based networking technology and exploit a common distributed control plane with lambda and/or TDM-based transport layer. Its OPEX and CAPEX are expected to be lower than MPLS.

The ITU-T recommendations specify the architectural approach of T-MPLS but are focused on the data plane aspects of T-MPLS. Many other areas require further work in order to develop a fully fledged suite of T-MPLS standards that can be applied to large-scale network deployments.

For T-MPLS (or its successor MPLS-TP) to succeed, it must ultimately address operator concerns about its interoperability with IP/MPLS, its support for multipoint-to-multipoint connections, its potential alignment with recent OAM standards work for Ethernet.

REFERENCES

1. ITU-T G.8110. MPLS layer network architecture; 2005.
2. IETF RFC 3031. Multiprotocol label switching architecture; 2001.
3. ITU-T G.8110.1. Architecture of transport MPLS (T-MPLS) layer network; 2006.
4. ITU-T G.8112. Interfaces for the transport MPLS (T-MPLS) Hierarchy (TMH); 2006.
5. ITU-T G.8121. Characteristics of multi-protocol label switched (MPLS) equipment functional blocks; 2006.
6. ITU-T G.8151. Management aspects of the T-MPLS network element; 2007.
7. ITU-T G.8113. Operations, administration and maintenance mechanism for MPLS-TP in Packet Transport Network (PTN); 2011.
8. ITU-T G.8114 (draft). Operation & Maintenance mechanisms for T-MPLS layer networks; 2008.
9. ITU-T Y.1711. Operation & Maintenance mechanism for MPLS networks; 2004.
10. ITU-T G.8131. Linear protection switching for transport MPLS (T-MPLS) networks; 2007.
11. ITU-T. Draft recommendation G.8132, T-MPLS Shared Protection Ring; 2008.
12. RFC 3270. Multi-Protocol Label Switching (MPLS) Support of Differentiated Services; 2002 May.
13. ITU-T Y.1731. OAM Functions and Mechanisms for Ethernet based Networks; 2006 May.
14. IEEE 802.1ag-2007. Virtual Bridged Local Area Networks—Amendment 5: Connectivity Fault Management; 2007 Dec.
15. ITU-T G.807. Requirements for an Automatic Switched Transport Networks (ASTN); 2004.
16. ITU-T G.808, (808.1). Generic protection switching-Linear trail and subnetwork protection; 2010.
17. RFC 3471. Generalized Multi-Protocol Label Switching (GMPLS) Signaling Functional Description; 2003 Jan.

Chapter 13

MPLS-TP (MPLS-Transport Profile)

13.1 INTRODUCTION

MPLS-TP or MPLS (Multiprotocol Label Switching) Transport Profile has been defined by the IETF to be used as a network-layer technology in transport networks. It is a continuation of T-MPLS (Transport MPLS). The objective is to achieve the transport characteristics of Synchronous Optical Network/Synchronous Digital Hierarchy (SONET/SDH) that are connection oriented, a high level of availability, and quality-of-service and extensive operations, administration, and maintenance (OAM) capabilities. MPLS-TP is expected to enable MPLS to be deployed in a transport network and operated in a similar manner to existing transport technologies and support packet transport services with a similar degree of predictability to that found in existing transport networks.

As discussed in Chapter 13, the ITU Study Group 15 took MPLS as a starting point and has labored to develop T-MPLS, a stripped down version suitable for connection-oriented transport. It is a packet-based transport network that will provide an evolution path for next-generation networks reusing a profile of existing MPLS and complementing it with transport-oriented OAM and protection capabilities. T-MPLS promises multiservice provisioning, multilayer OAM, and protection resulting in optimized circuit and packet resource utilization. The objective is to have an infrastructure for any client traffic type, in any scale, at the lowest cost per bit.

ITU-T approved the first version of T-MPLS Architecture in 2006. By 2008, some vendors started supporting T-MPLS in their optical transport products. At the same time, the IETF was working on a new mechanism called *Pseudo Wire Emulation Edge-to-Edge* (*PWE3*) that emulates the essential attributes of a service such as ATM, TDM, frame relay, or Ethernet over a packet switched network (PSN), which can be an MPLS network [1].

A Joint Working Group (JWT) was formed between the IETF and the ITU-T to achieve mutual alignment of requirements and protocols and came up with

Networks and Services: Carrier Ethernet, PBT, MPLS-TP, and VPLS, First Edition. Mehmet Toy.
© 2012 John Wiley & Sons, Inc. Published 2012 by John Wiley & Sons, Inc.

another approach. The T-MPLS is renamed to MPLS-TP to produce a converged set of standards for MPLS-TP. The MPLS-TP is a packet-based transport technology based on the MPLS Traffic Engineering (MPLS-TE) and pseudowire (PW) data plane architectures defined in References 2–4.

MPLS-TP is client agnostic that can carry L3, L2, L1 services and physical layer agnostic that can run over IEEE Ethernet PHYs, SONET/SDH and optical transport network (OTN), Wavelength Division Multiplexing (WDM), etc. Its OAM functions are similar to those available in traditional OTNs such as SONET/SDH and OTN. Several protection schemes are available at the data plane similar to those available in traditional OTNs.

With MPLS-TP, network provisioning can be achieved via a centralized Network Management System (NMS) and/or a distributed control plane. The Generalized Multiprotocol Label Switching (GMPLS) can be used as a control plane that provides a common approach for management and control of multilayer transport networks.

Networks are typically operated from a network operation center (NOC) using an NMS that communicates with the network elements (NEs). The NMS provides FCAPS management functions (i.e., fault, configuration, accounting, performance, and security management) as defined in Reference 5.

For MPLS-TP, NMS can be used for static provisioning while the GMPLS can be used dynamic provisioning of transport paths. The control plane is mainly used to provide restoration functions for improved network survivability in the presence of failures and facilitates end-to-end path provisioning across network or operator domains. The operator has the choice to enable the control plane or to operate the network in a traditional way without control plane by means of an NMS. The NMS also needs to configure the control plane and interact with the control plane for connection management purposes.

Static and dynamic provisioning models are possible. The static provisioning model is the simplified version commonly known as *static MPLS-TP*. This version does not implement even the basic MPLS functions, such as Label Distribution Protocol (LDP) and Resource Reservation Protocol–Traffic Extension (RSVP-TE), since the signaling is static. It does, however, implement support for GAL (Generic Associated Channel Label) and G-ACh (Generic Associated Channel), which is used in supporting OAM functions. There are several proposals for a dynamic provisioning version, one of which includes GMPLS as a signaling mechanism.

On one hand, MPLS-TP uses a subset of IP/MPLS standards where features that are not required in transport networks, such as IP forwarding, penultimate hop popping (PHP), and equal cost multiple path (ECMP), are not supported (as in T-MPLS) or made optional. On the other hand, MPLS-TP defines extensions to existing IP/MPLS standards and introduces established requirements from transport networks. Among the key new features are comprehensive OAM capable of fast detection, localization, troubleshooting and end-to-end SLA verification, linear and ring protection with sub-50-ms recovery, separation of control and data planes, and fully automated operation without control plane using NMS.

MPLS-TP allows for operation in the networks without requiring the nodes to implement IP forwarding. This means that the OAM operation has to be able to operate in IP and non-IP modes. The control plane is optional, and the protocols such as OAM and protection are structured to be able to operate fully without a control plane.

The essential features of MPLS-TP are listed [6]:

- A connection-oriented packet switching technology; its architecture based on ITU-T G.805 [7].
- No PHP.
- There is no modification to MPLS forwarding architecture. Existing PW and Label Switched Path (LSP) constructs will be used.
- Bidirectional T-MPLS LSPs are supported by pairing the forward and backward directions to follow the same path (i.e., the same nodes and links).
- TTL (Time to Live) is supported according to IETF RFC 3443 [8] for only the pipe and short-pipe models (as discussed in Chapter 13).
- Both E-LSP (EXP (Experimental Bits) Inferred LSP) and L-LSP (Label Only Inferred) are supported.
- Support QoS-guaranteed Multiservice by PWE3.
- EXP is supported for only the pipe and short-pipe models.
- Packet loss probability comparable to TDM transport with a single drop precedence (i.e., green frames) and statistical multiplexing gain with two drop precedence values (i.e., green and yellow frames) are supported. per-hop behavior (PHB) scheduling class (PSC) and drop precedence are inferred directly from the EXP field in the MPLS shim header.
- The model used for TTL and EXP is consistent, either both use the pipe or both use the short-pipe model.
- Both per-platform and per-interface label spaces are supported.
- OAM tools for operation and fault management are based on ITU-T Y.1711 [9] where the OAM function is responsible for monitoring the LSP/PWE and initiating path recovery actions.
- High reliability, equipment protection, network protection less than 50 ms based on ITU-T Y.1720 [10].
- LSP merging is not supported (e.g., no use of LDP multipoint-to-point signaling in order to avoid losing LSP head-end information).
- Point-to-multipoint multicast (i.e., not multipoint to multipoint) is supported. Similarly, only point-to-multipoint connection is supported, while multipoint to multipoint is not.
- The ECMP protocol is not supported.
- IP forwarding is not required to support OAM or data packets.

- Different options are supported for signaling communication network (SCN) links:
  - shared trail SCN links
  - shared hop SCN links
  - independent SCN links
- It can be used with static provisioning systems or with control plane. With static provisioning, there will be no dependency on routing or signaling (e.g., GMPLS or IGP, RSVP, BGP, LDP).
- It should be able to interoperate with existing MPLS and PWE control and forwarding planes.
- It should support high accuracy timing and clock.

13.2 FRAME FORMAT

MPLS-TP uses the MPLS frame format and MPLS label for packet forwarding. In addition, MPLS-TP defines G-ACh to support OAM.

Figure 13.1 illustrates the structure of the MPLS label. The MPLS label is a fixed 4-byte identifier added to the packet to switch the packet to its destination without the need for any routing table (Layer 3) lookups. The MPLS header is called the *shim header*. One or more labels are pushed on the packet at the ingress router forming a label stack. The first label is called the *top label* or the *transport label*, other labels are used by different MPLS applications if needed.

MPLS label is a short, fixed length, physically contiguous identifier used to identify a *Forwarding Equivalence Class (FEC)*, which aggregates individual flows.

In conventional IP forwarding, a particular router will typically consider two packets to be in the same FEC if there is some address prefix X in that router's routing tables such that X is the *longest match* for each packet's destination address. As the packet traverses the network, each hop in turn reexamines the packet and assigns it to an FEC. Three bits of EXP field represent eight packet priorities, called *Traffic Class*.

The Stack (S) field (1 bit) supports a hierarchical label stack. A packet with no label can be thought of as a packet whose label stack is empty (i.e., whose label stack has depth 0) (Fig. 13.2). If a packet's label stack is of depth m, the label at the bottom of the stack is called the *level 1 label* and the label at the top is called the *level m label*. The label stack is used for routing packets through LSP tunnels.

The TTL field (8 bits) provides conventional IP TTL functionality.

Figure 13.1   MPLS packet format.

Figure 13.2 MPLS-TP packet format for pseudowires.

Figure 13.3 Generic exception label and G-ACh for LSP monitoring and alarming.

To support FCAPS, a G-ACh is created that runs on PWE3 and carries OAM messages.

For PWs, the MPLS G-ACh uses the first 4 bits of the PW control word to provide the initial discrimination between data packets and packets belonging to the associated channel [11] (Fig. 13.3). The first 32 bits following the bottom of stack label then have a defined format called an Associated Channel Header (ACH), which further defines the content of the packet. The ACH is both a demultiplexer for G-ACh traffic on the PW and a discriminator for the type of G-ACh traffic.

When the control message is carried over a section or an LSP, rather than over a PW, a reserved label, *Transport Alert Label* as a *Label For yoU* (LFU) with a value of 13 at the bottom of the label stack (Fig. 13.3), indicates that the packet payload is something other than a client data packet. This reserved label is referred to as the *GAL*, as defined in Reference 12. When a GAL is found, it indicates that the payload begins with an ACH. The GAL is a demultiplexer for G-ACh traffic on the section or the LSP, and the ACH is a discriminator for the type of traffic carried on the G-ACh.

MPLS-TP forwarding follows the normal MPLS model, and thus, a GAL is invisible to an LSR unless it is the top label in the label stack. The only other circumstance in which the label stack may be inspected for a GAL is when the TTL has expired.

13.3 ARCHITECTURE

MPLS-TP architectural requirements are defined in Reference 6. The following is a brief description of them.

- MPLS-TP does not modify the MPLS forwarding architecture and is based on existing PW and LSP constructs.
- Point-to-point LSPs may be unidirectional or bidirectional. Congruent bidirectional LSPs can be constructed.
- MPLS-TP LSPs do not merge with other LSPs at an MPLS-TP Label Switched Router (LSR). If a merged LSP is created, it will be detected.

- Packets are forwarded solely based on switching the MPLS or PW label. LSPs and/or PWs can be established in the absence or presence of a dynamic control plane (i.e., dynamic routing or signaling).
- OAM and protection mechanisms, and forwarding of data packets operate without IP forwarding support.
- LSPs and PWs are monitored through the use of OAM in the absence of control plane or routing functions.

The MPLS-TP layered architecture supporting the requirements above is depicted in Figure 13.4. The LSP (PSN tunnel) can transport multiple PWs, each demultiplexed by a unique PW MPLS label. A 4-octet control word may be added to the MPLS payload field to identify different payload types encapsulated in PW, such as Ethernet, frame relay, ATM, Point-to-Point Protocol (PPP), PDH, and SDH.

For a PW server layer, the client signal is an *attachment circuit* (AC) (Figs 13.5 and 13.6). An AC may be an Ethernet port, an Ethernet VLAN, a frame relay DLCI, a frame relay port, etc. A PW forwarder binds an AC to a particular PW. The PW encapsulates service-specific PDUs or circuit data received from ACs, carries encapsulated data over a PSN tunnel, and manages the signaling, timing, order, or other aspects of the service at the boundaries of the PW. From

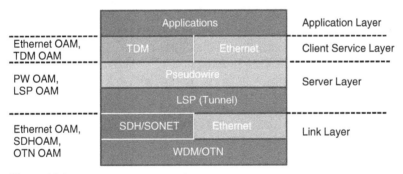

Figure 13.4   MPLS-TP network layering.

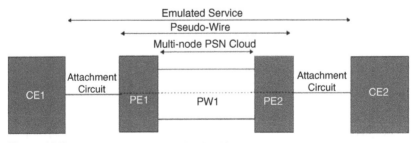

Figure 13.5   MPLS-TP network high level architecture.

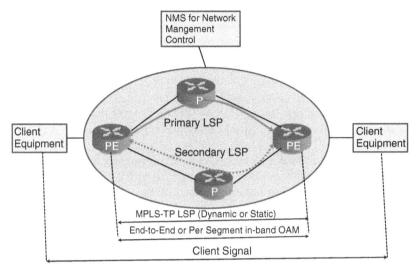

Figure 13.6 MPLS-TP network architecture with protected LSPs.

an OAM perspective, the PW provides status and alarm management for each MPLS-TP service instance.

The relationship between the client layer network and the MPLS-TP server layer network is defined by the MPLS-TP network boundary and the label context (Fig. 13.7). It is not explicitly indicated in the packet. In terms of the MPLS label stack, when the native service traffic type is itself MPLS labeled, then the S bits of all the labels in the MPLS-TP label stack carrying that client traffic are zero. Otherwise, the bottom label of the MPLS-TP label stack has the S bit set to 1.

The S bit is used to identify when MPLS label processing stops and network-layer processing starts. Only one label stack entry (LSE) contains the S bit (Bottom of Stack bit) set to 1.

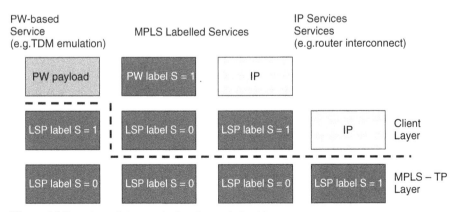

Figure 13.7 MPLS-TP layer and client layer relationship.

13.3.1 Data Plane

MPLS-TP uses a standard MPLS data plane as defined in Reference 2. Sections, LSPs, and PWs (Fig. 13.8) are the data transport entities providing a packet transport service. LSPs and PWs abstract the data plane of the client layer network from the MPLS-TP data plane by encapsulating the payload. The PWs can emulate services such as Ethernet, frame relay, or PPP/high level data link control (HDLC). The adaptation supports IP packets and MPLS-labeled packet services such as PWs, Layer 2 VPNs,_and Layer 3 VPNs.

In addition, the data plane provides MPLS-TP forwarding function based on the label that identifies the transport path (i.e., LSP or PW). The label value specifies the processing operation to be performed by the next hop at that level of encapsulation.

The lowest server layer provided by MPLS-TP is an MPLS-TP LSP. The client layers of an MPLS-TP LSP may be network-layer protocols, MPLS LSPs, or PWs. Therefore, the LSP payload can be network-layer protocol packets and PW packets.

The links traversed by a layer N+1 MPLS-TP LSP are layer N MPLS-TP sections. Such an LSP is referred to as a *client of the section layer*, and the section layer is referred to as the server layer with respect to its clients.

The MPLS label stack associated with an MPLS-TP section at layer N consists of N labels, in the absence of stack optimization mechanisms. In order for two LSRs to exchange non-IP MPLS-TP control packets over a section, an additional label, the GAL, must be at the bottom of the label stack.

A section provides a means of identifying the type of payload it carries. If the section is a data link, link-specific mechanisms such as a protocol type indication in the data-link header may be used. If the section is an LSP, the payload is implied by the LSP label. If the LSP payload is MPLS labeled, the payload is indicated by the setting of the S bit (Fig. 13.7).

LSP can be point-to-point unidirectional, point-to-point associated bidirectional, point-to-point corouted bidirectional, and point-to-multipoint

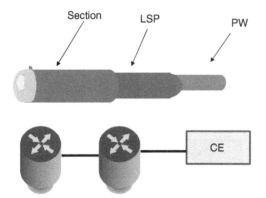

Section   LSP   PW

CE

Figure 13.8 LSP, section, and PW.

unidirectional. Point-to-point unidirectional LSPs are supported by the basic MPLS architecture. Multipoint-to-point and multipoint-to-multipoint LSPs are not supported.

A point-to-point associated bidirectional LSP between LSR A and LSR B consists of two unidirectional point-to-point LSPs, one from A to B and the other from B to A, which are regarded as a pair providing a single logical bidirectional transport path.

A point-to-point corouted bidirectional LSP is a point-to-point associated bidirectional LSP with the additional constraint that its two unidirectional component LSPs in each direction follow the same nodes and links (i.e., the same path).

In a point-to-multipoint unidirectional LSP functions, an LSR may have more than one (egress interface, outgoing label) pair associated with the LSP, and any packet on the LSP is transmitted out to all associated egress interfaces.

MPLS-TP LSPs use the MPLS label switching operations and TTL processing procedures defined in References 2, 13, and 14. In addition, MPLS-TP PWs use the single-segment pseudowires (SS-PWs) and optionally the multisegment pseudowires (MS-PWs) forwarding operations defined in References 3 and 4.

MPLS-TP supports quality-of-service capabilities via the MPLS Differentiated Services (Diffserv) architecture [15]. Both E-LSP (Exp-Inferred-LSP) and L-LSP (Label-Only-Inferred-LSP) MPLS Diffserv modes are supported. Data plane quality-of-service capabilities are included in the MPLS-TP in the form of traffic-engineered (TE) LSPs [16]. The uniform, pipe, and short-pipe Diff-Serv tunneling and TTL Processing in References 14 and 15 may be used for MPLS-TP LSPs.

ECMP load balancing is not performed on an MPLS-TP LSP. However, MPLS-TP LSPs may operate over a server layer that supports load balancing, which is transparent to MPLS-TP.

PHP is disabled by default on MPLS-TP LSPs.

13.3.2 MPLS-TP Router Types

An MPLS-TP LSR is either an MPLS-TP Provider Edge (PE) router or an MPLS-TP Provider (P) router for a given LSP. An MPLS-TP PE router is an MPLS-TP LSR that adapts client traffic and encapsulates it to be transported over an MPLS-TP LSP by pushing a label or using a PW. An MPLS-TP PE exists at the interface between a pair of layer networks. For an MS-PW, an MPLS-TP PE may be either an S-PE (Switching Provider Edge) or a T-PE (Terminating Provider Edge).

An MPLS-TP Label Edge Router (LER) is an LSR that exists at the end points of an LSP and therefore pushes or pops the LSP label.

A PE, which pushes or pops an LSP label, resides at the edge of a given MPLS-TP network domain of a service provider's network. It has links to another MPLS-TP network domain or to a CE (customer edge), except for the case of a PW S-PE router, which is not restricted to the edge of an MPLS-TP network domain.

An MPLS-TP Provider router is an MPLS-TP LSR that switches LSPs that carry client traffic, but it does not adapt client traffic and encapsulate it to be carried over an MPLS-TP LSP. A provider router does not have links to other MPLS-TP network domains.

A PW S-PE router is a PE capable of switching the control and data planes of the preceding and succeeding PW segments in an MS-PW. It terminates the PSN tunnels of the preceding and succeeding segments of the MS-PW and includes a PW switching point for an MS-PW. Furthermore, an S-PE can exist anywhere a PW must be processed or policy applied.

A PW Terminating Provider Edge (T-PE) router is a PE where ACs are bound to a PW forwarder. A terminating PE is present in the first and last segments of an MS-PW.

CE equipment on either side of the MPLS-TP network are peers and view the MPLS-TP network as a single link.

13.3.3 Service Interfaces

An MPLS-TP network consists of two layers, the transport service layer and the transport path layer. The transport service layer provides the interface between CE nodes and the MPLS-TP network. Each packet transmitted by a CE node for transport over the MPLS-TP network is associated with the receiving MPLS-TP PE node by a single logical point-to-point connection at the transport service layer between the ingress PE and the corresponding egress PE to which the peer CE is attached. Such a connection is called an *MPLS-TP Transport Service Instance*, and the set of client packets belonging to the native service associated with such an instance on a particular CE-PE link is called a *client flow*.

The transport path layer provides aggregation of Transport Service Instances over LSPs and aggregation of transport paths.

An MPLS-TP PE node can provide two types of interfaces to the transport service layer: the user–network interface (UNI) providing the interface between a CE and the MPLS-TP network (Fig. 13.9) and the network–network interface (NNI) providing the interface between two MPLS-TP PEs in different administrative domains (Fig. 13.10). A UNI service interface may carry only network-layer clients by MPLS-TP LSPs or both network-layer and non-network-layer clients where a PW is required to adapt the client traffic received over the service interface. An NNI service interface may be to an MPLS LSP or a PW.

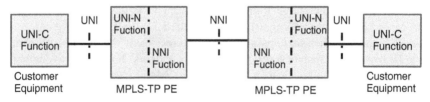

Figure 13.9 Transport service layer interfaces.

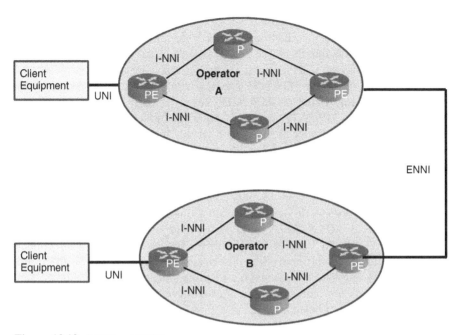

Figure 13.10 I-NNI and ENNI.

The UNI for a particular client flow may or may not involve signaling between the CE and PE, and if signaling is used, it may or may not traverse the same AC that supports the client flow. Similarly, the NNI for a particular Transport Service Instance may or may not involve signaling between the two PEs; if signaling is used, it may or may not traverse the same data link that supports the service instance.

13.3.4 IP Transport Service

MPLS-TP network can provide a point-to-point IP transport service between attached CE nodes. The IP packet is extracted from the link-layer frame and associated with a service LSP based on the source MAC address and the IP protocol version. The packet formed after encapsulation by adding the service LSP label to the frame is mapped to a tunnel LSP where the tunnel LSP label is pushed, and the packet is transmitted over the outbound interface associated with the LSP.

For packets received over a tunnel LSP carrying the service LSP label, the steps are performed in the reverse order.

MPLS-TP uses PWs to provide a Virtual Private Wire Service (VPWS), a Virtual Private Local Area Network Service (VPLS), a Virtual Private Multicast Service (VPMS), and an Internet Protocol Local Area Network Service (IPLS) [17–19]. PWs and their associated labels may be configured or signaled.

If the MPLS-TP network provides a layer 2 interface as a service interface to carry both network-layer and non-network-layer traffic then a PW is required to support the service interface. The PW is a client of the MPLS-TP LSP server layer. MS-PWs may be used to provide a packet transport service.

Figure 13.11 depicts an MPLS-TP network with SS-PWs, while Figure 13.12 shows the architecture for an MPLS-TP network when MS-PWs are used. The corresponding MPLS-TP protocol stacks including PWs are shown in Figure 13.13. In this figure, the transport service [6] is identified by the PW demultiplexer (Demux) label, and the transport path layer is identified by the LSP Demux label [6].

With network-layer adaptation, the MPLS-TP domain provides either a unidirectional or bidirectional point-to-point connection between two PEs in order to deliver a packet transport service to attached CE nodes. For example, a CE may be an IP, MPLS, or MPLS-TP node. As shown in Figures 13.14 and 13.15, there is an AC between the CE1 and PE1 providing the service interface, a bidirectional LSP across the MPLS-TP network to the corresponding PE node on the right, and an AC between that PE node and the corresponding CE node for this service.

Client packets are received at the ingress service interface of the PE. The PE pushes one or more labels onto the client packets that are then label switched

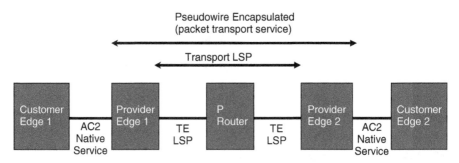

Figure 13.11 MPLS-TP architecture for single-segment PW.

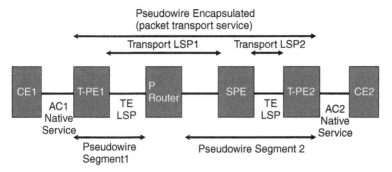

Figure 13.12 MPLS-TP architecture (multisegment PW).

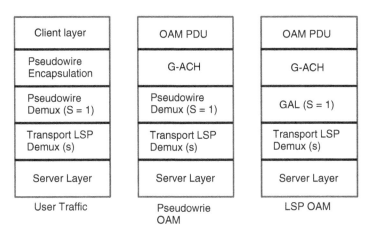

Figure 13.13 MPLS-TP label stack using pseudowires.

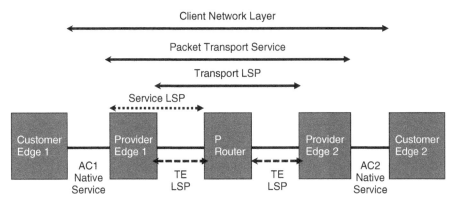

Figure 13.14 MPLS-TP architecture for network-layer clients.

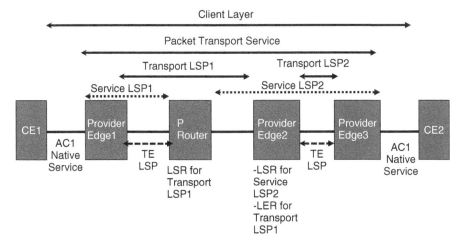

Figure 13.15 MPLS-TP architecture for service LSP switching.

over the transport network. Correspondingly, the egress PE pops any labels added by the MPLS-TP networks and transmits the packet for delivery to the attached CE via the egress service interface.

In Figure 13.16, the transport service layer [6] is identified by the Service LSP (SvcLSP) demultiplexer (Demux) label, and the transport path layer [6] is identified by the transport (Trans) LSP Demux label. The functions of the encapsulation label and the service label (SvcLSP Demux) shown above may alternatively be represented by a single LSE. The S bit is always zero when the client layer is MPLS labeled. It may be necessary to swap a service LSP label at an intermediate node (Fig. 13.7).

Within the MPLS-TP transport network, the network-layer protocols are carried over the MPLS-TP network using a logically separate MPLS label stack, which is the server stack. The server stack is entirely under the control of the nodes within the MPLS-TP transport network and not visible outside that network. Figure 13.16 shows how a client network protocol stack (which may be an MPLS label stack and payload) is carried over a network-layer client service over an MPLS-TP transport network.

A label may be used to identify the network-layer protocol payload type. Therefore, when multiple protocol payload types are to be carried over a single service LSP, a unique LSE needs to be present for each payload type. Such labels are referred to as *encapsulation labels* (Fig. 13.16). An encapsulation label may be either configured or signaled.

Both an encapsulation label and a service label should be present in the label stack when a particular packet transport service is supporting more than one network-layer protocol payload type.

Service labels are typically carried over an MPLS-TP transport LSP edge to edge. An MPLS-TP transport LSP is represented as an LSP transport Demux

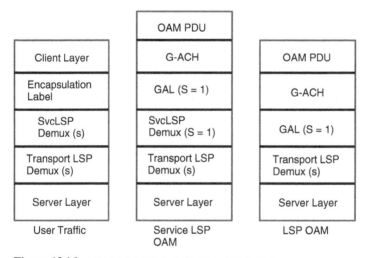

Figure 13.16 MPLS-TP label stack for IP and LSP clients.

label (Fig. 13.16). Transport LSP is commonly used when more than one service exists between two PEs.

13.3.5   Generic Associated Channel (G-ACh)

The MPLS G-ACh [12] is an auxiliary logical data channel supporting control, management, and OAM traffic associated with MPLS-TP transport entities, MPLS-TP sections, LSPs, and PWs in the data plane.

For correct operation of OAM mechanisms, OAM packets follow the same path that data frames use, which is called *fate sharing*. In addition, in MPLS-TP, it is necessary to discriminate between user data payloads and other types of payload such as packets associated with a Signaling Communication Channel (SCC) or a channel used for a protocol to coordinate path protection state. This is achieved by a generic control channel associated to the LSP, PW, or a section with or without IP encapsulation.

MPLS-TP makes use of G-ACh to support FCAPS functions by carrying packets related to OAM, a protocol used to coordinate path protection state, SCC, Management Communication Channel (MCC), or other packet types in-band over LSPs, PWs, or sections. The G-ACh is similar to the PW-Associated Channel [11], which is used to carry OAM packets over PWs. The G-ACh is indicated by an ACH, presenting for all sections, LSPs, and PWs that make use of FCAPS functions supported by the G-ACh.

Figure 13.17 shows the reference model depicting how the control channel is associated with the PW protocol stack. This is based on the reference model for Virtual Circuit Connection Verification (VCCV) in Reference 20.

PW-associated channel messages that control the PW PDUs are encapsulated using the PWE3 encapsulation, such that they are handled and processed in the same manner as the PW PDUs.

Figure 13.18 shows the reference model depicting how the control channel is associated with the LSP protocol stack.

13.3.6   Control Plane

The control plane is responsible for end-to-end segment LSPs and PWE-3 application labels, determining and defining primary and backup paths and configuring the OAM function along the path. OAM is responsible for monitoring and driving switches between primary and backup paths for the end-to-end path and path segments.

Figure 13.19 illustrates the interactions between the MPLS-TP control plane, the forwarding plane, the management plane, and the OAM for point-to-point MPLS-TP LSPs or PWs.

The use of a control plane protocol is optional in MPLS-TP. LSPs and PWs can be provisioned by an NMS similar to provisioning a SONET network. On

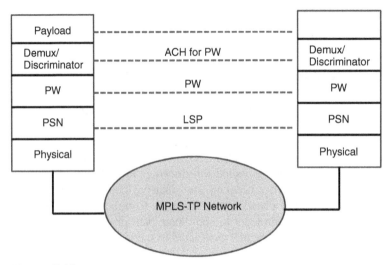

Figure 13.17 PWE3 protocol stack reference model showing the G-ACh [17].

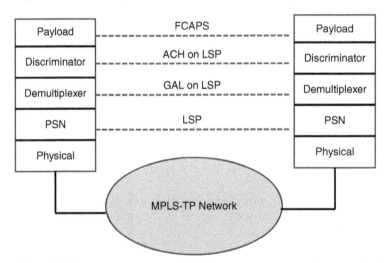

Figure 13.18 MPLS protocol stack reference model showing the LSP associated control channel [17].

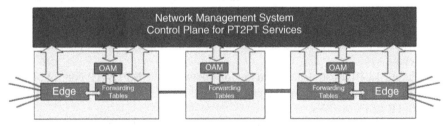

Figure 13.19 MPLS-TP control plane architecture context.

the other hand, they can be configured by the network using Generalized (G)-MPLS and Targeted Label Distribution Protocol (T-LDP), respectively. GMPLS is based on the TE extensions to MPLS (MPLS-TE). It may also be used to set up the OAM function and define recovery mechanisms. T-LDP is part of the PW architecture and is widely used these days to signal PWs and their status.

A distributed dynamic control plane may be used to enable dynamic service provisioning. NMS may be centralized or distributed while control plane is distributed. The control plane may be transported over the server layer, an LSP, or a G-ACh and capable of activating MPLS-TP OAM functions such as fault detection and localization.

The distributed MPLS-TP control plane may provide signaling, routing, traffic engineering, and constraint-based path computation. It is based on existing MPLS and PW control plane protocols, GMPLS signaling for LSPs, and T-LDP signaling for PWs. However, PW control and maintenance take place separately from LSP tunnel signaling.

In a multidomain environment, the MPLS-TP control plane supports different types of interfaces at domain boundaries or within the domains such as UNI, Internal Network–Network Interface (I-NNI), and External Network–Network Interface (ENNI).

The MPLS-TP control provides functions to ensure its own survivability and to enable it to recover gracefully from failures and degradations. These include graceful restart and hot redundant configurations. The control plane is logically decoupled from the data plane such that a control-plane failure does not imply a failure of the existing transport paths.

A PW or an LSP may be statically configured without the support of a dynamic control plane. Static operation is independent for a specific PW or LSP instance; therefore, it is possible to configure a PW statically while setting up the LSP by a dynamic control plane. When static configuration mechanisms are used, care must be taken to ensure that loops are not created.

In order to monitor, protect, and manage a segment or multisegments of an LSP, a hierarchical LSP [2], called a *Subpath Maintenance Element* (*SPME*), which does not carry user traffic is instantiated. For example, in Figure 13.20, two SPMEs are configured to allow monitoring, protection, and management of the LSP-concatenated segments. One SPME is defined between LER2 and LER3, and a second SPME is set up between LER4 and LER5. Each of these SPMEs may be monitored, protected, or managed independently [17].

13.3.7 Network Management

The network management architecture and requirements for MPLS-TP are based on ITU-T G.7710/Y.1701 [21].

Element Management System (EMS) can be used to manage MPLS-TP NEs such as LSR, LER, PE, S-PE, or T-PE. The MCC, realized by the G-ACh, provides a logical operations channel between NEs. Management at network

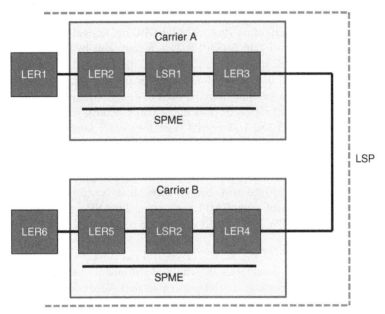

Figure 13.20   SPMEs in intercarrier network.

level can be provided by an NMS. Between EMS and NMS, the Netconf [22], SNMP, or CORBA (Common Object Request Broker Architecture) protocol may be used. With this architecture, FCAPS can be realized.

13.4 OAM (OPERATIONS, ADMINISTRATION, AND MAINTENANCE)

The MPLS-TP OAM architecture supports a wide range of OAM functions to verify connectivity, monitor path performance, and generate, filter, and manage local and remote defect alarms. These functions are applicable to any layers, MPLS-TP sections, LSPs, and PWs.

The MPLS-TP OAM functions can be performed without relying on a dynamic control plane or IP functionality in the data path. In an MPLS-TP network where IP functionality is available, all existing IP/MPLS OAM functions such as LSP ping, bidirectional forward detection (BFD), connectivity check, and connectivity verification may be used. The default mechanism for OAM demultiplexing in MPLS-TP LSPs and PWs is the G-ACh [12]. This uses a combination of an ACH and a GAL to create a control channel associated to an LSP, section, or PW. Forwarding based on IP addresses for OAM or user data packets is not required for MPLS-TP.

RFC4379 [23] and BFD for MPLS LSPs [24] define mechanisms that enable an MPLS LSR to identify and process MPLS OAM packets when the OAM packets are encapsulated in an IP header. These mechanisms are based on TTL expiration and/or use an IP destination address in the range 127/8 for IPv4.

OAM and monitoring in MPLS-TP are based on the concept of maintenance entities, where a maintenance entity (ME) can be viewed as the association of two Maintenance Entity Group End Points (MEPs). A Maintenance Entity Group (MEG) is a collection of one or more MEs that belongs to the same transport path and that are maintained and monitored as a group. The MEPs limit OAM responsibilities of an OAM flow to the domain of a transport path or segment in the layer network being monitored and managed. An MEG may also include a set of Maintenance Entity Group Intermediate Points (MIPs). MEP and MIP support continuity checks (CCs) and connectivity verification (CC) messages between the two end points, allowing the detection of lost connectivity and connection misconfiguration. The CC/CV messages can be exchanged with a period that can be set from 3.3 ms to 10 min, thus allowing for very fast failure detection. In summary, service OAM concepts described in Chapter 7 are pretty much carried over to MPLS-TP.

MPLS-TP OAM also supports performance monitoring, including delay and loss measurements (LMs) defined in Reference 25 to detect performance degradations.

OAM packets follow the same path as traffic data using the G-ACh. A G-ACh packet may be directed to an individual MIP along the path of an LSP or MS-PW by setting the appropriate TTL in the LSE for the G-ACh packet. When the location of MIPs along the LSP or PW path is not known by the MEP, the LSP path tracing may be used to determine the appropriate setting for the TTL to reach a specific MIP. Any node on an LSP and PW can send an OAM packet on that LSP. An OAM packet can only be received to be processed at an LSP end point, a PW end point, or on the expiry of the TTL in the LSP or PW LSE.

The MPLS-TP OAM framework is applicable to sections, LSPs, MS-PWs, and Subpath Maintenance Entities (SPMEs). It supports corouted and associated bidirectional or unidirectional point-to-point transport paths and point-to-multipoint transport paths. Management, control, and OAM protocol functions may need response packets to be delivered from the receiver back to the originator of a message exchange.

Let us assume U is the upstream LER (Fig. 13.21), D is the downstream LER, and Y is an intermediate LSR along the LSP. Return path packet transmission can be from D to U, from Y to U, or from D to Y. In-band or out-of-band return paths can be used for the traffic from a downstream node D to an upstream node U. The LSP between U and D is bidirectional, and therefore, D has a path via the MPLS-TP LSP to return traffic back to U, or D has an out-of-band mechanism for directing traffic back to U.

An arbitrary part of a transport path that can be monitored via OAM independent of the end-to-end monitoring (OAM) is called a *Tandem Connection* [7]. Tandem Connection Monitoring (TCM) allows monitoring of a subset or a segment of LSP in terms of connectivity, fault, quality, and alarms. In MPLS-TP, TCM can be implemented with Path Segment Tunnels (PSTs), enabling a

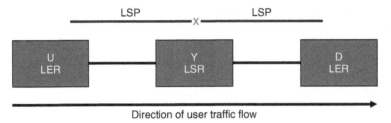

Figure 13.21 Return path reference model.

subset of the segments of LSP or MS-PWs to be monitored independent of any end-to-end OAM.

OAM packets are subject to the same forwarding treatment as the user data packets; therefore, they have the same PSC in both E-LSP and L-LSP.

13.4.1 OAM Hierarchy

The OAM hierarchy for MPLS-TP maybe expressed as follows (Fig. 13.22):

- OAM for physical layer involved in power level and signal quality, including alarms such as loss of light, loss of signal (LoS), loss of frame (LoF), errored second (ES), severely errorred second (SES);
- OAM at link layer such as 802.3 ah functions (i.e., facility loopback, AIS, RDI, and Dying Gasp);
- OAM at LSP layer, that is, the OAM for end-to-end and segment LSPs;
- OAM for PW layer.

The MPLS-TP OAM framework builds on the concept of an MEG, and its associated MEPs and MIPs, as described above.

MEs define a relationship between two points (MEPs) of a transport path to which maintenance and monitoring operations apply. The collection of one or more MEs that belongs to the same transport path and that are maintained and monitored as a group are known as an *MEG*. Intermediate points in between

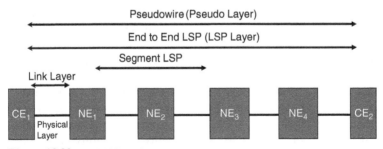

Figure 13.22 OAM hierarchy.

MEPs are called *MIPs*. MEPs and MIPs are associated with the MEG and can be shared by more than one ME in an MEG.

In case of unidirectional point-to-point transport paths, a single unidirectional ME is defined to monitor it. In case of associated bidirectional point-to-point transport paths, two independent unidirectional MEs are defined to independently monitor each direction. In case of corouted bidirectional point-to-point transport paths, a single bidirectional ME is defined to monitor both directions congruently. In case of unidirectional point-to-multipoint transport paths, a single unidirectional ME for each leaf is defined to monitor the transport path from the root to that leaf.

OAM functionalities are available not only on a transport path granularity (e.g., LSP or MS-PW) but also on arbitrary parts of transport paths, defined as Tandem Connections, between any two arbitrary points along a transport path. Subpath Maintenance Elements (SPMEs), are hierarchical LSPs instantiated to provide monitoring of a portion of a set of transport paths (LSPs or MS-PWs) that follow the same path between the ingress and the egress of the SPME. In the TCM, there is a direct correlation between all fault management and performance monitoring information gathered for the SPME and the monitored path segment of the end-to-end transport path.

MEPs are responsible for originating almost all of the proactive and on-demand monitoring OAM functionality for the MEG. An MEP is capable of originating and terminating OAM packets for fault management and performance monitoring. These OAM packets are carried within the G-ACh with the proper encapsulation and an appropriate channel type as defined in Reference 12.

A server MEP is an MEP of an MEG that encapsulates and transports the MPLS-TP layer network or the sublayer being referenced. It can run appropriate OAM functions for fault detection within the server sublayer network and provides a fault indication to its client MPLS-TP layer network, such as lock report (LKR) and alarm indication signal (AIS) that are originated by intermediate nodes and triggered by server layer events.

A policing function is normally colocated with an MEP at UNI or NNI.

In the context of an MPLS-TP LSP, only LERs can implement MEPs. In the context of an SPME, any LSR of the MPLS-TP LSP can be an LER of SPMEs, monitoring infrastructure of the transport path. Regarding PWs, only Terminating Pseudowire End Points (T-PEs) can implement MEPs, while for SPMEs supporting one or more PWs, both T-PEs and S-PEs can implement SPME MEPs. Any MPLS-TP LSR can implement an MEP for an MPLS-TP section. A node hosting an MEP can either support per-node MEP or per-interface MEPs.

An MIP that is a point between the MEPs of an MEG for a PW, an LSP, or an SPME reacts with some OAM packets and forwards all the other OAM packets. It can generate OAM packets only in response to OAM packets that it receives from the MEG it belongs to. The OAM packets generated by the MIP are sent to the originating MEP. An MIP can be a per-node MIP in an unspecified location within the node or a per-interface MIP with two or more MIPs per node on both sides of the forwarding engine.

The following MPLS-TP MEGs are specified in Reference 26.

- A Section Maintenance Entity Group (SMEG), allowing monitoring and management of MPLS-TP sections between MPLS LSRs.
- An LSP Maintenance Entity Group (LMEG), allowing monitoring and management of an end-to-end LSP between LERs.
- A PW Maintenance Entity Group (PMEG), allowing monitoring and management of an end-to-end SS-/MS-PWs (between T-PEs). A PMEG can be configured on any SS-PW or MS-PW. It is intended to be deployed in scenarios where it is desirable to monitor an entire PW between a pair of MPLS-TP-enabled T-PEs rather than monitoring the LSP aggregating multiple PWs between PEs.
- An LSP SPME MEG (LSMEG), allowing monitoring and management of an SPME between a given pair of LERs and/or LSRs along an LSP. An LSMEG is an MPLS-TP SPME with an associated MEG intended to monitor an arbitrary part of an LSP between the MEPs instantiated for the SPME. An LSMEG can monitor an LSP path segment and may support the forwarding engine(s) of the node(s) at the edge(s) of the path segment.
- A PW SPME MEG (PSMEG), allowing monitoring and management of an SPME between a given pair of T-PEs and/or S-PEs along an MS-PW to monitor an arbitrary part of an MS-PW between the MEPs instantiated for the SPME independent of the end-to-end monitoring (PMEG).

A PSMEG may include the forwarding engine(s) of the node(s) at the edge(s) of the path segment. A PSMEG is no different than an SPME, it is simply named so to discuss SPMEs specifically in a PW context.

Hierarchical LSPs are also supported in the form of SPMEs. In this case, each LSP in the hierarchy is a different sublayer network that can be monitored, independent from higher and lower level LSPs in the hierarchy, on an end-to-end basis (from LER to LER) by an SPME. It is possible to monitor a portion of a hierarchical LSP by instantiating a hierarchical SPME between any LERs/LSRs along the hierarchical LSP.

Figure 13.23 depicts these MEGs in a network path between two customer equipment (i.e., CE1 and CE2). Domains 1 and 2 may represent the section of the path within carriers 1 and 2, respectively.

T-PE1 is adjacent to LSR1 via the MPLS-TP section TL1, and LSR1 is adjacent to S-PE1 via the MPLS-TP section LS1. Similarly, in domain 2, T-PE2 is adjacent to LSR2 via the MPLS-TP section TL2 and LSR2 is adjacent to S-PE2 via the MPLS-TP section LS2.

Figure 13.23 also shows a bidirectional MS-PW (MS-PW12) between AC1 and AC2. The MS-PW consists of three bidirectional PW path segments: (i) PW1 path segment between T-PE1 and S-PE1 via the bidirectional LSP1-1, (ii)

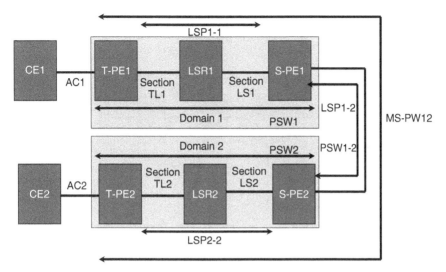

Figure 13.23   The MPLS-TP OAM MEGs.

PW1-2 path segment between S-PE1 and S-PE2 via the bidirectional LSP1-2, and (iii) PW2 path segment between S-PE2 and T-PE2 via the bidirectional LSP2-2.

Let us consider an end-to-end LSP traversing two carrier networks in Fig. 13.24. OAM is end to end or per segment where a segment is between MEPs. The OAM in each segment is independent of any other segment. Similarly, protection or restoration is always between MEPs that are on a per-segment or an end-to-end basis.

OAM monitoring for a PW with switching and without switching is depicted in Figures 13.25 and 13.26 and the related packet format is depicted in Figure 13.3. At the end points or PW stitch point, the process verifies the operational status of the PW and works with the native AC technology by transporting and acting on native AC OAM.

In Figure 13.25, PW over LSP monitoring is depicted where the PW has no switching. End-to-end LSP OAM is used since PW OAM cannot create MIPs at the intercarrier boundary without a PW switching function.

In Figure 13.26, PW over LSP monitoring is depicted where the PW has switching. End-to-end LSP OAM is not required since the PW switching points can support an MIP.

13.4.2   OAM Functions for Proactive Monitoring

Proactive monitoring for an MEG is usually configured at transport path creation time and performed "in-service." Such transactions are universally MEP to MEP in operation while notifications can be node to node, such as MS-PW transactions, or node to MEPs, such as AIS. They are CV, LM, and delay measurements configured at the MEPs and typically reported outside to a management system.

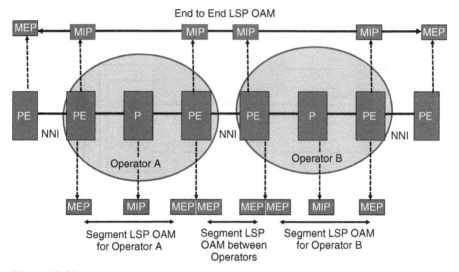

Figure 13.24 An LSP monitoring.

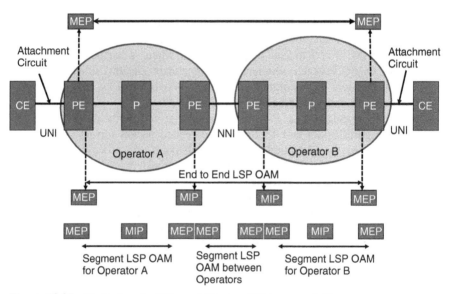

Figure 13.25 Monitoring of a PW over LSP where PW has no switching.

CC and CV functions are used to detect a loss of continuity (LOC) defect between two MEPs in an MEG; an unexpected connectivity defect between two MEGs, such as mismerging or misconnection; and an unexpected connectivity within the MEG with an unexpected MEP. Each Continuity Check and Connectivity Verification (CC-V) OAM packet also includes a globally unique Source

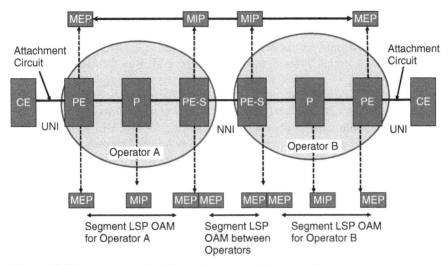

Figure 13.26 Monitoring of a PW over LSP where PW has switching.

MEP identifier, whose value needs to be configured on the source and peer sink MEP(s). The packet is transmitted at a regular, operator configurable rate. Default transmission periods for fault management, performance management, and protection switching are 1 s, 100 ms, and 3.3 ms, respectively.

If protection switching with CC-V defect entry criteria of 12 ms is required in conjunction with the requirement to support 50-ms recovery time, then a transmission period of 3.33 ms is necessary. For a 50-ms recovery time associated with the requirement for a CC-V defect entry criteria period of 35 ms, a transmission period of 10 ms can be used.

When proactive CC-V is enabled, a sink MEP detects an LOC defect when it fails to receive proactive CC-V OAM packets from the source MEP. A sink MEP identifies a misconnectivity defect (e.g., mismerge, misconnection, or unintended looping) when the received packet carries an unexpected globally unique Source MEP identifier.

If proactive CC-V OAM packets are received with the expected Source MEP identifier and a transmission period different than the locally configured reception period then a CC-V period misconfiguration defect is detected.

If proactive CC-V OAM packets are received with the expected globally unique Source MEP identifier but with an unexpected encapsulation then a CC-V unexpected encapsulation defect is detected. When the sink MEP does not receive any CC-V OAM packets with the expected MEP identifier and an unexpected encapsulation for an interval equal at least to 3.5 times the longest transmission period of the pro-active CC-V OAM packets, a mis-connectivity defect is raised. It blocks all the traffic (also including the user data packets) that it receives from the misconnected transport path.

LM is used to exchange periodically (i.e., proactive LM) counter values for ingress and egress packets transmitted and received by the transport path.

Each MEP counts its transmitted and received user data packets. These counts are then correlated in real time with the peer MEP in the ME to calculate packet loss.

For an MEP, near-end packet loss refers to packet loss associated with incoming data packets from the far-end MEP, while far-end packet loss refers to packet loss associated with egress data packets (toward the far-end MEP).

Proactive LM can be operated one way and two way. One-way LM is applicable to both unidirectional and bidirectional (corouted or associated) transport paths, while two-way LM is applicable only to bidirectional (corouted or associated) transport paths.

MIPs, as well as intermediate nodes, do not process the LM information and forward these proactive LM OAM packets as regular data packets.

Proactive DM is used to measure the long-term packet delay and packet delay variation in the transport path monitored by a pair of MEPs. Periodic DM OAM packets are sent from an MEP to a peer MEP. Proactive DM can also be operated one way or two way.

MIPs, as well as intermediate nodes, do not process the DM information and forward these proactive DM OAM packets as regular data packets.

Out-of-service throughput testing can be performed on-demand verifying the bandwidth/throughput of an LSP or a PW before it is put in service, in one-way or two-way mode between MEPs and between MEP and MIP.

When configured to perform such tests, a source MEP inserts OAM test packets with a specified packet size and transmission pattern at a rate to exercise the throughput.

13.4.3 Data Plane Loopback, RDI, AIS

Data plane loopback is an out-of-service function that places a transport path, at either an intermediate or terminating node, into a data plane loopback state such that all traffic received on the looped back interface is sent on the reverse direction of the transport path without modification.

The MEG is locked such that user data traffic is prevented from entering/exiting that MEG, and test traffic is inserted at the ingress of the MEG. This test traffic can be generated from an internal process residing within the ingress node or injected by external test equipment connected to the ingress node. The MEG should be put into a locked state before the diagnostic test is started.

OAM packets are sent to the MIPs or MEPs in the data plane loopback mode via TTL expiry. MIPs can be addressed with more than one TTL value on a corouted bidirectional path set into data plane loopback. If the loopback function is to be performed at an intermediate node, it is only applicable to corouted bidirectional paths. If the loopback is to be performed end to end, it is applicable to both corouted bidirectional or associated bidirectional paths.

When there is a signal fail condition, remote defect indication (RDI) is transmitted by a sink MEP to communicate the condition to its source MEP. In case of corouted and associated bidirectional transport paths, the RDI can be piggybacked onto the CC-V packet. In case of unidirectional transport paths, the RDI can be sent only using an out-of-band return path if it exists and its usage is enabled by policy actions.

When the signal fail condition clears, the MEP should stop transmitting the RDI to its peer MEP. When RDI was piggybacked onto CC-V, it will be cleared from subsequent transmission of CC-V packets. An MEP should clear the RDI defect on reception of an RDI indicator cleared.

The Alarm Reporting relies on an AIS packet to suppress alarms following detection of defective conditions at the server layer. When a server MEP has a signal fail condition, it notifies that to the colocated MPLS-TP client/server adaptation function, which then generates OAM packets with AIS information in the downstream direction to allow the suppression of secondary alarms at the MPLS-TP MEP in the client layer. MIPs and intermediate nodes forward these AIS OAM packets as regular data packets.

The AIS condition is cleared if no AIS packet has been received in 3.5 times the AIS transmission period, which is traditionally 1/s.

Similarly, the Lock Reporting function upon an LKR packet is used to suppress alarms following administrative locking action in the server layer.

When a server MEP is locked, the MPLS-TP client layer adaptation function generates packets with LKR information to allow the suppression of secondary alarms at the MEPs in the client layer. MIPs, as well as intermediate nodes, do not process the LKR information and forward these LKR OAM packets as regular data packets.

Locked condition is cleared if no LKR packet has been received for 3.5 times the transmission period. The LKR transmission period is traditionally 1/s.

Lock Instruct (LKI) command instructs the peer MEP(s) to put the MPLS-TP transport path into a locked condition. It is single-side provisioning for administratively locking (and unlocking) an MPLS-TP transport path. It is also possible to administratively lock (and unlock) an MPLS-TP transport path using two-side provisioning, where the NMS administratively puts both MEPs into an administrative lock condition. In this case, the LKI function is not required/used. MIPs, as well as intermediate nodes, treat LKI OAM packets as regular data packets.

An MEP, on receiving a single-side administrative unlock command from NMS, sends an LKI removal request OAM packet to its peer MEP(s).

13.4.4 Client Failure Indication (CFI), Route Tracing, and Lock Instruct

The CFI is used to help process client defects and propagate a client signal defect condition from the local AC, where the defect was detected, to the far-end AC, in case the client of the transport path does not support a native defect/alarm indication mechanism such as AIS. MIPs, as well as intermediate nodes, do not process the CFI information and forward these proactive CFI OAM packets as regular data packets.

It is often necessary to trace a route covered by an MEG from a source MEP to its peer MEP(s) including all the MIPs in between for troubleshooting purposes.

Route tracing discovers the full list of MIPs and peer MEPs. In case a defect exists, the route trace function will only be able to trace up to the defect and needs to be able to return the incomplete list of OAM entities, allowing localization of the fault.

Route tracing is a basic means to perform CV. For this function to work properly, a return path must be present.

13.5   PROTECTION SWITCHING

MPLS-TP also supports sub-50-ms fully automated protection switching without any control plane and by relying on fast OAM messages and the Protection State Coordination (PSC) mechanism to synchronize the both ends in case of failure. The OAM message can be exchanged at rates up to 3.3-ms intervals to achieve protection switching in sub-50-ms time.

The fast, accurate, and coordinated protection switching time is achieved for LSP and PW by linear protection switching that uses a PSC protocol very similar to that in ITU-T G.8131 [29]. When deployed in ring topologies, a ring protection mechanism similar to that in ITU-T G.8132 [27] can be used. MPLS-TP not only supports various modes of protection (i.e., 1 + 1, 1 : 1, N : 1) but also provides protection on every layer (i.e., PW, LSP, and section). All these protection switching mechanisms are very similar to the current mechanisms being used by SONET/SDH and OTN networks.

MPLS-TP uses existing GMPLS and PW control-plane-based restoration mechanisms applicable to bidirectional paths. Traditional PW redundancy can be used for PE/AC failure protection. In addition, 1 + 1 and 1 : 1 LSP protection are supported, as is full LSP reroute mechanism (as is common in MPLS networks).

Different protection schemes of two concurrent traffic paths (1 + 1) can be provided: one active and one standby path with guaranteed bandwidth on both paths (1 : 1) and one active path and a standby path the resources of which are shared by one or more other active paths (shared protection).

MPLS-TP recovery schemes are applicable to all levels in the MPLS-TP domain (i.e., section, LSP, and PW) providing segment and end-to-end recovery. These mechanisms, which can be based on data plane, control plane, or management plane, support the coordination of protection switching at multiple levels to prevent race conditions occurring between a client and its server layer.

13.6 SECURITY CONSIDERATIONS

Protection against possible attacks on data plane and control plane is necessary. Both G-ACh and its signaling need to be secured. The MPLS-TP control plane security is discussed in Reference 28.

A peer MPLS-TP node can be flooded with G-ACh messages to deny services to others. G-ACh packets can be intercepted as well. To protect against these potential attacks, G-ACh message throttling mechanisms can be used. Messages between line cards and the control processor will be limited.

The contents of G-ACh messages need to be protected. OAM traffic can reveal sensitive information such as performance data and details about the current state of the network. Therefore, authentication, authorization, and encryption should be in place. This will prevent both unauthorized access to vital equipment and third parties from learning about sensitive information about the transport network.

In case of misconfiguration, some nodes can receive OAM packets that they cannot recognize. In such a case, these OAM packets should be silently discarded in order to avoid malfunctions whose effect may be similar to malicious attacks (e.g., degraded performance or even failure).

13.7 CONCLUSION

MPLS-TP is a continuation of T-MPLS. It is client agnostic that can carry L3, L2, L1 services and physical layer agnostic that can run over IEEE Ethernet PHYs, SONET/SDH and OTN, WDM, etc. Its OAM functions are similar to those available in traditional OTNs and Metro Ethernet networks. However, there are substantial differences between Metro Ethernet OAM PDUs and MPL-TP OAM PDUs. In order for them to interoperate, mapping between them is necessary.

MPLS-TP uses a subset of IP/MPLS standards. PHP and ECMP are not supported or made optional. Control and data planes are separated. The operation is fully automated without control plane using NMS.

Alignment between Carrier Ethernet OAM and MPLS-TP OAM is very helpful in providing end-to-end OAM where access is formed by Carrier Ethernet networks while the backbone is formed by MPLS-TP networks.

REFERENCES

1. RFC 3916. Requirements for Pseudo-Wire Emulation Edge-to-Edge; 2004 Sept.
2. RFC 3031. Multiprotocol label switching architecture; 2001 Jan.
3. RFC 3985. Pseudowire emulation edge-to-edge (PWE3) architecture; 2005 Mar.
4. RFC5659. An architecture for multi-segment pseudowire emulation edge-to-edge; 2009 Oct.
5. ITU-T. Recommendation M.3400 TMN management functions; 1997 April.
6. RFC 5654. Requirements of an MPLS transport profile; 2009 Sept.
7. ITU-T G.805. Generic functional architecture of transport networks; 2000 Mar.
8. IETF RFC 3443. Time To Live (TTL) processing in Multi-Protocol Label Switching (MPLS) networks; 2003 Jan.

9. ITU-T Y.1711. Operation & maintenance mechanism for MPLS networks; 2004 Feb.
10. ITU-T Y.1720. Protection switching for MPLS networks; 2002.
11. RFC 4385. Pseudowire emulation edge-to-edge (PWE3)Control Word for Use over an MPLS PSN; 2006 Feb.
12. RFC 5586. MPLS generic associated channel; 2009 June.
13. RFC 3032. MPLS label stack encoding; 2001 Jan.
14. RFC 3443. Time To Live (TTL) Processing in Multi-Protocol Label Switching (MPLS) Networks; 2003.
15. RFC 3270. Multi-Protocol Label Switching (MPLS) Support of Differentiated Services; 2002 May.
16. RFC 3209. RSVP-TE: extensions to RSVP for LSP tunnels; 2001 Dec.
17. RFC 5921. A framework for MPLS in transport networks; 2010 July.
18. RFC 4664. Framework for Layer 2 Virtual Private Networks (L2VPN); 2006 Sept.
19. R. Aggarwal, et. al, BGP based Virtual Private Multicast Service Auto-Discovery and Signaling, draft-raggarwa-l2vpn-p2mp-pw-02.txt; 2009 July.
20. RFC 5085. Pseudowire Virtual Circuit Connectivity Verification (VCCV): a control channel for pseudowires; 2007 Dec.
21. ITU-T G.7710. Common equipment management function requirements; 07/07.
22. RFC 4741. NETCONF configuration protocol; 2006 Dec.
23. RFC 4379. Detecting Multi-Protocol Label Switched (MPLS) data plane failures; 2006.
24. RFC 5884. Bidirectional Forwarding Detection (BFD) for MPLS Label Switched Paths (LSPs); 2010.
25. ITU-T Y.1731. OAM functions and mechanisms for Ethernet based Networks; 2006 May.
26. I. Busi, et. al, Operations, Administration and Maintenance Framework for MPLS-based Transport Networks, draft-ietf-mpls-tp-oam-framework-11.txt; 2011 Feb.
27. ITU-T. Draft recommendation G.8132, T-MPLS Shared Protection Ring; 2008.
28. RFC 5920. Security framework for MPLS and GMPLS networks; 2010 July.
29. ITU-T G.8131. Linear protection switching for transport MPLS (T-MPLS) networks; 2007.

Chapter 14

Virtual Private LAN Services (VPLS)

14.1 INTRODUCTION

Multiprotocol Label Switching (MPLS) is generally accepted as the common convergence technology and facilitates the deployment and management of Virtual Private Networks (VPNs). MPLS-based VPN can be classified as either a layer 2 [1, 2] or a layer 3 point-to-point service or multipoint service:

- Layer 3 multipoint VPNs or Internet Protocol (IP) VPNs that are often referred to as *Virtual Private Routed Networks* (VPRNs),
- Layer 2 point-to-point VPNs, which basically consist of a collection of separate Virtual Leased Lines (VLLs) or pseudowires (PW), and
- Layer 2 multipoint VPNs, or Virtual Private LAN Services (VPLSs).

VPLS is a multipoint service, but unlike IP VPNs, it can transport non-IP traffic and leverages advantages of Ethernet. VPLS is also used within a service provider's (SP) network to aggregate services for delivery to residential and enterprise customers.

Regardless of how the VPN service is used, most enterprise customers use routers at the LAN/WAN boundary. However, VPLS is a layer 2 VPN service and allows the use of layer 2 switches as the customer edge (CE) device.

VPLS is introduced by RFC 2764 [1], as a VPN service, which emulates a LAN segment using IP-based facilities (Fig. 14.1) for a given users. In other words, VPLS creates a layer 2 broadcast domain that is fully capable of learning and forwarding on Ethernet Media Access Control (MAC) addresses. Multiple CEs such as bridges, router, layer 2 switches across a shared Ethernet and IP/MPLS SP network infrastructure are interconnected via transparent LAN service (TLS).

Two VPLS solutions are proposed (Table 14.1):

- RFC 4761 [3]—VPLS using BGP that uses BGP for signaling and discovery. The signaling is similar to RFC 2547 [4]. The encapsulation is similar

Networks and Services: Carrier Ethernet, PBT, MPLS-TP, and VPLS, First Edition. Mehmet Toy.
© 2012 John Wiley & Sons, Inc. Published 2012 by John Wiley & Sons, Inc.

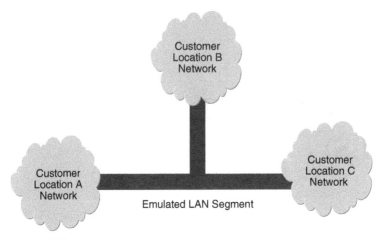

Figure 14.1   VPLS emulating a LAN segment.

Table 14.1   VPLS Solutions

VPLS Implementation Model	Discovery	Signaling
RFC 4761 (BGP-based VPLS)	BGP	BGP
RFC 4762 (LDP-based VPLS)	None	LDP

to Martini encapsulation. BGP signaling mechanism is used in BGP VPNs, therefore, operators can use the same signaling for VPLS.

- RFC 4762 [5]—VPLS using label distribution that uses Label Distribution Protocol (LDP) signaling and basically an extension to the Martini draft [6, 7]. Broadcast mode is used.

Both approaches assume tunnel LSPs between PEs (Fig. 14.2). Pseudowires (PWE3s) are setup over tunnel LSPs (i.e., Virtual Connection (VC) LSPs). The ID for a PWE3 is VCID, which is the VPN ID. RFC 4762 [5] is more widely implemented.

VPLS as described in [5] is a class of VPN that allows the connection of multiple sites in a single bridged domain over a provider-managed IP/MPLS network (Fig. 14.3).

All customer sites that are in one or more metro areas in a VPLS instance appear to be on the same LAN, regardless of their location. VPLS uses an Ethernet interface as the customer handoff, simplifying the LAN/WAN boundary and allowing rapid and flexible service provisioning.

Use of IP/MPLS routing protocols and procedures instead of the Spanning Tree Protocol (STP) and MPLS labels instead of Virtual LAN (VLAN) IDs improves the scalability of the VPLS service.

In VPLS, customers maintain complete control over their routing, since all the customer routers in the VPLS are part of the same subnet (LAN). The result is

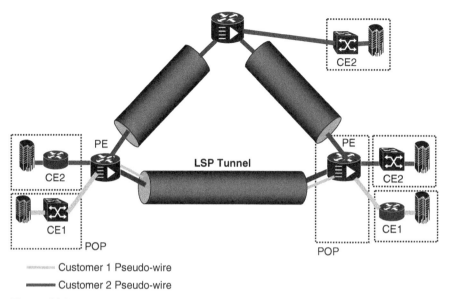

Figure 14.2 Tunnel LSPs between PEs for VPLS.

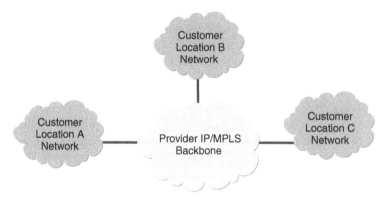

Figure 14.3 VPLS-single bridge domain over IP/MPLS backbone.

a simplified IP addressing plan, especially when compared to a mesh constructed from many separate point-to-point connections. The SP also benefits from reduced complexity to manage the VPLS service, as it has no awareness or participation in the customer's IP addressing space and routing.

A VPLS-capable network consists of CEs, provider edges (PEs), and a core MPLS network:

- The CE device is a router or switch located at the customer's premises. It can be owned and managed by the customer or SP. It is connected to the PE via an attachment circuit (AC). Ethernet is the interface between the CE and PE.

- The PE device is where all the VPN intelligence resides, the VPLS origi-
  nates and terminates, and all the necessary tunnels are set up to connect to
  all the other PEs. As VPLS is an Ethernet layer 2 service, the PE must be
  capable of MAC learning, bridging and replication on a per-VPLS basis.
- The IP/MPLS core network interconnects the PEs. It does not really participate in the VPN functionality. Traffic is simply switched based on the
  MPLS labels.

The basis of any multipoint VPN service (IP VPN or VPLS) is the full mesh
of MPLS tunnels that are Label Switched Paths (LSPs) set up between all the
PEs participating in the VPN service. For each VPN instance, an Ethernet PW
runs inside tunnels. CEs appear to be connected by a single LAN (Fig. 14.4).
An Ethernet bridge determines where PE needs to send a frame. As a result,
VPLS can be defined as a group of virtual switch instances (VSIs) that are
interconnected using Ethernet over MPLS circuits in a full mesh topology to
form a single logical bridge.

In addition to its MPLS functions, A VPLS-enabled PE has VPLS code module based on IETF RFCs and Bridging module based on IEEE 802.1D learning
bridge, as shown in Figure 14.4. Service provider network inside rectangle looks
similar to a single Ethernet bridge. If CE is a router then PE only sees one MAC
per customer location.

Depending on the exact VPLS implementation, when a new PE or VPLS
instance is added, the amount of effort to establish this mesh of LSPs can vary
dramatically. Once the LSP mesh is built, the VPLS instance on a particular PE
is now able to receive Ethernet frames from the customer site and, based on the
MAC address, switch those frames into the appropriate LSP. This is possible
because VPLS enables the PE router to act as a learning bridge with one MAC
table per VPLS instance on each PE. In other words, the VPLS instance on the
PE router has a MAC table that is populated by learning the MAC addresses, as
the Ethernet frames enter specific physical or logical ports, exactly the same way
that an Ethernet switch does.

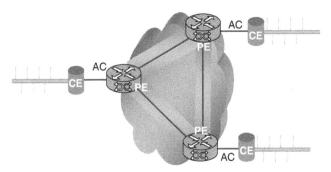

Figure 14.4   Full mesh of PEs.

Once an Ethernet frame enters via a customer-facing ingress port, the destination MAC address is looked up in the MAC table and the frame is sent unaltered as long as the MAC table contains the MAC address into the LSP that will deliver it to the correct PE attached to the remote site. If the MAC address is not in the MAC address table, the Ethernet frame is replicated and flooded to all logical ports associated with that VPLS instance, except the ingress port where it just entered. Once the PE hears back from the host that owns that MAC address on a specific port, the MAC table is updated in the PE. Similar to a switch, the MAC addresses that have not been used for a certain amount of time are aged out.

Each MPLS LSP consists of two unidirectional LSPs, providing bidirectional transport for each bidirectional PW. A VPLS network is established by first defining the MPLS LSPs that will support the VPLS PW tunnels. The paths are determined using the Open Path Shortest First (OSPF) link-state protocol, which determines the shortest path for the LSP to the target destination. A full bidirectional mesh of LSPs needs to be established between all participating VPLS PEs. LDP or Resource Reservation Protocol (RSVP) is then used to distribute the label information to allow label swapping at each node. After that, the PWE3 tunnels are established over the existing LSPs. Either LDP or BGP is used to exchange PWE3 labels.

In order to establish MPLS LSPs, OSPF-TE and Resource Reservation Protocol-Traffic Engineering (RSVP-TE) can be used as well where OSPF-TE will take bandwidth availability into account when calculating the shortest path, while RSVP-TE allows reservation of bandwidth.

The MPLS mechanisms and tunnels are merely used as transport. Pseudowires are generic and designed to be implemented on any suitable packet-switched network. It is therefore possible to envisage a VPLS-like approach based on other tunnel mechanisms, such as MPLS-TP and PBB-TE.

There are two key components of VPLS, PE discovery and signaling. PE discovery can be via provisioning application, BGP, and Remote Authentication Dial-In User Service (RADIUS). Signaling can be targeted through LDP and BGP.

The advantages of VPLS may be summarized as follows:

- Complete customer control over their routing where there is a clear demarcation of functionality between SP and customer that makes troubleshooting easier.
- Ability to add a new site without configuration of the SP's equipment or the customer equipment at existing sites.
- Faster provisioning, with potential for customer provisioned bandwidth-on-demand.
- Minimizes MAC address exposure, improving scaling by having one MAC address per site (i.e., one MAC per router) or per service.
- Improves customer separation by having CE router to block unnecessary broadcast or multicast traffic from customer LANs.

- MPLS core network emulates a flat LAN segment that overcomes distance limitations of Ethernet-switched networks and extends Ethernet broadcast capability across WAN.
- Point-to-multipoint connectivity connects each customer site to many customer sites.
- A single CE–PE link transmits Ethernet packets to multiple remote CE routers.
- Fewer connections required to get full Connectivity among customer sites.
- Adding, removing, or relocating a CE router requires configuring only the directly attached PE router. This results in substantial OpEx savings.

14.2   DATA PLANE

Data plane of VPLS deals with frame encapsulation, classification, MAC address learning and aging, and forwarding. They are described in the following sections.

14.2.1   VPLS Encapsulation

In a VPLS, a customer Ethernet frame without preamble is encapsulated with a header as defined in RFC 4448 [8]:

- If the frame, as it arrives at the PE, has an encapsulation that is used by the local PE as a service delimiter to identify the customer and/or the particular service of that customer then that encapsulation may be stripped before the frame is sent into the VPLS. As the frame exits the VPLS, the frame may have a service-delimiting encapsulation inserted.
- If the frame, as it arrives at the PE, has an encapsulation that is not service delimiting then its encapsulation should not be modified by the VPLS.

According to rules mentioned above, if a customer frame arrives at a customer-facing port with a VLAN tag that identifies the customer's VPLS instance, the tag would be stripped before it is encapsulated in the VPLS. At egress, the frame may be tagged again, if a service-delimiting tag is used, or it may be untagged, if none is used. At both ingress and egress, dealing with service delimiters is a local action that neither PE has to signal to the other.

If a customer frame arrives at a customer-facing port with a VLAN tag that identifies a VLAN domain in the customer L2 network then the tag is not modified or stripped.

14.2.2   Classification and Forwarding

VPLS packets are classified as belonging to a given service instance and associated forwarding table based on the interface over which the packet is received.

After that they are forwarded based on the destination MAC address. The former mapping is determined by configuration.

In order to offer different Classes of Service within a VPLS, 802.1p bits in a customer Ethernet frame with a VLAN tag is mapped to EXP bits in the pseudowire and/or tunnel label. This mapping can be different for each VPLS, as each VPLS customer may have its own view of the required behavior for a given setting of 802.1p bits.

When a bridge receives a packet to a destination that is not in its Forwarding Information Base (FIB), a PE will flood packets to an unknown destination to all other PEs in the VPLS. An Ethernet frame whose destination MAC address is the broadcast MAC address is sent to all stations in that VPLS.

In forwarding a broadcast Ethernet frame or one with an unknown destination MAC address, **split horizon rule** is used to prevent loops. When a PE receives a broadcast Ethernet frame or one with an unknown destination MAC address from an attached CE, PE must send a copy of the frame to every other attached CE, as well as to all other PEs participating in the VPLS. When the PE receives the frame from another PE, this PE must send a copy of the packet only to attached CEs and must not send the frame to other PEs, as the other PE would have already done so.

14.2.3 MAC Address Learning and Aging

VPLS is a multipoint service, and therefore, the entire SP network appears as a single logical learning bridge for each VPLS. The logical ports of this SP bridge are the customer ports as well as the pseudowires on a VPLS edge (VE), which is the designated multihomed PE router acting as the end point for the VPLS pseudowire from the remote PE router. The SP bridge learns MAC addresses at its VEs, while a learning bridge learns MAC addresses on its ports. Source MAC addresses of packets with the logical ports on which they arrive are associated in the FIB to forward packets. If a VE learns a source MAC address S-MAC on logical port P, and later sees S-MAC on a different port R, then the VE updates its FIB to reflect the new port R.

Ethernet MAC learning based on the 6-octet MAC address is called *unqualified learning*. If the key for learning includes the VLAN tag when present, it is called *qualified learning*.

Choosing between qualified and unqualified learning mainly involves in whether one wants a single global broadcast domain (unqualified) or a broadcast domain per VLAN (qualified). The latter makes flooding and broadcasting more efficient, but requires larger MAC tables.

In unqualified learning, MAC addresses need to be unique and nonoverlapping among customer VLANs else they cannot be differentiated within the VPLS instance. An application of unqualified learning is port-based VPLS service for a given customer where all the traffic on a physical port, which may include multiple customer VLANs, is mapped to a single VPLS instance.

In qualified learning, each customer VLAN is assigned to its own VPLS instance. MAC addresses among customer VLANs may overlap with each other, but they will be handled correctly, as each customer VLAN has its own FIB. As a result, the qualified learning can result in large FIB table sizes, as the logical MAC address is now a VLAN tag + MAC address.

For STP to work in qualified learning mode, a VPLS-PE must be able to forward STP BPDUs over the proper VPLS instance.

VPLS PEs have an aging mechanism to remove a MAC address associated with a logical port, similar to learning bridges. This is needed so that a MAC address can be relearned if it moves from a logical port to another logical port. In addition, aging reduces the size of a VPLS MAC table to only active MAC addresses.

The age of a source MAC address S-MAC on a logical port P is the time since it was last seen as a source MAC on port P. If the age exceeds the aging time T, S-MAC is flushed from the FIB that keeps track of the mapping of customer Ethernet frame addressing and the appropriate PW to use. Every time S is seen as a source MAC address on port P, S-MAC's age is reset.

PEs that learn remote MAC addresses should be able to remove unused entries associated with a PW label via an aging mechanism such as an aging timer. The aging mechanism conserves memory and increases efficiency of operation. The aging timer for MAC address S-MAC should be reset when a packet with the source MAC address S-MAC is received.

14.3 LDP-BASED VPLS

An interface participating in a VPLS must be able to flood, forward, and filter Ethernet frames. Each PE will form remote MAC address to PW associations and associate directly attached MAC addresses to local-customer-facing ports.

Connectivity between PEs can be via MPLS transport tunnels as well as other tunnels over PWs such as GRE, L2TP, and IPsec. The PE runs the LDP signaling protocol and/or routing protocols to set up PWs, setting up transport tunnels to other PEs and delivering traffic over PWs.

The AC providing access to a customer site could be a physical Ethernet port, a VLAN, an ATM PVC carrying Ethernet frames, an Ethernet PW, etc.

14.3.1 Flooding, Forwarding, and Address Learning

One of the capabilities of an Ethernet service is that frames sent to broadcast addresses and to unknown destination MAC addresses are flooded to all ports. To achieve flooding within the SP network, all unknown unicast, broadcast, and multicast frames are flooded over the corresponding PWs to all PE nodes participating in the VPLS, as well as to all ACs.

To forward a frame, a PE associates a destination MAC address with a PW. MAC addresses on both ACs and PWs are dynamically learned. Packets are forwarded and replicated across both ACs and PWs.

Unlike BGP VPNs [9], reachability information is not advertised and distributed via a control plane. Reachability is obtained by standard learning bridge (Fig. 14.5) functions in the data plane. Standard learning, filtering, and forwarding actions [10] are performed when a PW or AC state changes.

It may be desirable to remove or unlearn MAC addresses that have been dynamically learned for faster convergence. This is accomplished by sending an LDP Address Withdraw message with the list of MAC addresses to be removed to all other PEs over the corresponding LDP sessions. An LDP Address Withdraw message contains a new TLV, the MAC List TLV specifying a list of MAC addresses that can be removed or unlearned using the LDP Address Withdraw message. Its format is shown in Figure 14.6.

The Address Withdraw message with MAC List TLVs expedites removal of MAC addresses as the result of a topology change (e.g., failure of the primary link for a dual-homed VPLS-capable switch). PEs that do not understand the message can continue to participate in the VPLS.

14.3.2 Tunnel Topology

Tunnels are set up between PEs to aggregate traffic. PWs are signaled to demultiplex encapsulated Ethernet frames from multiple VPLS instances that traverse the transport tunnels.

The topology of a VPLS must be loop-free topology such as full mesh of PWs. In a full mesh topologic PWs established between PEs, there is no need to relay packets between PEs. A simpler loop-breaking rule: the *split horizon* rule, whereby a PE must not forward traffic from one PW to another in the same VPLS mesh, is instantiated.

To prevent forwarding loops, *split horizon* rule is used where a PE must not forward traffic from one PW to another in the same VPLS mesh. The fact that there is always a full mesh of PWs between the PE devices ensures that every destination within the VPLS will be reached by a broadcast packet.

VPLS and PW relationships are shown in Figures 14.7 and 14.8.

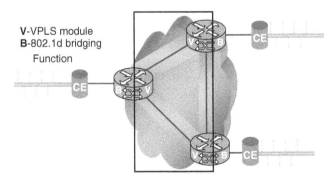

Figure 14.5 PE consisting of VPLS module (V) and IEEE 802.1d bridge (B).

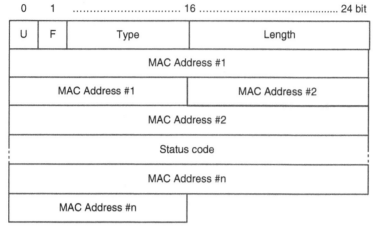

- U bit: Unknown bit. This bit MUST be set to 1. If the MAC address format is not understood, then the TLV is not understood and MUST be ignored.

- F bit: Forward bit. This bit MUST be set to 0. since the LDP mechanism used here is targeted, the TLV MUST NOT be forwarded.

- Type: Type field. This field MUST be set to 0 x 0404. This identifies the TLV type as MAC List TLV.

- Length: Length field. This field specifies the total length in octets of the MAC addresses in the TLV. The length MUST be a multiple of 6.

- MAC Address: The MAC address(es) being removed.

Figure 14.6 MAC list TLV address.

14.3.3 Discovery

A PE either can be configured with the identities of all the other PEs in a given VPLS or can use some protocol to discover the other PEs. The latter is called *autodiscovery*.

The former approach is fairly configuration intensive, as the PEs participating in a given VPLS are fully meshed. When a PE is added to, or removed from the VPLS, the VPLS configuration on all PEs in that VPLS must be changed. Therefore, autodiscovery is critical to keep the operational costs low, in automatically creating the LSP mesh. PEs can be manually configured when needed.

Multiple VPLS services can be offered over the same set of LSP tunnels. Signaling as specified in References [5] and [4] is used to negotiate a set of ingress and egress virtual VC labels on a per service basis. The VC labels are used by the PE routers for demultiplexing traffic arriving from different VPLS services over the same set of LSP tunnels.

In RFC 4762 [5], the LDP is used to set up these tunnels. Alternatively, the RSVP-TE or a combination of LDP and RSVP-TE can be used. Multipoint

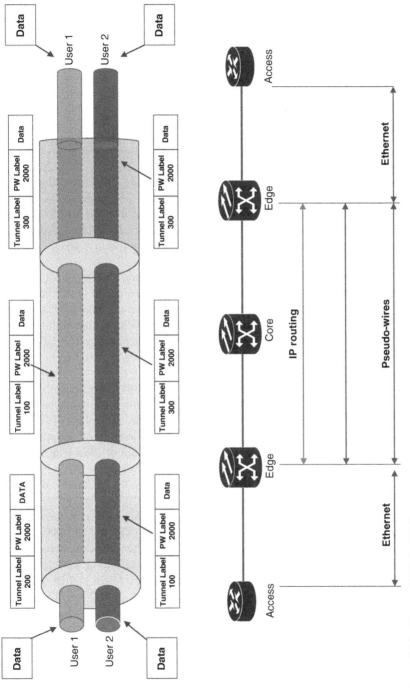

Figure 14.7 Usage of PWs in VPLS.

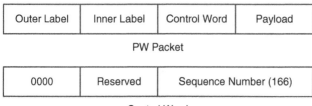

Outer Label	Inner Label	Control Word	Payload

PW Packet

0000	Reserved	Sequence Number (166)

Control Word

Figure 14.8   Format of PW packet and its control word.

VPNs can be created on top of this full mesh, hiding the complexity of the VPN from the backbone routers.

For every VPLS instance, a full mesh of inner tunnels that are called *PW* is created between all the PEs that participate in the VPLS instance. An autodiscovery mechanism locates all the PEs participating in a given VPLS instance. The SP can either configure the PE with the identities of all the other PEs in a given VPLS or can select its preferred autodiscovery mechanism, such as the (RADIUS).

A PW consists of a pair of point-to-point single hop unidirectional LSPs in opposite directions, each identified by a PW label, also called a *VC* label. PW labels are exchanged between a pair of PEs using the targeted LDP signaling protocol. The VPLS identifier is exchanged with the labels, so that both PWs can be linked and be associated with a particular VPLS instance. The PW labels only have local significance between each pair of PEs. Each PE will form a MAC address to PW associations.

In VPLS, unlike BGP, VPNs as defined in RFC 4364 [9], reachability information is not advertised and distributed via a control plane. Reachability is obtained by standard learning bridge functions in the data plane. PE routers need to implement a virtual bridge (Fig. 14.5) functionality to support MAC learning, packet replication, and forwarding. The creation of PWs with a pair of LSPs enables a PE to participate in MAC learning. When the PE receives an Ethernet frame with an unknown source MAC address, the PE knows on which VC it was sent.

When a packet arrives on a PW, if the source MAC address is unknown, it needs to be associated with the PW, so that outbound packets to that MAC address can be delivered over the associated PW. Similarly, when a packet arrives on an access port, if the source MAC address is unknown, it needs to be associated with the access port, so that outbound packets to that MAC address can be delivered over the associated access port.

Unknown packets (i.e., the destination MAC address has not been learned) are replicated and forwarded on all LSPs to the PE routers participating in that service until the target station responds, and the MAC address is learned by the PE routers associated with that service.

As LDP-based VPLS [5] does not specify autodiscovery, the SP must know explicitly which PEs are part of the VPLS instance. For every VPLS instance present on a PE, the SP will have to configure that PE with the addresses of all other PEs that are part of that VPLS instance. This process is operationally intensive and subject to human error.

Every time a PE joins a VPLS domain, the SP must manually look up the other PEs that are part of that VPLS domain. Once this information is attained, a full mesh of LDP sessions between that PE and every other PE that is a part of the VPLS domain is built. This overhead of a full mesh of LDP sessions is required because LDP does not have the advantage of BGP's route reflector (RR) architecture. For a VPLS instance with a small number of sites, the burden of LDP may not be that noticeable. However, it becomes more and more significant with the growth of the service, especially when LDP signaling sessions are authenticated via MD5. With a full LDP mesh, MD5 keys need to be configured on either end of every LDP session.

If the VPLS instance spans multiple autonomous systems (ASs), the globally significant 32-bit VCID used by LDP signaling requires manual coordination between ASs. For a VPLS instance spanned two ASs, the providers would need to use the same LDP VCID for that VPLS.

14.3.4 LDP-Based Signaling

A full mesh of LDP sessions is used to establish the mesh of PWs. Once an LDP session has been formed between two PEs, all PWs between these two PEs are signaled over this session. A hierarchical topology can be used to minimize the size of the VPLS full mesh when there is a large number of PWs.

Label mapping messages are exchanged between participating PEs to create tunnels, while label withdrawal messages are used to tear down the tunnels. The message has FEC TLV (Fig. 14.9) with PWid FEC element or Generalized ID FEC element.

Both ends of a pseudowire need to be configured with the PWid FEC, which is to be unique only within the targeted LDP session between the two PEs, and IP addresses. If pseudowires are the members of layer-2 VPN and a discovery mechanism is used, or where switched on-demand PWs are used or when distributed VPLS is used, it is desirable that only one end of a pseudowire is configured with the configuration information. That information can be combined with the information carried in the discovery mechanism to successfully complete the signaling procedures. The PWid FEC needs to be globally unique.

The generalized ID FEC (Fig. 14.9) tries to address all these cases. One can view the Generalized ID FEC as one signaling scheme that covers the superset of cases, including the PWid provisioning model.

PWid FEC contains PW type, Control bit (indicates presence of the control word), Group ID, PW ID, and interface parameters Sub-TLV. Generalized PW FEC contains [11] only PW type, Control bit, Attachment Group Identifier (AGI), Source Attachment Individual Identifier (SAII), and Target Attachment Individual Identifier (TAII) that replace the PW ID.

The Group ID and the interface Parameters are contained in separate TLVs, called *the PW Grouping TLV* and *the interface Parameters TLV*.

Either of these types of PW FEC may be used for the setup of TDM PWs with the appropriate selection of PW types and interface parameters. LDP signaling is designed for setting up point-to-point connections and used in Martini pseudowire services. After the label exchange, traffic parameters and OAM messages need to be negotiated for efficient signaling of each PW information.

The PWid FEC element can be used if a unique 32-bit value has been assigned to the PW, and if each endpoint has been provisioned with that value. The generalized PWid FEC element requires that the PW endpoints be uniquely identified. In addition, the end point identifiers are structured to support applications where the identity of the remote end points needs to be autodiscovered rather than statically configured.

The generalized PWid FEC Element is FEC type 0x81. Its fields are depicted in Figure 14.9:

- Control bit (C): this bit is used to signal the use of the control word as specified in Reference [12].
- PW type: the allowed PW types are Ethernet (0x0005) and Ethernet tagged mode (0x004), as specified in Reference [13].
- PW info length as specified in Reference [12].
- AGI, Length, Value: the unique name of this VPLS. The AGI identifies a type of name, and Length denotes the length of Value, which is the name of the VPLS. We use the term AGI interchangeably with VPLS identifier.

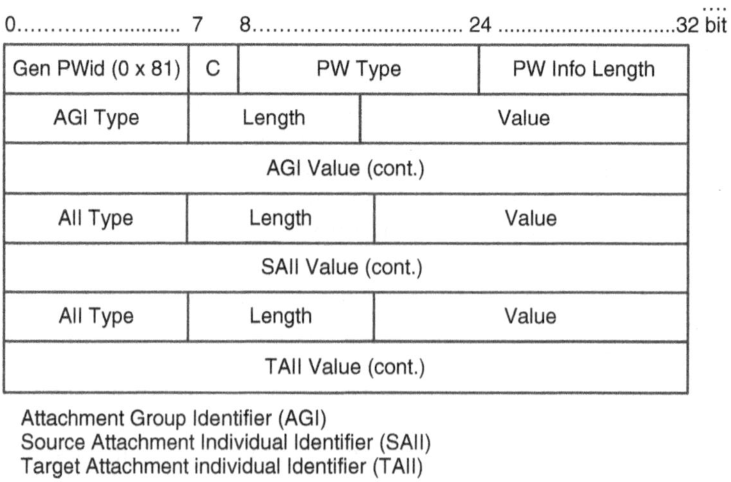

0....................... 7	8............................... 24 ....32 bit		
Gen PWid (0 x 81)	C	PW Type	PW Info Length
AGI Type	Length	Value	
AGI Value (cont.)			
All Type	Length	Value	
SAll Value (cont.)			
All Type	Length	Value	
TAll Value (cont.)			

Attachment Group Identifier (AGI)
Source Attachment Individual Identifier (SAII)
Target Attachment individual Identifier (TAII)

Figure 14.9   Generalized ID FEC element [12].

- TAII, SAII: these are null because the mesh of PWs in a VPLS terminates on MAC learning tables, rather than on individual ACs. The use of nonnull TAII and SAII is reserved for future enhancements.

14.3.5   Data Forwarding on an Ethernet PW

Let us consider a customer with three sites (Fig. 14.10), through CE1, CE2, CE3, and CE4, respectively. Assume that this configuration was determined using an unspecified autodiscovery mechanism. CE1, CE2, and CE3 are end stations at different customer sites and their ACs to their respective PE devices have been configured in the PEs. A VPLS has been set up between PE1, PE2, and PE3 such that the PEs belong to a particular VPLS identifier (AGI), 100.

Three PWs between PE1, PE2, and PE3 need to be created, each consisting of a pair of unidirectional LSPs or VCs. For VC-label signaling between PEs, each PE initiates a targeted LDP session to the peer PE and communicates to the peer PE about what VC label to use when sending packets for the considered VPLS. In this example, PE1 signals PW label 102 to PE2 and 103 to PE3, and PE2 signals PW label 201 to PE1 and 203 to PE3.The specific VPLS instance is identified in the signaling exchange using a service identifier.

Assume a packet from CE1 is bound to CE2. When it leaves CE1, say it has a source MAC address of MAC1 and a destination MAC of MAC2. If PE1 does not know where MAC2 is, it will flood the packet by sending it to PE2 and

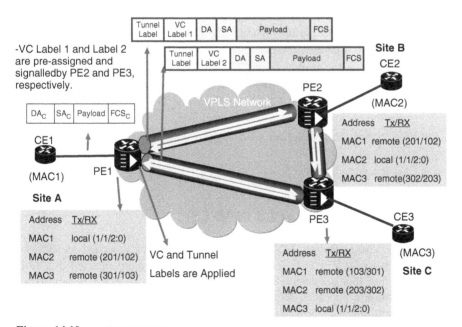

Figure 14.10   A VPLS example.

PE3. When PE2 receives the packet, it will have a PW label of 201. PE2 can conclude that the source MAC address MAC1 is behind PE1, as it distributed the label 201 to PE1. It can therefore associate MAC address MAC1 with PW label 102.

Once the VPLS instance with Svc-id 100 has been created, the first packets can be sent and the MAC learning process starts. Assuming CE2 is sending a packet to PE2 destined for CE1:

- PE2 receives the packet and learns from the source MAC address that CE2 can be reached on local port 1/1/2:0, stores this information in the FIB for Svc-id 100.
- PE2 does not yet know the destination MAC address CE1, so it floods the packet to PE1 with VC label 201 on the corresponding MPLS outer tunnel and to PE3 with VC label 203 on the corresponding MPLS outer tunnel.
- PE1 learns from VC label 201 that CE2 is behind PE2; it stores this information in the FIB for Svc-id 100.
- PE3 learns from VC label 203 that CE2 is behind PE2; it stores this information in the FIB for Svc-id 100.
- PE1 strips off label 201, does not know the destination CE1 and floods the packet on ports 1/1/2:0. PE1 does not flood the packet to PE3 because of the split horizon rule.
- PE3 strips off label 203. PE3 does not know the destination M1 and sends the packet on port 1/1/2:0, therefore, does not flood the packet to PE1 because of the split horizon rule.
- CE1 receives the packet. When CE1 receives the packet from CE2, it replies with a packet to CE2.
- PE1 receives the packet from CE1 and learns that CE1 is on local port 1/1/2:0. It stores this information in the FIB for Svc-id 100.
- PE1 already knows that CE2 can be reached via PE2 and therefore only sends the packet to PE2 using VC label 102.
- PE2 receives the packet for CE2. It knows that CE2 is reachable on port 1/1/2:0.
- CE2 receives the packet.

14.3.6 Hierarchical VPLS (H-VPLS)

Hierarchical Virtual Private LAN Service (H-VPLS) was designed to address scalability issues in VPLS. In VPLS, all PE nodes are interconnected in a full mesh to ensure that all destinations can be reached. In H-VPLS, a new type of node is introduced called *the multitenant unit* (MTU) (Figure 14.11), which aggregates multiple CE connections into a single PE, to reduce the number of PE-to-PE connections.

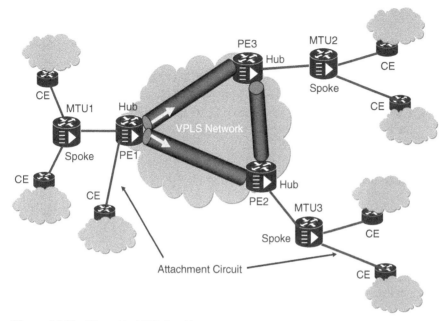

Figure 14.11 Hierarchical VPLS architecture.

The MTU is connected to the nearest PE using a *spoke* PW. PE-to-PE PWs are now referred to as *hub* PWs. Full connectivity is only required between PEs, which allows more linear scaling of the network.

Spoke PWs are created between the MTUs and the PE routers. Connection for the spoke PW implementation can be either an IEEE 802.1Q tagged connection or an MPLS LSP with LDP signaling. Spoke connections are generally created between layer 2 switches placed at the MTU and the PE routers placed at the SP's point of presence (POP). This considerably reduces both the signaling and replication overhead on all devices.

It is often beneficial to extend the VPLS service tunneling techniques into the access switch domain. This can be accomplished by treating the access device as a PE and provisioning PWs between this device and other edge devices, as a basic VPLS.

In H-VPLS, a CE is attached to an MTU via an AC. An AC from a specific customer is associated with a virtual bridge that is dedicated to that customer within the considered MTU.

An AC may be a physical or a VLAN tagged logical port. In the basic scenario, an MTU has one uplink to a PE. This uplink contains one spoke PW for each VPLS served by the MTU. The end points of this spoke PW are an MTU and a PE. Spoke PWs can be implemented using LDP-signaled MPLS PWs, if the MTU is MPLS enabled. Alternatively, they can be implemented using service provider Virtual LANs (S-VLANs) whereby every VLAN on the MTU-PE uplink of an Ethernet aggregation network identifies a spoke PW.

Let us consider the operation in Figure 14.11:

- Ethernet frames with known MAC addresses are switched accordingly within the VPLS.
- Frames with unknown or broadcast MAC addresses that are received from the PW are replicated and sent to all attached CE devices within the VPLS.
- Frames with unknown or broadcast MAC addresses that are received from a CE device are sent over the PW to the PE and to all other attached CE devices within the VPLS.
- Frames coming from the PW and CE devices with unknown MAC addresses are learned and aged within the VPLS.

The PE device needs to implement one VB for each VPLS served by the PE-attached MTUs. A particular spoke PW is associated with the PE VB dedicated to the considered VPLS instance. In the core network, the PE has a full mesh of PWs to all other PEs that serve the VPLS, as in the normal VPLS scenario.

H-VPLS also enables VPLS services to span multiple metro networks. A spoke connection that can be a simple tunnel is used to connect each VPLS service between the two metros. A set of ingress and egress VC labels is exchanged for each VPLS service instance to be transported over this LSP. The PE routers at each end treat this as a virtual spoke connection for the VPLS service in the same way as the PE-MTU connections. This architecture minimizes the signaling overhead and avoids a full mesh of VCs and LSPs between the two metro networks.

Spoke connectivity between MTU and PE depends on PE capabilities. PE that supports bridging functions, routing, and MPLS encapsulation is called *PE-rs*. In Figure 14.13, each MTU-s has one PW to PE-rs. The PE-rs devices are connected in a basic VPLS full mesh. For each VPLS service, a single spoke PW is set up between the MTU-s and the PE-rs based on [12]. Unlike traditional PWs that terminate on a physical port or a VLAN, a spoke PW terminates on a VSI on the MTU-s and the PE-rs devices (Fig. 14.12). The VSI is a virtual MAC bridging instance with *split horizon* learning that is unknown MAC addresses received from a PW tunnel will not be forwarded to other PW tunnels, to avoid learning loops.

The MTU-s and the PE-rs treat each spoke connection similar to an AC of the VPLS service. The PW label is used to associate the traffic from the spoke to a VPLS instance.

If PE does not support bridging function (i.e., PE-r), for every AC participating in the VPLS service, PE-r creates a point-to-point PW that terminates on the VSI of PE1-rs as shown in Figure 14.13.

For every port that is supported in the VPLS service, a PW is set up from the PE-r to the PE-rs. Once the PWs are set up, there is no learning or replication function required on the part of the PE-r. Traffic between CEs that are connected to PE-r is switched at PE1-rs and not at PE-r.

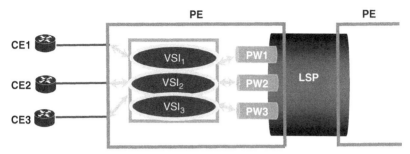

Figure 14.12 A diagram of MTU with VSI and PE-rs with VSI.

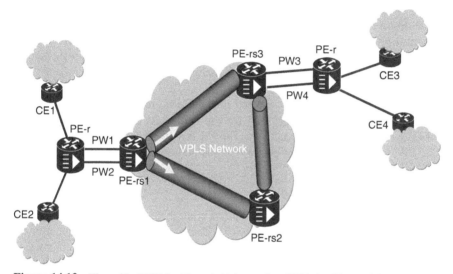

Figure 14.13 Hierarchical VPLS with nonbridging spokes (PW1 should go to PE_rs3).

If PE-r devices use Provider VLANs (P-VLAN) as demultiplexers instead of PWs, PE1-rs can treat them as such and map these "circuits" into a VPLS domain to provide bridging support between them.

This approach adds more overhead than the bridging-capable (MTU-s) spoke approach, since a PW is required for every AC that participates in the service versus a single PW required per service (regardless of ACs) when an MTU-s is used. However, this approach offers the advantage of offering a VPLS service in conjunction with a routed Internet service without requiring the addition of new MTU-s.

Virtual Private Wire Service (VPWS)

The VPWS-PE (Virtual Private Wire Service) and VPLS-PE are functionally very similar, in that they both use forwarders to map ACs to pseudowires. The only difference is that while the forwarder in a VPWS-PE does a one-to-one mapping between the AC and pseudowire, the forwarder in a VPLS-PE is a VSI that maps multiple ACs to multiple pseudowires.

In a VPWS, each CE device is presented with a set of point-to-point virtual circuits. The other end of each virtual circuit is another CE device. Frames transmitted without affecting their content by a CE on such a virtual circuit are received by the CE device at the other end point of the virtual circuit. The PE thus acts as a virtual circuit switch.

As stated earlier, an MTU-s supports layer-2 switching and bridging on all its ports including the spoke. It is treated as a virtual port. Packets to unknown destinations are replicated to all ports in the service including the spoke. Once the MAC address is learned, traffic between CEs that are connected to the spoke will be switched locally by the MTU-s, while traffic between a CE and any remote destination is switched directly onto the spoke and sent to the PE-rs over the point-to-point PW. As the MTU-s is bridging capable, only a single PW is required per VPLS instance for any number of access connections in the same VPLS service. This further reduces the signaling overhead between the MTU-s and PE-rs.

If the MTU-s is directly connected to the PE-rs, other encapsulation techniques, such as Q-in-Q, can be used for the spoke. At PE-rs, the spoke from the MTU-s is treated as a virtual port. The PE-rs will switch traffic between the spoke PW, hub PWs, and ACs once it has learned the MAC addresses.

Protection via Dual Homing

Protection against PW link failure and PE-rs equipment failure, MTU-s or PE-r, can be dual homed into two PE-rs devices. These two PE-rs devices must be a part of the same VPLS service instance. The MTU-s sets up two PWs connected to each PE-rs for each VPLS instance. If we use 1 + 1 protection scheme, one of the two PWs can be designated as primary that carries active traffic, while the other can be designated as secondary which is a standby. The MTU-s negotiates the PW labels for both the primary and secondary PWs, but does not use the secondary PW unless the primary PW fails.

The MTU-s should control the usage of the spokes to the PE-rs devices. If the spokes are PWs then LDP signaling is used to negotiate the PW labels. The hello messages of the LDP session could be used to detect failure of the primary PW.

On failure of the primary PW, MTU-s immediately switches to the secondary PW. At this point, the PE-rs3 that terminates the secondary PW starts learning MAC addresses on the spoke PW. All other PE-rs nodes in the network think that CE1 and CE2 are behind PE-rs1 and may continue to send traffic to PE1-rs until they learn that the devices are now behind PE-rs3. The unlearning process can take a long time and may adversely affect the connectivity of higher level protocols from CE1 and CE2.

To enable faster convergence, the PE3-rs where the secondary PW got activated may send out a flush message, using the MAC List TLV, to all PE-rs nodes.

On receiving the message, PE-rs nodes flush the MAC addresses associated with that VPLS instance.

Hierarchical VPLS Model Using Ethernet Access Network

Ethernet-based access networks are currently deployed by SPs to offer VPLS services to their customers. Tagged or untagged Ethernet traffic of customers can be tunneled via an additional VLAN tag (S-Tag) to the customer's data. Therefore, there is a one-to-one correspondence between an S-Tag and a VPLS instance.

The PE-rs needs to perform bridging functionality over the standard Ethernet ports toward the access network, as well as over the PWs toward the network core. In this model, the PE-rs may need to run STP toward the access network, in addition to split-horizon over the MPLS core. The PE-rs needs to map a S-VLAN to a VPLS instance and its associated PWs, and vice versa.

As each P-VLAN corresponds to a VPLS instance, the total number of VPLS instances supported is limited to 4K. This 4K limit applies only within an Ethernet access network and not to the entire network.

14.4 BGP APPROACH

VPLS control plane functions mainly autodiscovery and provisioning of PWs are accomplished with a single BGP update advertisement. In the autodiscovery, each PE discovers other PEs that are a part of a given VPLS instance via BGP. When a PE joins or leaves a VPLS instance, only the affected PE's configuration changes, while other PEs automatically find out about the change and adapt.

BGP Route Target (RT) community (or extended communities) [3, 13, 14] is used to identify the members of a VPLS. If VPLS is fully meshed, a single RT is adequate to identify a given VPLS instance. A PE announces usually via IBGP that it belongs to a specific VPLS instance by annotating its network layer reachability information (NLRI) for that VPLS instance with RT, and acts on this by accepting NLRIs from other PEs that have RT. A PE announces that it no longer participates in that specific VPLS instance by withdrawing all NLRIs that it had advertised with RT. RT format is given in Figure 14.14. Detailed description of RT is given in RFC 4360 [4].

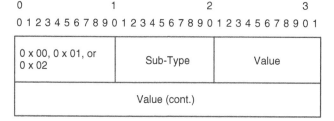

Figure 14.14   RT format.

14.4.1 Autodiscovery

In order to understand autodiscovery in BG-based VPLS, let us assume a new PE is added by the SP. As in Reference [3], a single BGP session is established between the new PE and an RR. The new PE then joins a VPLS domain when the VPLS instance is configured on that PE. Each VPLS instance is identified by a particular RT BGP extended community, which is configured as a part of configuring a VPLS instance.

Once this occurs, the PE advertises that it is a part of the VPLS domain via the RR to other PEs joined in that VPLS instance.

The advertisement carries the BGP RT extended community that is configured for that VPLS. This Community identifies the advertisement with a particular VPLS. For the purpose of redundancy, the PE may establish BGP sessions with more than one RR.

When a PE is added, only a BGP session between it and the RR is established. If the session is to be authenticated with MD5 then only keys for the two end points of that BGP session are configured. When a new VPLS instance is configured on that PE, it then advertises its availability via the RR, making all other relevant PEs aware of its presence. At the same time, BGP signaling automatically builds the mesh of LSPs for that VPLS instance.

If the VPLS instance needs to span multiple ASs, use of the RT for identifying a particular VPLS simplifies operations, as each AS can assign a particular RT to that VPLS on its own. This is possible because RT extended community embeds the AS number, and these numbers are globally unique.

14.4.2 Signaling

Once discovery is done, each pair of PEs in a VPLS must be able to establish pseudowires to each other, transmit certain characteristics of the pseudowires that a PE sets up for a given VPLS, and tear down the pseudowires when they are no longer needed. This mechanism is called *signaling*.

Multiple streams are multiplexed in a tunnel where each stream is identified by an MPLS label if MPLS is used, called *demultiplexer*. The label determines which VPLS instance a packet belongs to and the ingress PE. Each stream may represent a service.

A distinct BGP update message can be used to send a demultiplexer to each remote PE. This would require the originating PE to send N such messages for N remote PEs. It is also possible for PE to send one common update message that contains demultiplexers for all the remote PEs. Doing this reduces the control plane load both on the originating PE and on the BGP RRs (Figure 14.15) that may be involved in distributing this update to other PEs.

To accomplish this, *label blocks* are introduced. A label block, defined by a label base LB and a VE block size VBS, is a contiguous set of labels {LB, LB + 1, ..., LB + VBS − 1}. All PEs within a given VPLS are assigned unique

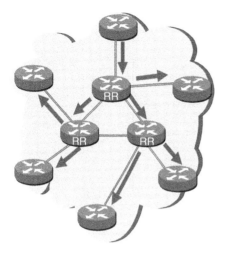

- Route reflectors can pass on iBGP updates to clients
- Each RR passes along only best routes

Figure 14.15   Route reflector.

VE IDs as a part of their configuration. A PE*x* wishing to send a VPLS update sends the same label block information to all other PEs. Each receiving PE infers the label intended for PE*x* by adding its unique VE ID to the label base. In this manner, each receiving PE gets a unique demultiplexer for PE*x* for that VPLS.

VPLS BGP NLRI

A VPLS BGP NLRI (Fig. 14.16) consists of a VE ID, a VE Block Offset, a VE block size, and a label base. The Address Family Identifier (AFI) is the L2VPN AFI (25), and the Subsequent Address Family Identifier (SAFI) is the VPLS SAFI (65).

A PE participating in a VPLS must have at least one VE ID. If the PE is the VE, it typically has one VE ID. If the PE is connected to several u-PEs (Fig. 14.19), it has a distinct VE ID for each u-PE. It may additionally have a VE ID for itself, if it itself acts as a VE for that VPLS.

VE IDs are typically assigned by the network administrator. A given VE ID should belong to only one PE, unless a CE is multihomed.

The *Layer 2 Info extended community* (Fig. 14.17) is used to signal control information about the pseudowires to be set up for a given VPLS. The extended community value, 0x800A, is allocated by IANA. This information includes the *Encaps Type*, which is the type of encapsulation on the pseudowires; *Control Flags* (Fig. 14.18) which is the control information regarding the pseudowires; and the *Maximum Transmission Unit*, which is used on the pseudowires. The Encaps Type for VPLS is 19.

| Length (2 octets) |
| Route Distinguisher (octets) |
| VE ID (2 octets) |
| VE Block Offset (2 octets) |
| VE Block Size (2 octets) |
| Label Base (3 octets) |

Figure 14.16   BGP NLRI for VPLS information.

| Extended Community Type (2 octets) |
| Encaps Type (1 octet) |
| Control Flags (1 octet) |
| L2 MTU (2 octets) |
| Reserved (2 octets) |

Figure 14.17   Layer 2 info extended community.

0	1	2	3	4	5	6	7
		MBZ				C	S

MBZ: Must be zero
C: A Control word [7] MUST or MUST NOT be present when sending
VPLS packets to this PE, depending on whether C is 1 or 0, respectively
S: Sequenced delivery of frames MUST or MUST NOT be used when sending
VPLS packets to this PE, depending on whether S is 1 or 0, respectively

Figure 14.18   Control Flags bit vector.

14.4.3   BGP VPLS Operation

To create a new VPLS, VPLSnew, a network administrator must pick an RT
for VPLSnew, RTnew. This will be used by all PEs that serve VPLSnew. To
configure a given PE, PE1, to be part of VPLSnew, the network administrator
only has to choose a VE ID, V, for PE1.

   If PE1 is connected to u-PEs (Fig. 14.19), PE1 may be configured with more
than one VE ID. The PE1 may also be configured with a route distinguisher (RD),

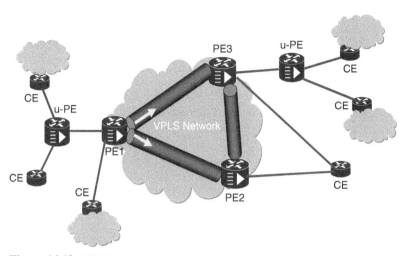

Figure 14.19 VPLS operation.

RDnew. Otherwise, it generates a unique RD for VPLSnew. PE1 then generates an initial label block and a remote VE set for V, defined by VE Block Offset VBO, VE block size VBS, and label base LB.

PE1 then creates a VPLS BGP NLRI with RD RDnew, VE ID V, VE Block Offset VBO, VE block size VBS and label base LB. It attaches a *Layer 2 Info extended community* and an RT, RTnew. It sets the BGP Next Hop for this NLRI as itself and announces this NLRI to its peers. The network layer protocol associated with the Network Address of the Next Hop for the combination <AFI = L2VPN AFI, SAFI = VPLS SAFI> is IP [15]. If the value of the length of the Next Hop field is four, then the Next Hop contains an IPv4 address. If this value is 16, then the Next Hop contains an IPv6 address.

If PE1 hears from another PE, for example, PE2, a VPLS BGP announcement with RTnew and VE ID Z, then PE1 knows via autodiscovery that PE2 is a member of the same VPLS. PE1 then has to set up its part of a VPLS pseudowire between PE1 and PE2. Similarly, PE2 will have discovered that PE1 is in the same VPLS, and PE1 must set up its part of the VPLS pseudowire. Thus, signaling and pseudowire setup are also achieved with the same update message.

If PE1's configuration is changed to remove VE ID V from VPLSnew then PE1 withdraws all its announcements for VPLSnew that contain VE ID V. If all PE1's links to its CEs in VPLSnew goes down then PE1 either withdraws all its NLRIs for VPLSnew or let other PEs in the VPLSnew know that PE1 is no longer connected to its CEs.

14.4.4 Multi-AS VPLS

As in References [13] and [3], autodiscovery and signaling functions are typically announced via IBGP. This assumes that all the sites in a VPLS are connected to

PEs in a single AS. However, sites in a VPLS may connect to PEs in different ASs. In this case, IBGP connection between PEs and PE-to-PE tunnels between the ASs must be established.

There are three methods for signaling interprovider VPLS:

- *VPLS-to-VPLS Connections at the* autonomous system border routers (ASBRs) requiring an Ethernet interconnect between the ASs, and both VPLS control and data plane state on the ASBRs. This method is easy to deploy.

  An ASBR (ASBR1) acts as a PE for all VPLSs that span AS1 and an AS to which ASBR1 is connected to, such as AS2 (Fig. 14.20). The ASBR on the neighboring AS (ASBR2) is viewed by ASBR1 as a CE for the VPLSs that span AS1 and AS2. Similarly, ASBR2 acts as a PE for this VPLS for AS2 and views ASBR1 as a CE.

  This method does not require MPLS protocol on the ASBR1–ASBR2 link, but requires L2 Ethernet. A VLAN ID is assigned to each VPLS traversing this link. Furthermore, ASBR1 performs the PE operations (discovery, signaling, MAC address learning, flooding, encapsulation, etc.) for all VPLSs that traverse ASBR1. This imposes a significant load on ASBR1, both on the control plane and the data plane, which limits the number of multi-AS VPLSs.

- *EBGP Redistribution of VPLS Information between ASBRs* requiring VPLS control plane state on the ASBRs and MPLS on the AS–AS interconnect. This method requires IBGP peerings between the PEs in AS1 and ASBR1, an EBGP peering between ASBR1 and ASBR2 in AS2, and IBGP peerings between ASBR2 and the PEs in AS2.

- *Multihop EBGP Redistribution of VPLS Information between ASs* requiring MPLS on the AS–AS interconnect, but no VPLS state on the ASBRs.

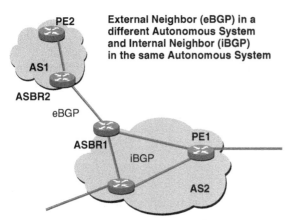

Figure 14.20 Inter-AS VPLS.

In this method, there is a multihop EBGP peering between the PEs (or preferably, an RR) in AS1 and the PEs (or RR) in AS2.

In the multi-AS configuration, a range of VE IDs identifying VPLS spanning multiple ASs is assigned for each AS. For example, AS1 uses VE IDs in the range 1–100, AS2 from 101 to 200, etc. If there are 10 sites attached to AS1 and 20 to AS2, the allocated VE IDs could be 1–10 and 101–120, respectively. This minimizes the number of VPLS NLRIs that are exchanged while ensuring that VE IDs are kept unique.

There will be no overlap between VE ID ranges among ASs, except when there is multihoming. When a VPLS site is connected to multiple PEs in the same AS or PEs in different ASs, the PEs connected to the same site can be configured either with the same VE ID or with different VE IDs. When PEs are in different ASs, it is mandatory to run STP on the CE device, and possibly on the PEs, to construct a loop-free VPLS topology.

14.4.5 Hierarchical BGP VPLS

The purpose of hierarchical BGP VPLS to scale the VPLS control plane when using BGP. The hierarchy

a. alleviates the full mesh connectivity requirement among VPLS BGP speakers,

b. limits BGP VPLS message passing to just the interested speakers rather than all BGP speakers, and

c. simplifies the addition and deletion of BGP speakers.

The basic technique for the hierarchy is to use BGP RRs [6] that helps to achieve objectives in (1) and (2). A designated small set of RRs are fully meshed. A BGP session between each BGP speaker and one or more RRs is established.

In this approach, there is no need for a full mesh connectivity among all the BGP speakers. If a large number of RRs is needed then this method will be used recursively.

The use of RRs introduces no data plane state and no data plane forwarding requirements on the RRs, and does not change the forwarding path of VPLS traffic. This is in contrast to the technique of H-VPLS defined in Reference [8].

One can define several sets of RRs, for example, a set to handle VPLS, another to handle IP VPNs, and another for Internet routing. Another partitioning could be to have some subset of VPLSs and IP VPNs handled by one set of RRs, and another subset of VPLSs and IP VPNs handled by another set of RRs. The use of Route Target Filtering (RTF) [4] can make this simpler.

Limiting BGP VPLS message passing to just the interested BGP speakers is addressed by the use of RTF. RTF is also very effective in inter-AS VPLS. More details on how RTF works and its benefits are provided in [4].

14.5 SECURITY

Data in a VPLS is only distributed to other nodes in that VPLS and not to any external agent or other VPLS. However, VPLS does not offer confidentiality, integrity, or authentication capabilities. VPLS packets are sent in the clear packet-switched network. A device-in-the-middle can eavesdrop and may be able to inject packets into the data stream. If security is desired, the PE-to-PE tunnels can be IPsec tunnels. The end systems in the VPLS sites can use encryption to secure their data even before it enters the SP network.

14.6 EXTERNAL NETWORK–NETWORK INTERFACE

An SP can choose VPLS to transport Ethernet service over its domain independent from handoffs to other provider. The handoff, which is an ENNI (External Network–Network Interface), can be 802.1ad (Fig. 14.21).

802.1ad packets can be mapped onto to the appropriate tunnel in the provider domain. Each provider network determines the best path *within* that network optimized within each domain. QoS is honored within each provider network. At interprovider boundaries, EXP-PCP mappings are done to preserve end-to-end QoS.

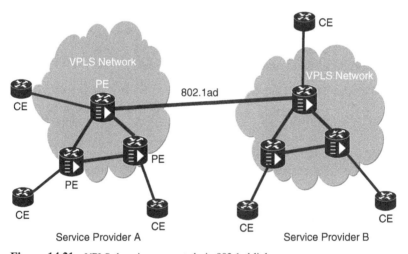

Figure 14.21 VPLS domains connected via 802.1ad links.

14.7 CONCLUSION

VPLS has received widespread industry support from both vendors and SPs and has been deployed widely. It offers flexible connectivity and a cost-effective solution for enterprise customers and carriers.

We have described two VPLS architectures, namely, LDP and BGP approaches, to emulate LAN over an MPLS network. VPLS offers enterprise customers intersite connectivity, protocol transparency and scalability, and a simplified LAN/WAN boundary. VPLS simplifies the network management and reduces the involvement of the SP in the customer IP scheme. Instead of requiring customers to connect to an IP network, with the complexity of IP routing protocols, they connect with simple and low cost Ethernet. All these are provisioned using standards-based Ethernet and MPLS gear.

REFERENCES

1. RFC 2764, A Framework for IP Based Virtual Private Networks, 2000.
2. L. Andersson, E. Rosen: Framework for Layer-2 Virtual Private Networks, http://www.ietf.org/internet-drafts/draft-ietf-L2vpn-l2-framework-05.txt, 2004.
3. RFC 4761, Virtual Private LAN Service (VPLS) Using BGP for Auto-Discovery and Signaling, 2007.
4. RFC 4360, BGP Extended Communities Attribute, 2006.
5. RFC 4762, Virtual Private LAN Service (VPLS) Using Label Distribution Protocol (LDP) Signaling, 2007.
6. RFC 4448, Encapsulation Methods for Transport of Ethernet over MPLS Networks, 2006.
7. RFC 4364, BGP/MPLS IP Virtual Private Networks (VPNs), 2006.
8. IEEE 802.1ad-2005- Local and Metropolitan Area Networks Virtual Bridged Local Area Networks.
9. RFC 2587, Internet X.509 Public Key Infrastructure LDAPv2 Schema, 1999.
10. RFC 4447, Pseudowire Setup and Maintenance Using the Label Distribution Protocol (LDP, 2006).
11. RFC 4446, IANA Allocations for Pseudowire Edge to Edge Emulation (PWE3), 2006.
12. M. Lasserre, V. Kompella: "Virtual Private LAN Services over MPLS", http://www.ietf.org/internet-drafts/draft-ietf-l2vpn-vpls-ldp-07.txt, 2005.
13. L. Martini, et. al, Encapsulation Methods for Transport of Ethernet Over MPLS Networks, draft-ietf-pwe3-ethernet-encap-11.txt, 2005.
14. L. Martini, et. al, Transport of Layer 2 Frames Over MPLS, draft-martini-l2circuittrans-mpls-12.txt, Nov 2003.
15. J. Witters, G. van Kersen, J. De Clerq, S. Khandekar: "Keys to Successful VPLS Deployment", Alcatel Telecommunications Review, 4th Quarter 2004.

Index

Absolute mode, for RTP header, 241
Access, with Carrier Ethernet, 110
Access control rules, 19
Access EPL Service attributes, 153–156. *See also* Ethernet private line (EPL)
Access EVPL Service attributes, 153–156. *See also* Ethernet Virtual Private Line (EVPL)
Access links, for Carrier Ethernet, 123, 124, 125, 126
Access-located PRC, 58
Access networks, xii, xiii. *See also* Copper access network
service availability and, 222
synchronization and, 34
Access points (APs)
Ethernet protocol stack and, 116
in ETH layer network, 118
Accuracy. *See also* Frequency accuracy
with precision time protocols, 45–46
in synchronization, 35–36, 39, 42, 43
AC reference model, 75. *See also* Attachment circuits (ACs)
Active EVC Status, ELMI protocol and, 249, 251
Active hubs, 17
Active LACP packet exchange mode, 104
Active Provider Edge (PE), multisegment pseudowire setup and, 78
Adaptation function headers, for CESoETH, 238–241
Adaptation functions, Ethernet protocol stack and, 117

Adaptation sink function, Ethernet protocol stack and, 117
Adaptation source function, Ethernet protocol stack and, 117
Adapter cards, 28
Adapters, 19
Adaptive methods
in network synchronous operation, 54, 55
packet network impairments and, 55–56
stabilization period and, 56–57
Adaptive timing, 242
IWF synchronization function and, 57
Address Family Identifier (AFI), in VPLS BGP NLRI, 355
Addressing modes, in OAM addressing, 194
Address learning, in LDP-based VPLS, 340–341
Address Withdraw message, in LDP-based VPLS, 341
Administrative domains, 188
Aggregation, with Carrier Ethernet, 110
Aggregation control, in link aggregation, 105
Aggregator, in link aggregation, 105
AIS frames, 194, 210–213. *See also* Alarm Indication Signal (AIS)
AIS generation, 190
AIS PDU, 210, 212. *See also* Protocol data units (PDUs)

Networks and Services: Carrier Ethernet, PBT, MPLS-TP, and VPLS, First Edition. Mehmet Toy.
© 2012 John Wiley & Sons, Inc. Published 2012 by John Wiley & Sons, Inc.